Certified Solidworks Professional Advanced Surface Modeling Exam Preparation

ISBN: 978-0-620-90676-0

Black and White Print version

.

Acknowledgements

Special thanks to my work colleagues and friends for their encouragement and support in writing this Book. I would also like to thank my wife Judith and kids Charlotte and Malcolm for their support and giving me the space I needed to write this book.

TABLE OF CONTENTS

V

TABLE OF FIGURES

CHAPTER 1

COMPLETING A PART DESIGN USING SURFACE MODELING

In this Chapter, an incomplete part designed using surface modeling techniques in Solidworks is made available to you to complete the design. The surfaces have been split and end faces trimmed. You will be tasked to recreate these faces using Solidworks surfacing techniques and functionality. Images below show the finished part and images overleaf show the part at different stages of its creation.

The Unit System is MMGS (millimeter, gram, second), Decimal Places: 2, Material = None and the part is symmetric about the YZ Plane.

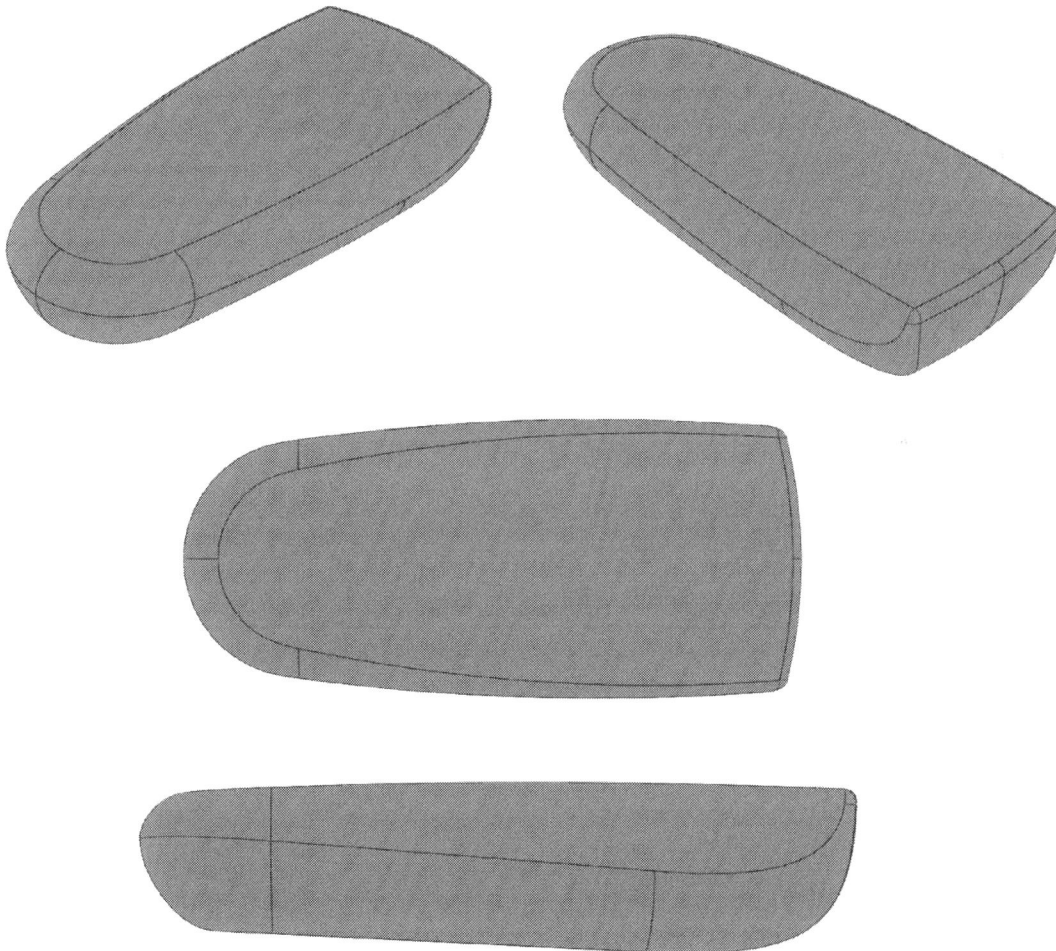

Figure 1 - Complete Part Designed using Surface Modeling Techniques and Functionality

Figure 2 – Complete Part Designed using Surface Modeling Techniques and Functionality

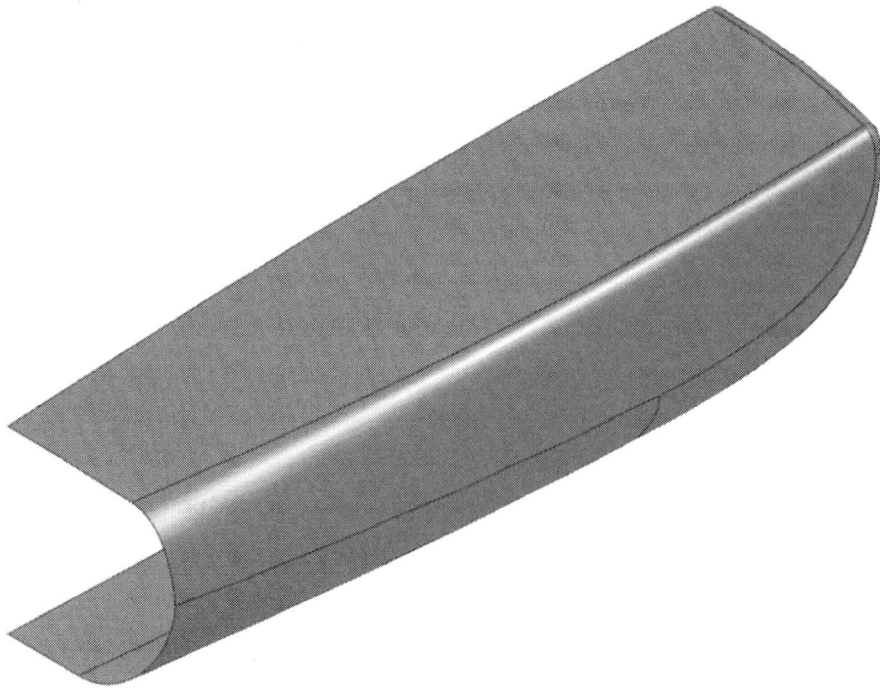

Figure 3 - Incomplete Part to be completed using Solidworks Surface Modeling Techniques and Functionality

Download the part RF_Start.sldprt from this Google drive location - *http://bit.ly/CSWPA-SU* or Scan the QR Code shown below:

If you experience any problems with downloading any files you may send an email to *cswpasmebook@gmail.com* with the title of the book indicated in your email subject. Open and save the downloaded part to your PC.

The part looks as shown in the following image - it consists of an imported surface - Surface-Imported1 and three sketch images - AAA, BBB and CCC as you can see in the part's Feature Manager Design Tree in the following image.

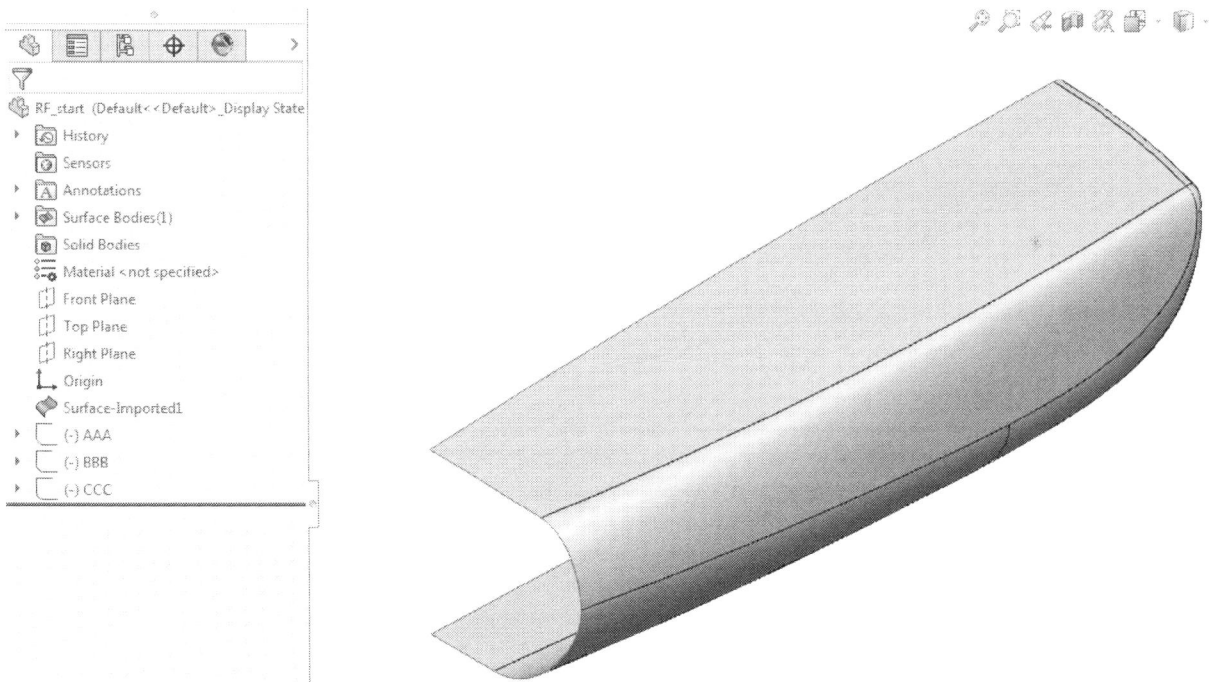

Figure 4 - Downloaded part to be completed using Solidworks Surface Modeling

Sketch AAA contains a sketch image of the missing faces as viewed from the side as shown in the following image.

3

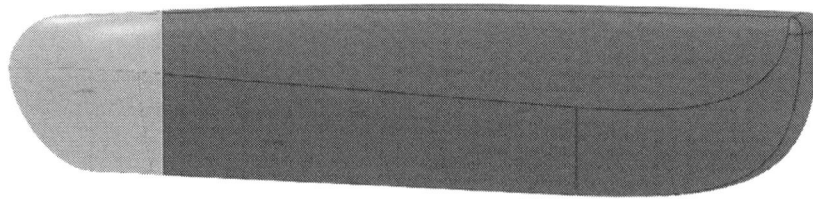

AAA

Figure 5 - Sketch image as viewed from the side view - Sketch AAA

Sketch BBB contains a sketch image of the missing faces as viewed from the top looking down as shown in the following image.

BBB

Figure 6 - Sketch image as viewed from the top looking down - Sketch BBB

Sketch CCC contains a sketch image of the missing faces as viewed from the bottom looking up as shown in the following image.

CCC

Figure 7 - Sketch image as viewed from the top looking down - Sketch CCC

Using the sketches and sketch pictures in the downloaded part RF_Start.sldprt, you must extend and trim the upper surface closely approximating the boundary shown in sketch BBB to get either of the results shown in the following images - thus an extended / trimmed surface that is either homogenous or made up of two or more faces after the trim / extend. Both results are acceptable. You must then measure the total surface area of the recreated upper surface in square millimeters as your answer to make sure you have executed this task correctly.

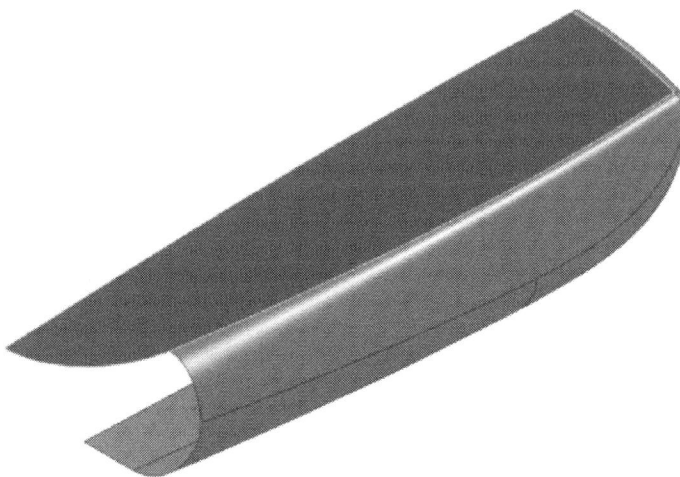

Figure 8 - Extended / trimmed surface that is homogenous

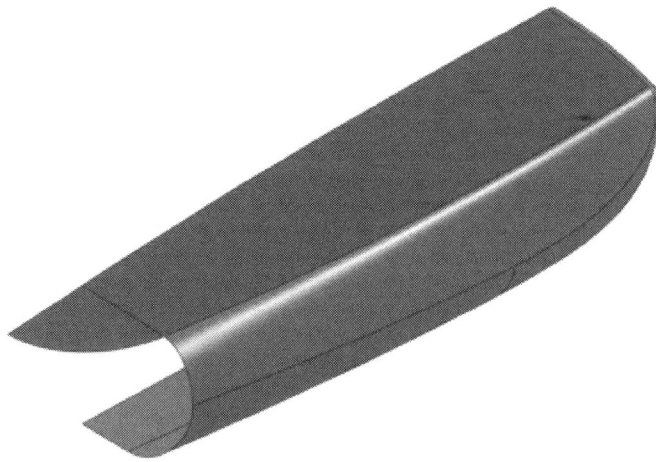

Figure 9 - Extended / trimmed surface that is made up of two or more faces

EXTENDING A SURFACE

Click Extend Surface (Surfaces toolbar) or Insert > Surface > Extend. In the PropertyManager:

Under Edges/Faces to Extend, select the edge in the graphics area as shown in the following image. Under End Condition, select distance and enter a dimension of 40mm. Under Extension Type, select Same Surface.

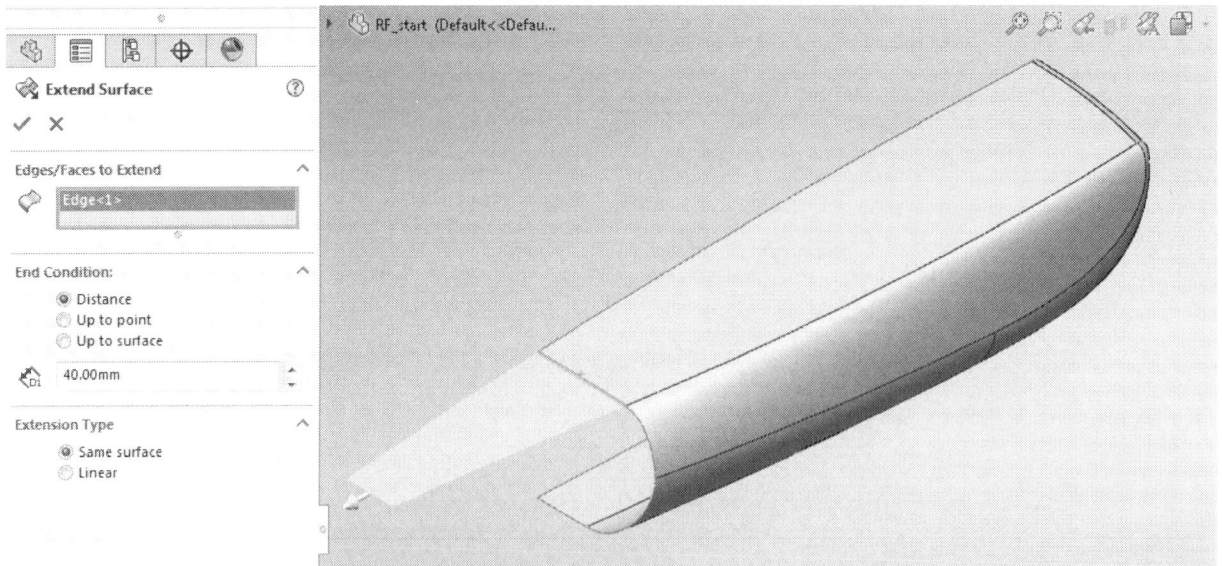

Figure 10 - Extending a surface through an en edge selection

Click OK to close the Property Manager and your part should now look as shown in the following image.

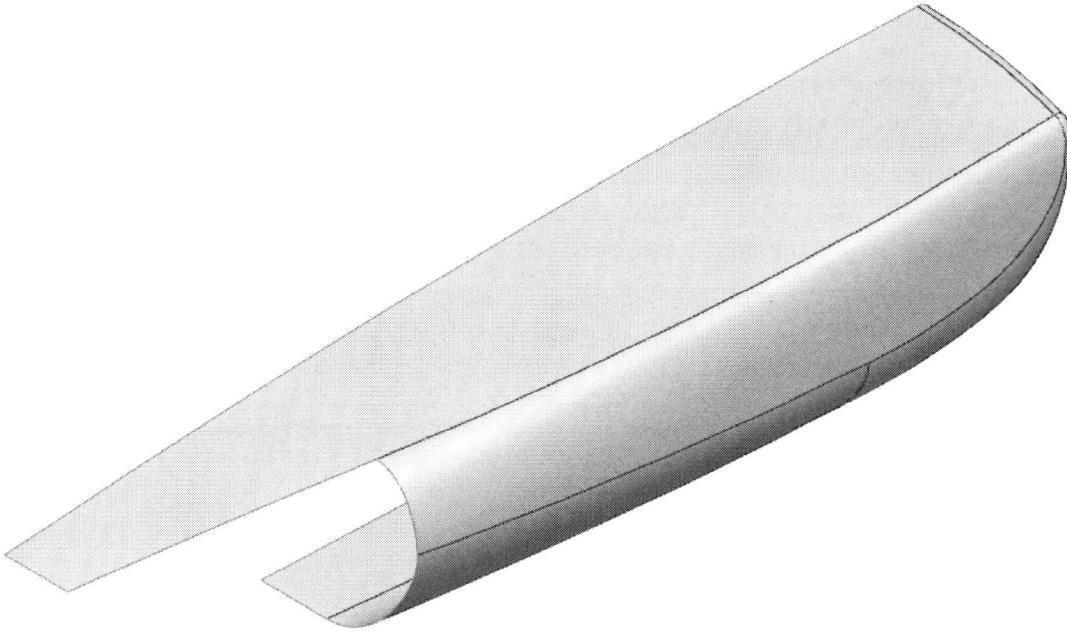

Figure 11 - Surface Model Current Status

UNDERSTANDING EXTEND SURFACE - EXTENSION TYPE OPTIONS

To appreciate the difference between the two options in the Extend Surface PropertyManager namely *Same Surface* and *Linear* under *Extension Type* you need to start a new solidworks part and save it as SURFACE EXTENSION TYPE.sldprt.

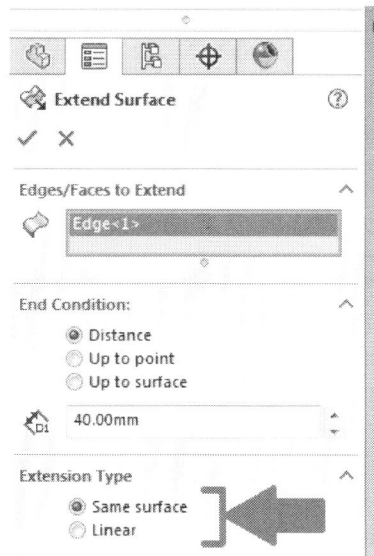

Figure 12 - Extension Type Options in the Extend Surface PropertyManager

Start a new Solidworks Part and Save it as SURFACE EXTENSION TYPE.sldprt.

CREATING A CENTERPOINT ARC

In this new part, under the Sketch Tab, click the Arc CommandManager to select the Centerpoint Arc from the Arc Flyout Tool or Click Tools > Sketch Entities > Centerpoint Arc then select the Right Plane in the Graphics Area. The Arc PropertyManager appears where you can still change the Arc Type if required that is before you click anywhere in the graphics area to start sketching.

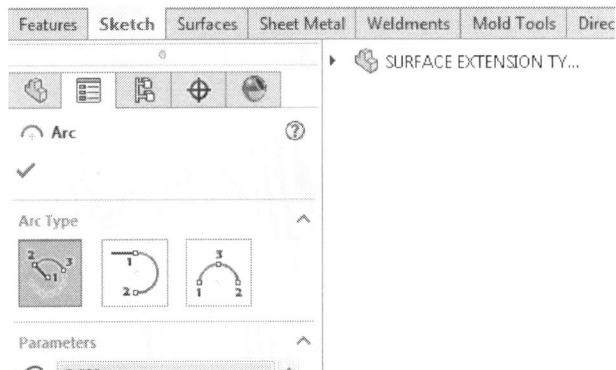

Figure 13 - Arc PropertyManager

With the Centerpoint Arc selected, click on the origin to place the center of the arc. Release and drag to set the radius and the start angle. Click to place a start point. Release, drag, and click to set an end point as shown in the following image.

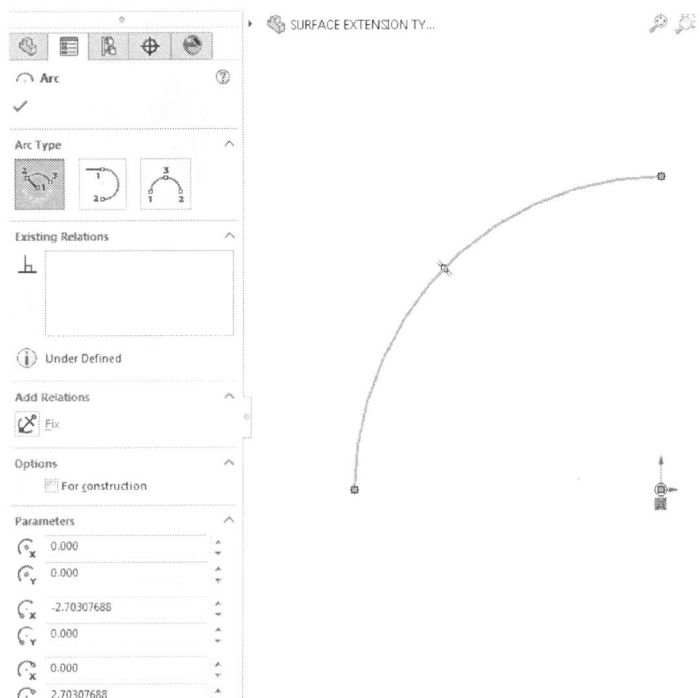

Figure 14 - Centerpoint Arc

8

Click Ok. Dimension and add relations as shown in the following image - dimensions are in millimeters.

TIP: You can change the unit system without opening Document Properties < Units - in the status bar *(Bottom Right Corner)*, Click Unit System, then Select and Click the preferred unit system as shown in the following image.

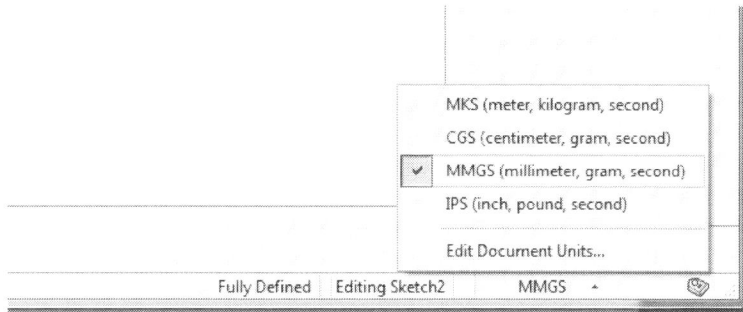

Figure 15 - Changing the Unit System without opening Document Properties

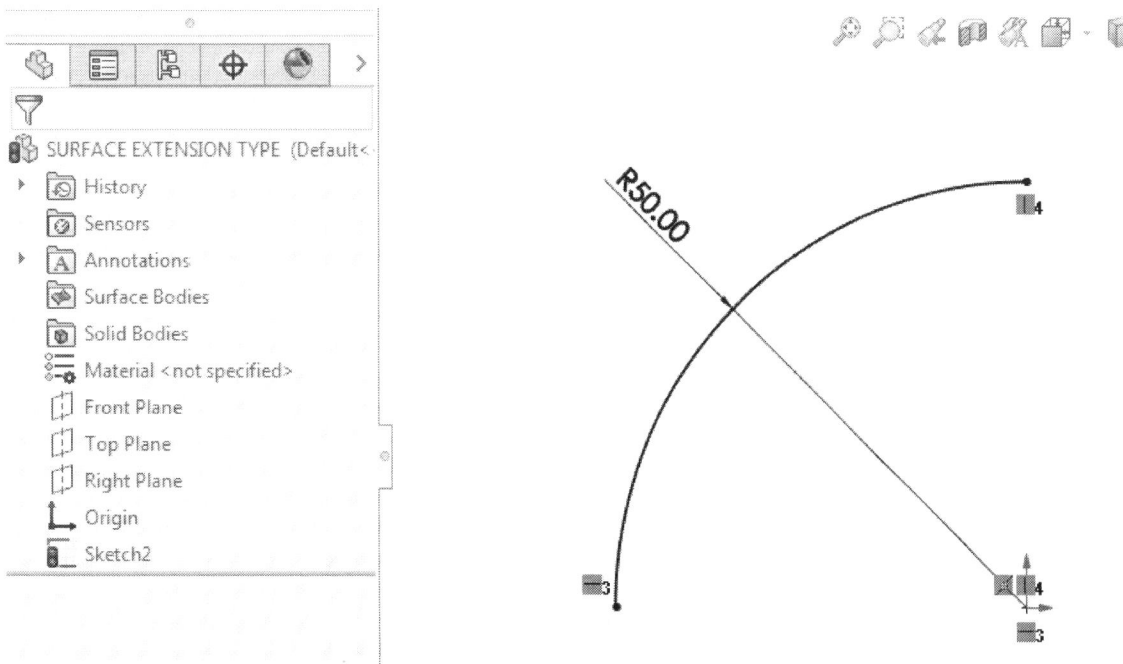

Figure 16 - Fully Defined Centerpoint Arc Sketch

Add a center line with relations *(a tangent and horizontal relation)* and dimensions as shown in the following image.

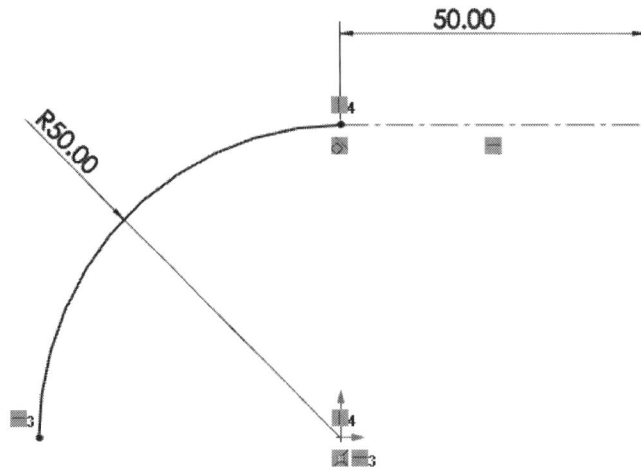

Figure 17 - Fully Defined Sketch

Add an arc *(construction)* with dimensions *(length of the arc)* and relations as shown in the following image.

TIP: You can dimension the true length of an arc in an open sketch by Clicking Tools > Dimensions > Smart or Click Smart Dimension on the Sketch Tab then Click on the Arc *(avoiding clicking or selecting the midpoint or midpoint relation)* in the Graphics Area followed by Clicking on the two End points of the Arc. Move the pointer to show the dimension preview then click to place the dimension. Set the value *(50mm)* in the Modify dialog box and click OK.

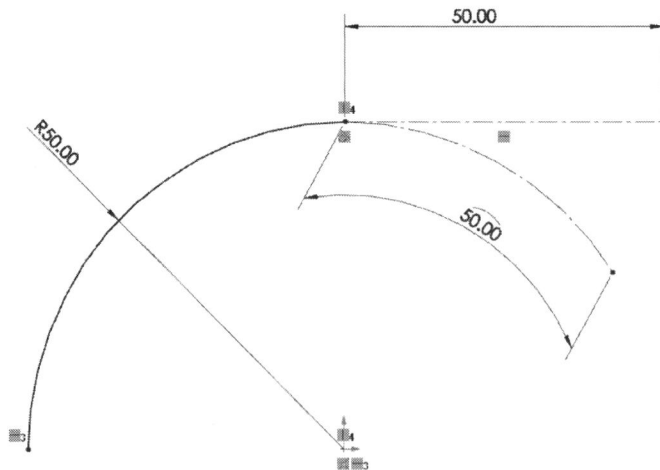

Figure 18 - Fully Defined Centerpoint Arc Sketch

EXTRUDED SURFACE

Without Exiting the Sketch, Click Extruded Surface (Surfaces Toolbar) or Insert > Surface > Extrude then set the PropertyManager options as shown in the following image.

10

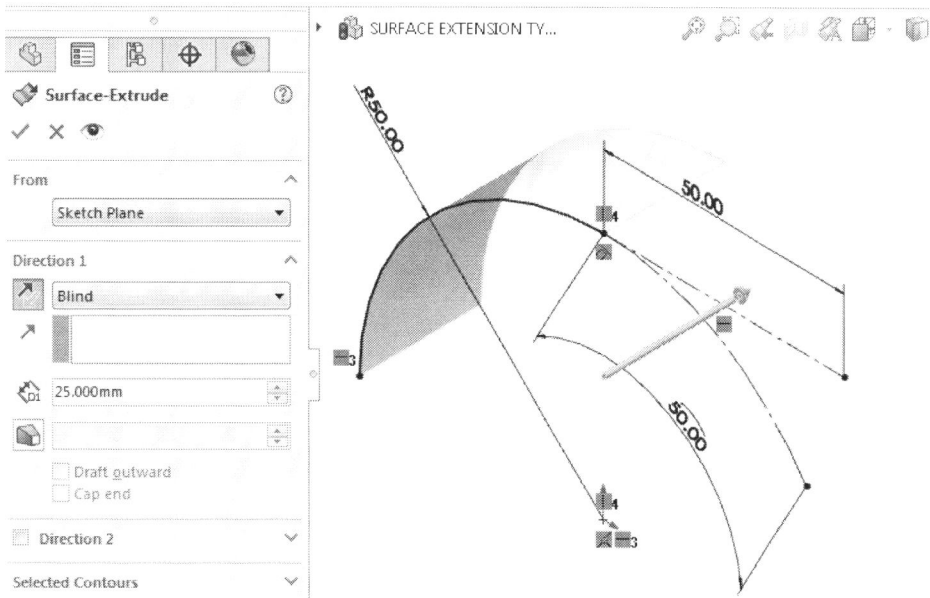

Figure 19 - Fully Defined Centerpoint Arc Sketch

Click OK.

HIDING AND SHOWING SKETCHES

Right-click the sketch in the Feature Manager Design Tree and select Show as shown in the following image *(the sketch entities highlight in the graphics area when you point over the sketch name in the Feature Manager Design Tree).*

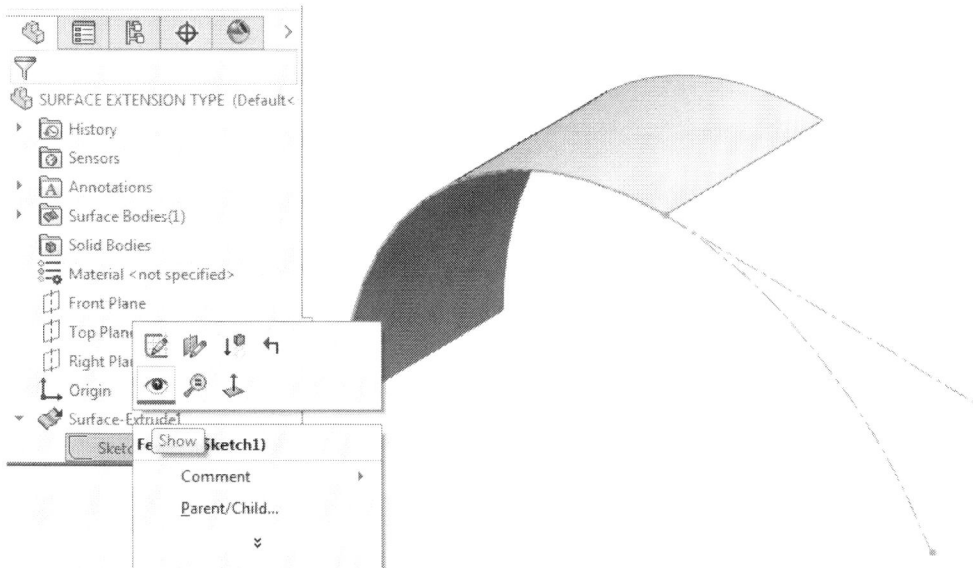

Figure 20 - Hiding and showing sketches

The resulting surface with the sketch shown should now look as shown in the following image.

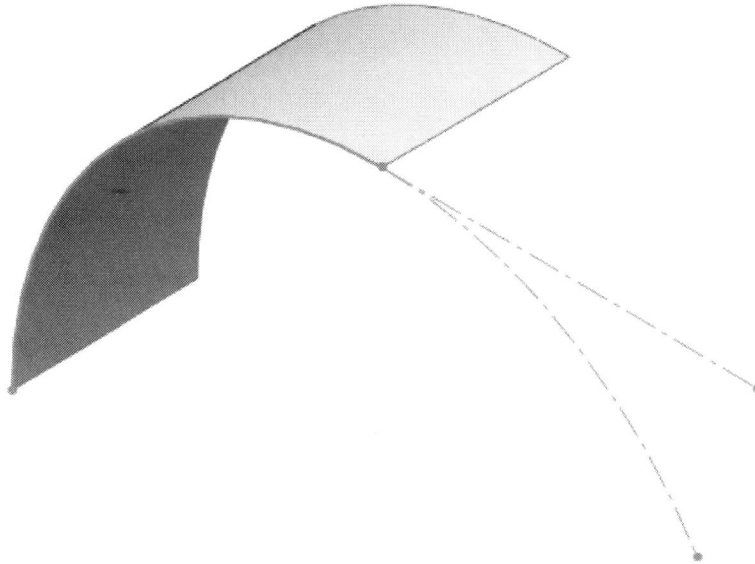

Figure 21 - Extruded Surface and absorbed sketch shown

EXTEND SURFACE - SAME SURFACE EXTENSION TYPE

Click Extend Surface (Surfaces toolbar) or Insert > Surface > Extend. In the PropertyManager:

Under Edges/Faces to Extend, select the edge in the graphics area as shown in the following image. Under End Condition, select distance and enter a dimension of 50mm. Under Extension Type, select Same Surface as shown in the following image then Click OK.

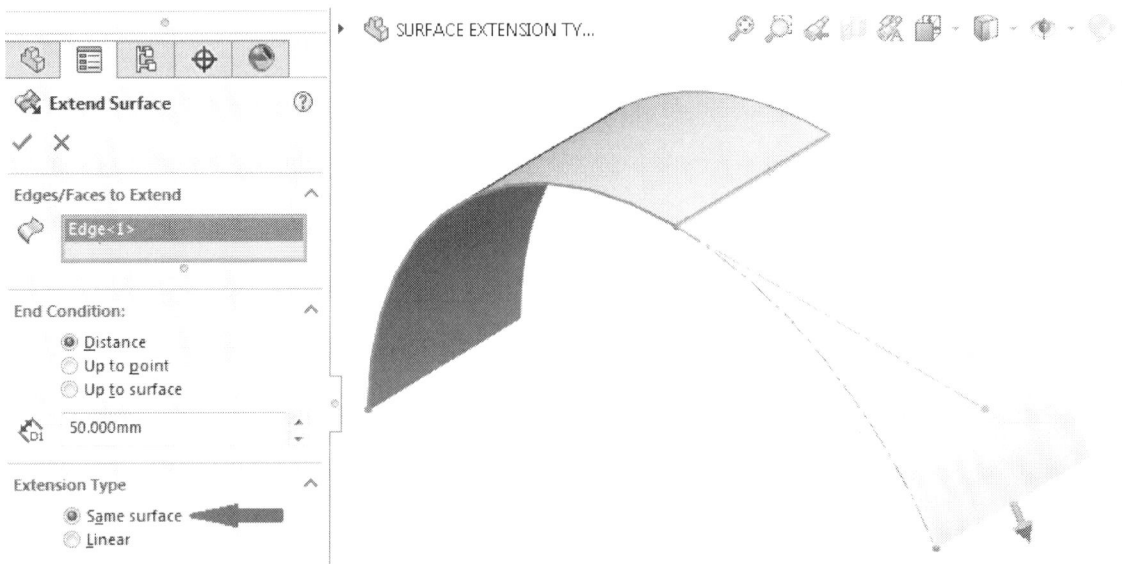

Figure 22 - Extend Surface - Same surface Extension Type

So the Same surface Extension Type Extends the surface **along** the geometry of the surface as you can see in the following image.

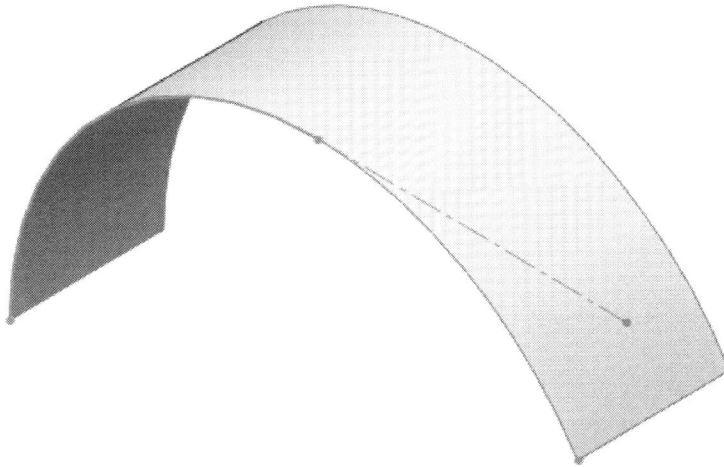

Figure 23 - Extend Surface Same surface Extension Type

Rename the Surface Extension Feature in the FeatureManager Design Tree to Same surface extension type and suppress it as shown in the following image.

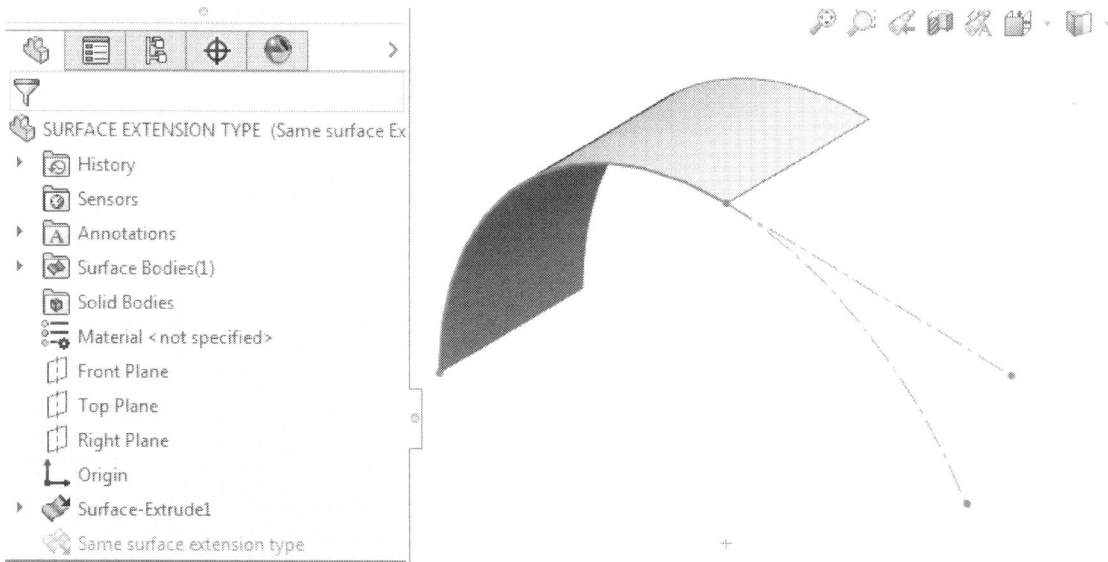

Figure 24 - Feature renaming and suppression

EXTEND SURFACE - LINEAR EXTENSION TYPE

Click Extend Surface (Surfaces toolbar) or Insert > Surface > Extend. In the PropertyManager:

Under Edges/Faces to Extend, select the edge in the graphics area as shown in the following image. Under End Condition, select distance and enter a dimension of 50mm. Under Extension Type, select Linear as shown in the following image then Click OK.

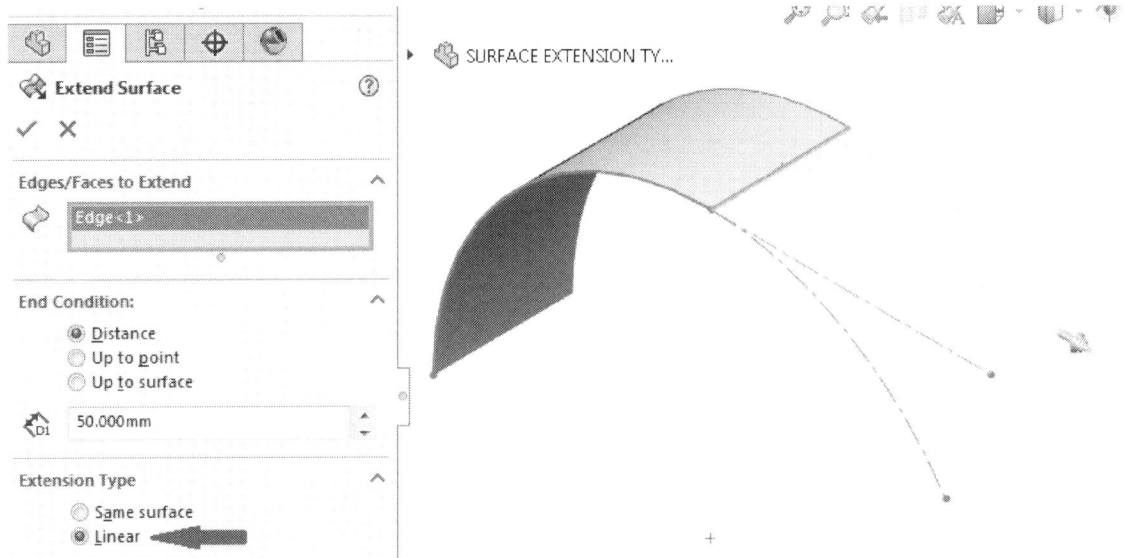

Figure 25 - Extend Surface - Linear Extension Type

So the Linear Extension Type Extends the surface **tangent** to the original surface along the edges as you can see in the following image.

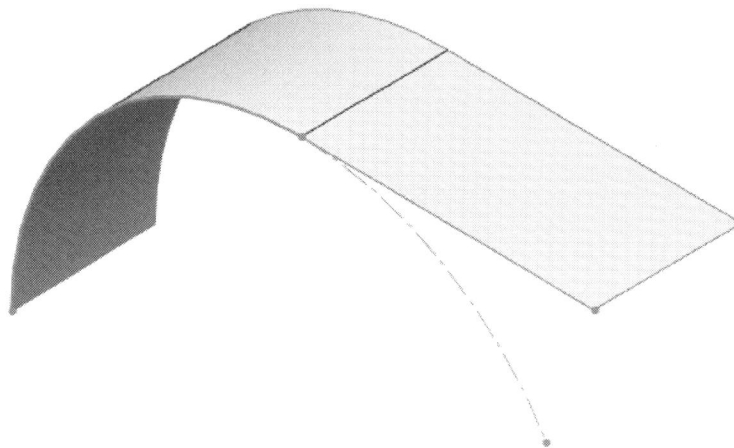

Figure 26 - Extend Surface Linear Extension Type

Rename the Surface Extension Feature in the FeatureManager Design Tree to Linear extension type as shown in the following image.

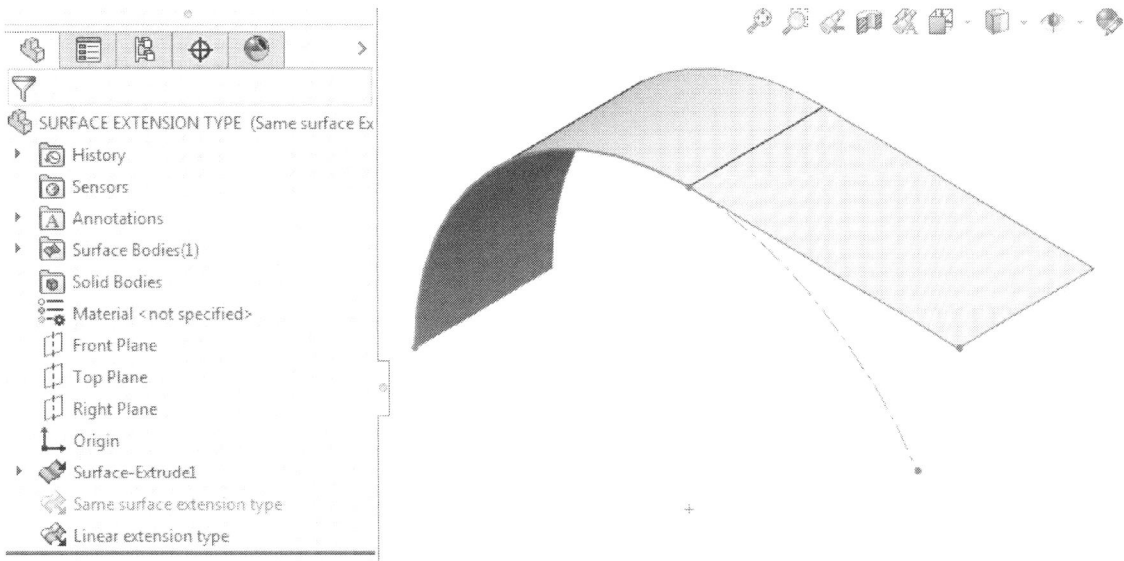

Figure 27 - Extend Surface Linear Extension Type - Feature renaming

Hopefully this clarifies the difference between the two Extension Types - Same surface vs Linear.

Save and close the Part.

Now let's refocus on the Downloaded part which we were working on.

TRIMMING A SURFACE USING A SKETCH

Click Trim Surface on the Surfaces toolbar, or click Insert > Surface > Trim. In the PropertyManager, under Trim Type, select Standard. Under Selections, choose sketch BBB in the Feature Manager Design Tree or in the Graphics Area. Select the Remove selections Radio Button and in the graphics area, select the area shown in purple in the following image. Under Surface Split Options, uncheck the Split all Checkbox and select the Natural Radio Button.

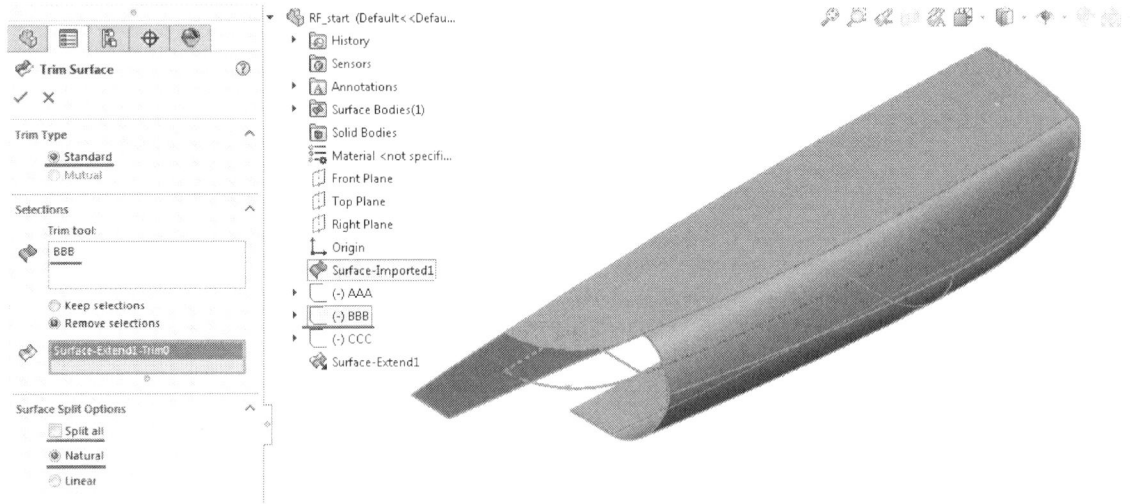

Figure 28 - Trimming an extended surface using a sketch

Click OK and your part should now appear as shown in the following image.

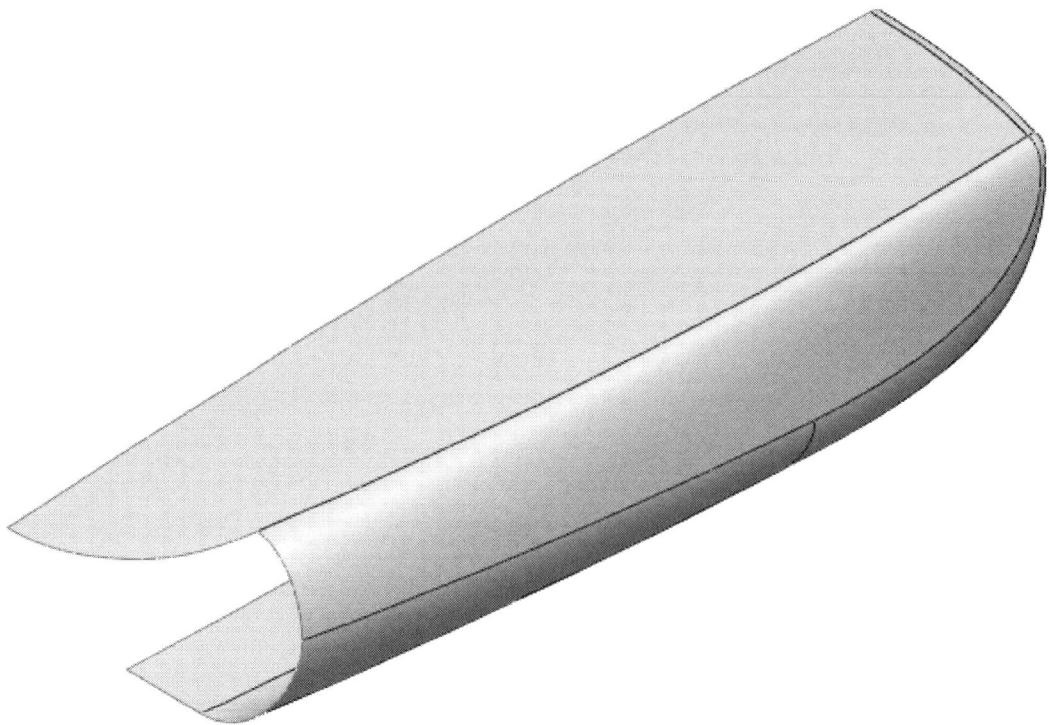

Figure 29 - Surface Model current status

MEASURING THE AREA OF A SURFACE

In the Graphics Area, click on the surface shown in the following image.

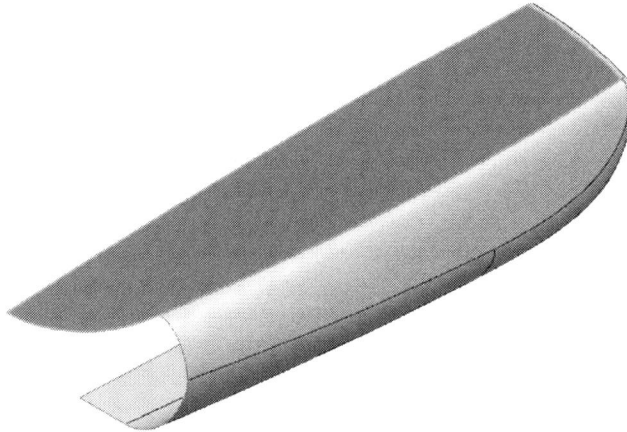

Figure 30 - Select a surface on the Surface Model

While the surface is still highlighted, Click Measure (Tools toolbar) or Tools > Evaluate > Measure. The Measure Dialog Box appears as shown in the following image. The surface area is underlined in the following image and thus the answer is 2608.58mm². Save your part and continue to Chapter 2.

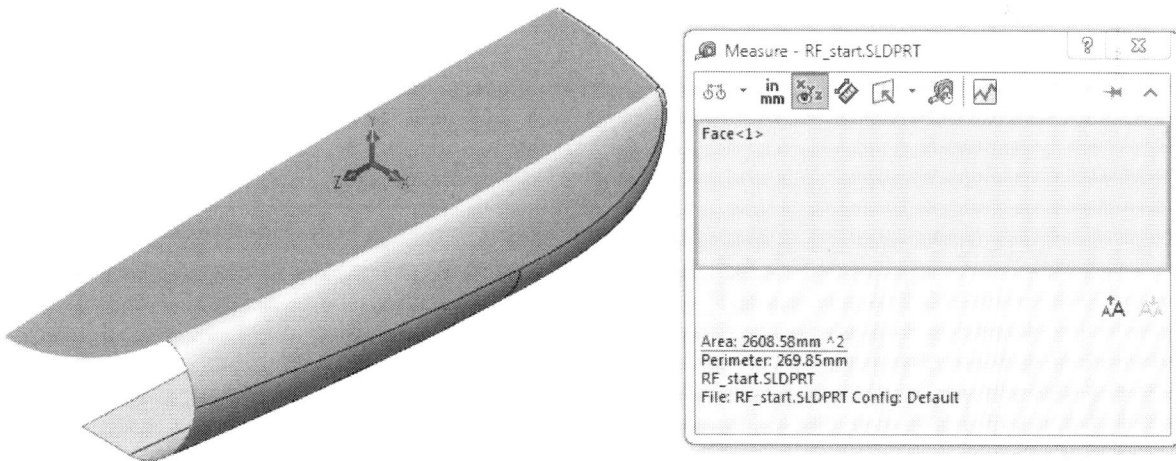

Figure 31 - Chapter One Answer - Measuring Surface Area

USING EXTEND AND TRIM SURFACE CONTROLS

In this Chapter, we will continue working on the same part from Chapter One.

Using the included sketches and sketch pictures, you are required to extend and trim the lower surface closely approximating the boundary shown in sketch CCC to get the resulting part or surface model as shown in the following image.

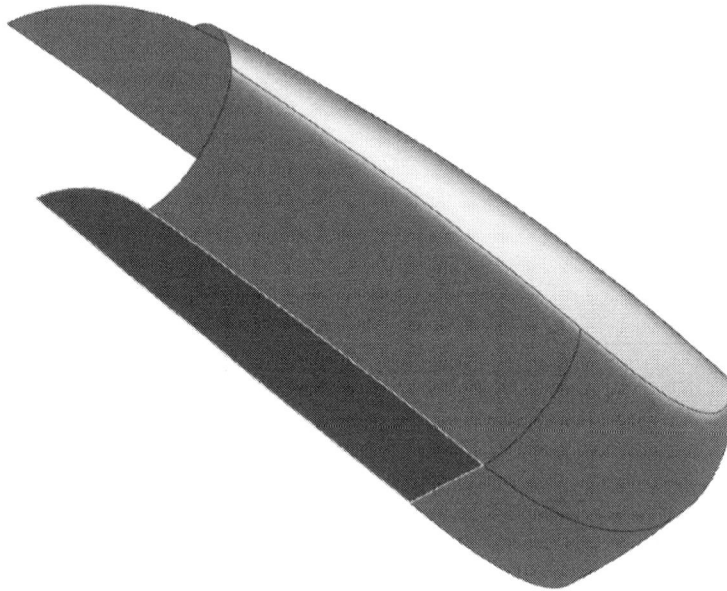

Figure 32 - Chapter Two - Completed surface model

You are then required to measure and provide the total surface area of the recreated lower surface and give your answer in square millimeters to two decimal places.

EXTENDING A SURFACE

Click Extend Surface (Surfaces toolbar) or Insert > Surface > Extend. In the PropertyManager: Under Edges/Faces to Extend, select the edge in the graphics area as shown in the following image. Under End Condition, select distance and enter a dimension of 40mm. Under Extension Type, select Same Surface.

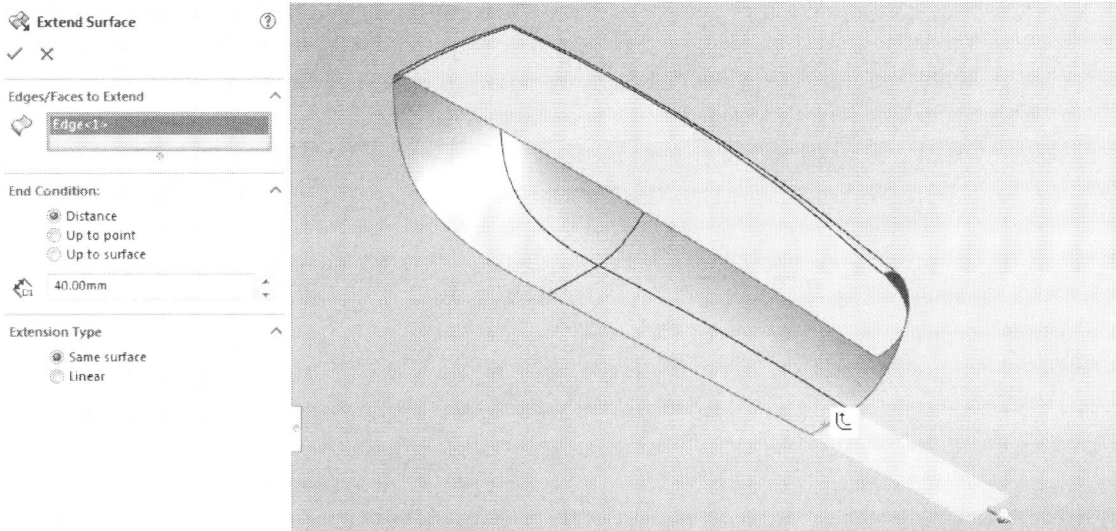

Figure 33 - Extending a Surface

Click OK to close the Property Manager and your part should now look as shown in the following image.

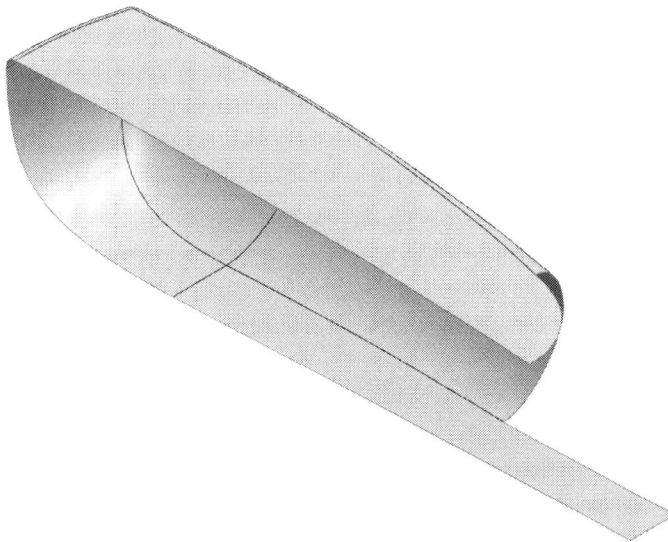

Figure 34 - Surface Model Current Status

19

TRIMMING A SURFACE USING A SKETCH PICTURE

Click Trim Surface on the Surfaces toolbar, or Click Insert > Surface > Trim. In the PropertyManager, under Trim Type, select Standard. Under Selections, choose sketch CCC in the Feature Manager Design Tree. Select the Remove selections Radio Button and in the graphics area, try to select the area shown in purple in the following image and you will notice that you can't select that area. Why? Now that is because sketch CCC has an image only in it and no other sketch entities in it hence we can't use it as a Trimming Tool. So we need to add some sketch entities in sketch CCC by tracing the sketch image in CCC. Click Cancel to close the Trim Surface Property Manager.

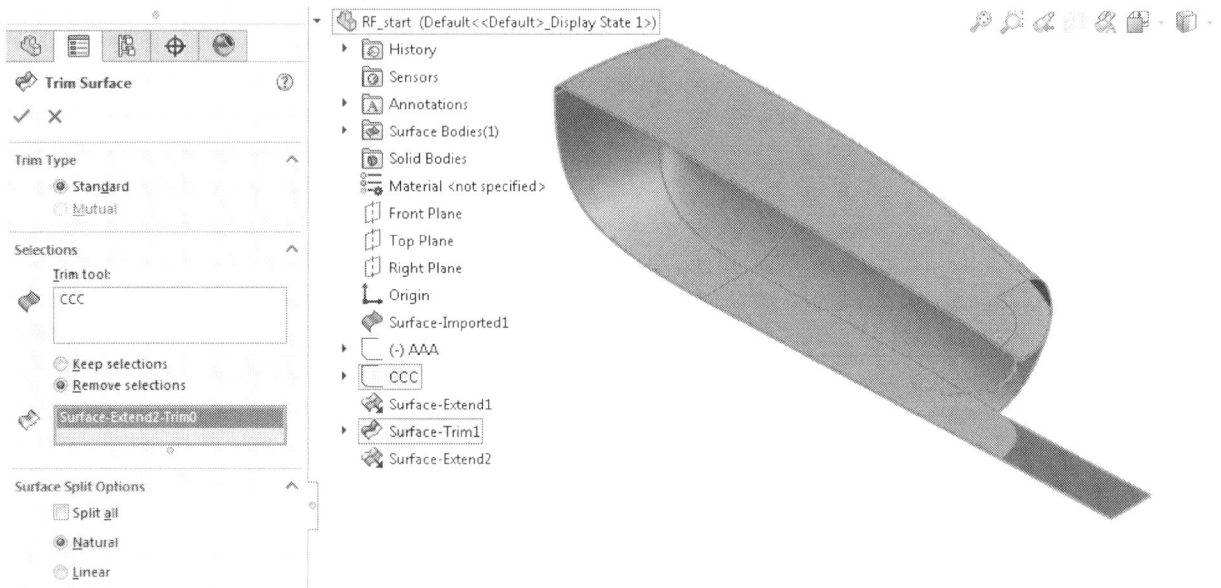

Figure 35 - Trimming a surface using a sketch

USING A SKETCH PICTURE TO CREATE A SKECTH

Right Click or Click on Sketch CCC and select Edit Sketch as shown in the following image.

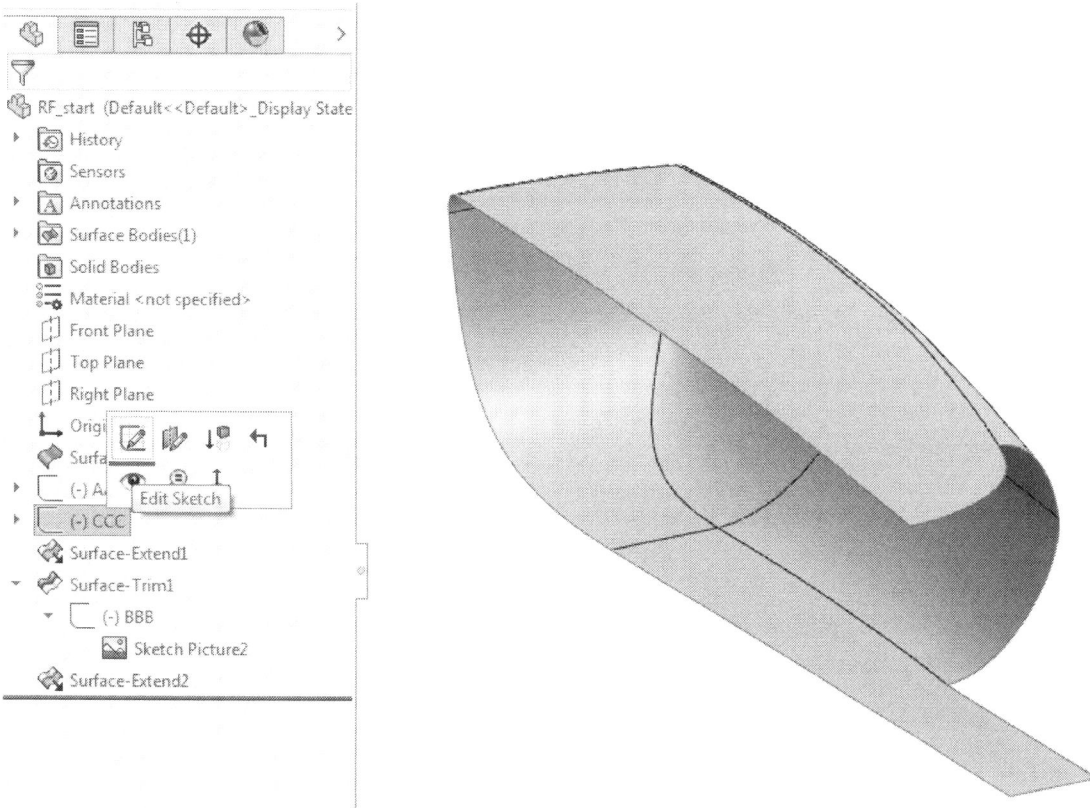

Figure 36 - Editing an existing Sketch

In Edit Sketch Mode, zoom to the area shown in the following image.

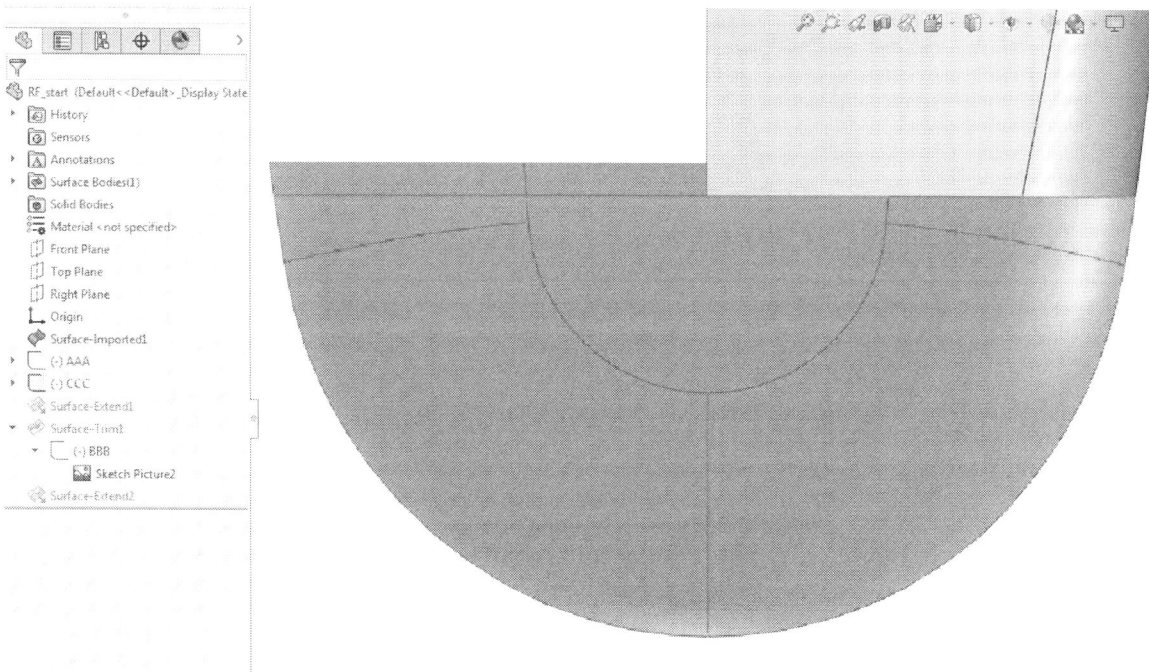

Figure 37 - Editing an existing Sketch

SKETCHING A 3 POINT ARC

Under the Sketch tab, Click the Arc Command Manager to select a 3 Point Arc from the Arc flyout tool as shown in the following image. You may also Click Tools > Sketch Entities > 3 Point Arc or you may also change to a different arc tool from the Arc PropertyManager.

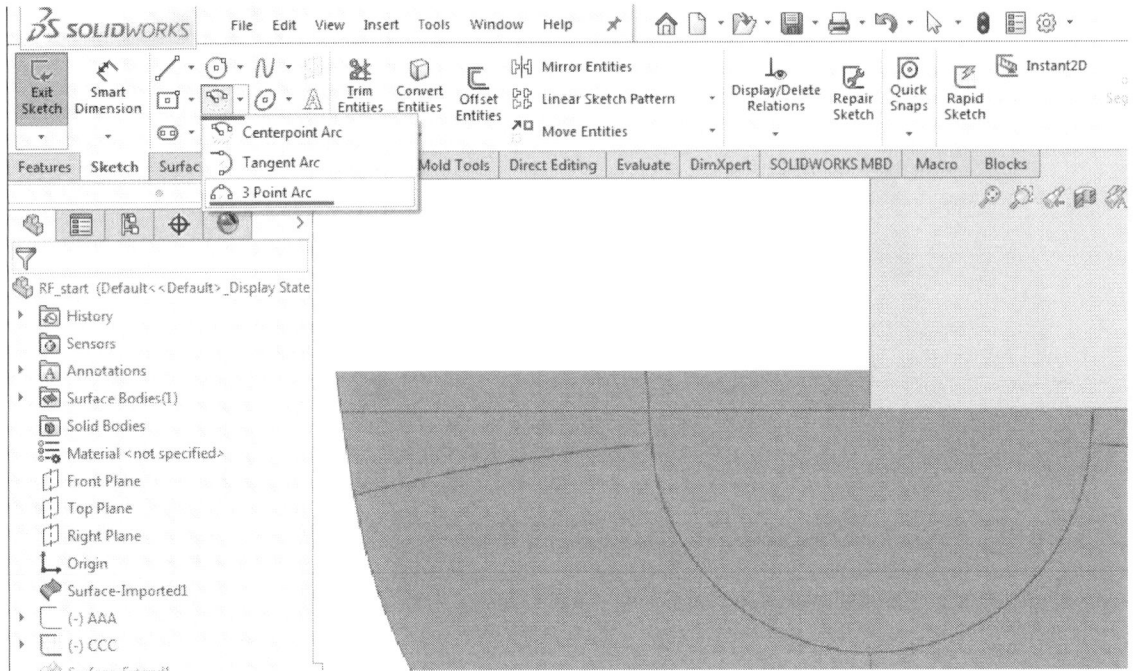

Figure 38 - Sketching a 3 Point Arc

Zoom to point 1 circled in the following image and Click to set a start point. Drag the pointer, then zoom and click to set an end point on point 2 circled in the following image. Drag to set the radius. Click to set the arc on point 3. Under Add Relation, click Fix. Then Click OK.

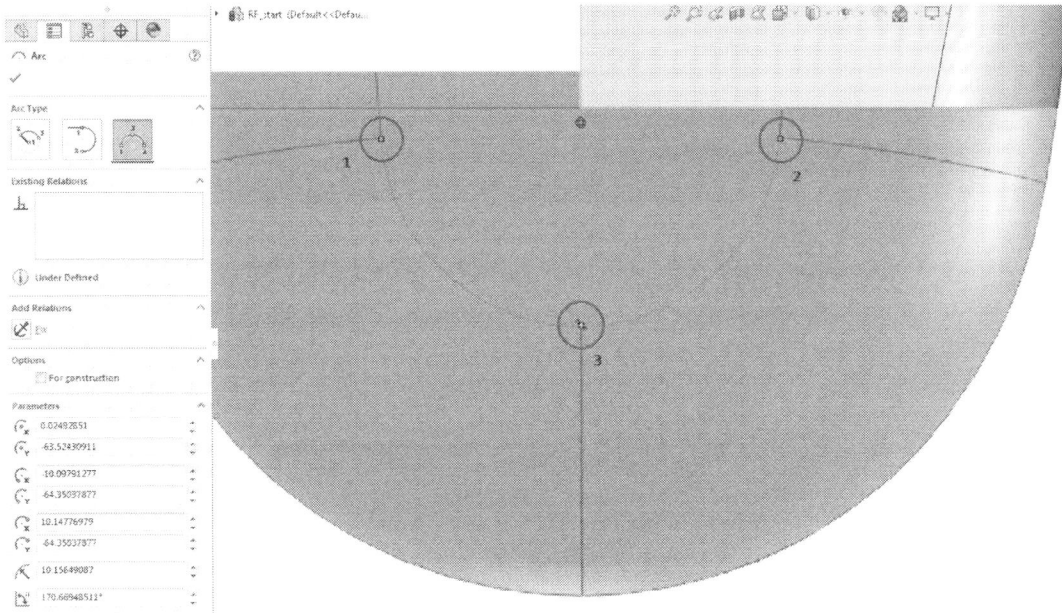

Figure 39 - Sketching a 3 Point Arc

Press the Shift Key on your Keyboard and select the two end points of the Arc then Click Fix under Add Relations in the Property Manager as shown in the following image. Click OK .

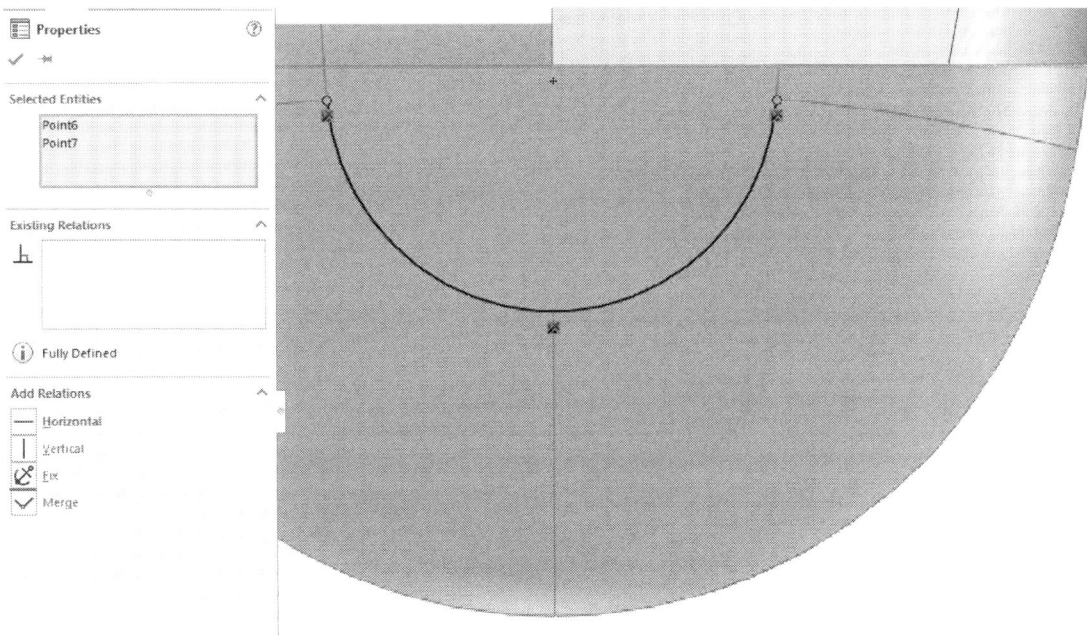

Figure 40 - Adding Sketch Relations

SKETCHING A SPLINE

Click Spline (Sketch toolbar) or Tools > Sketch Entities > Spline. In the graphics area, click the second endpoint of the arc and zoom to the area shown in the following image to click the second point which must also have a coincident relation with the edge of the existing surface - see the following image.

Figure 41 - Sketching a Spline

Press the Escape key on your Keyboard to complete the Spline. Add a Fixed Relation on the End Point of the Spline and a Tangent Relation between the spline and the Fully Defined Arc as shown in the following image.

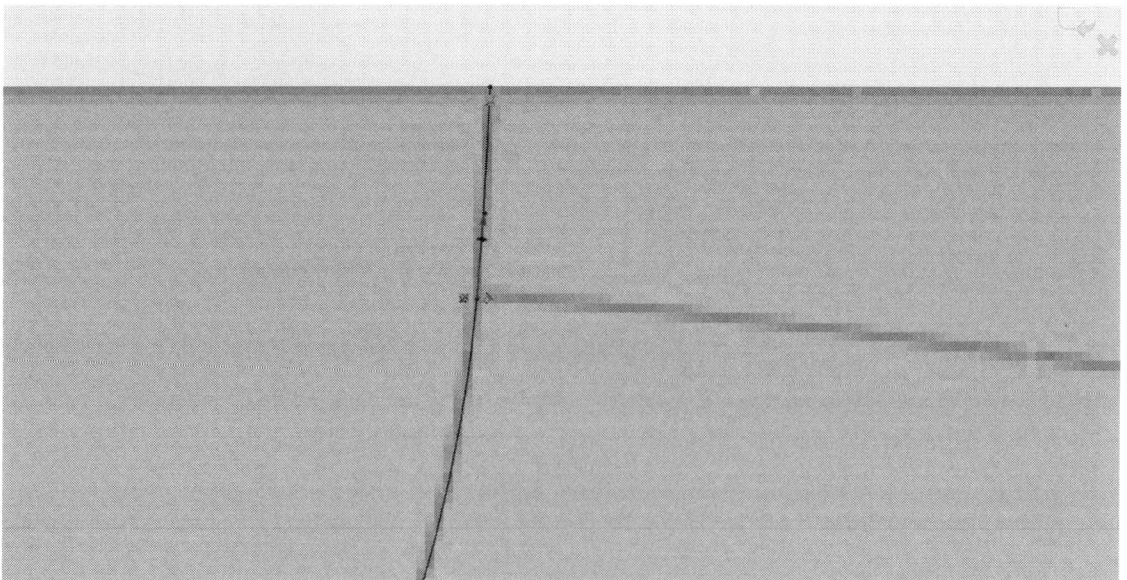

Figure 42 - Adding Sketch Relations

Exit the Sketch Mode. Save your part.

Now let's go back to Trimming our extended surface using Sketch CCC which now contains the sketch entities we created in the section above.

TRIMMING A SURFACE USING A SKETCH

Click Trim Surface on the Surfaces toolbar, or click Insert > Surface > Trim.   In the PropertyManager, under Trim Type, select Standard. Under Selections, choose sketch CCC in the Feature Manager Design Tree or in the Graphics Area. Select the Remove selections Radio Button and in the graphics area, select the area shown in purple in the following image. Under Surface Split Options, uncheck the Split all Checkbox and select the Natural Radio Button.

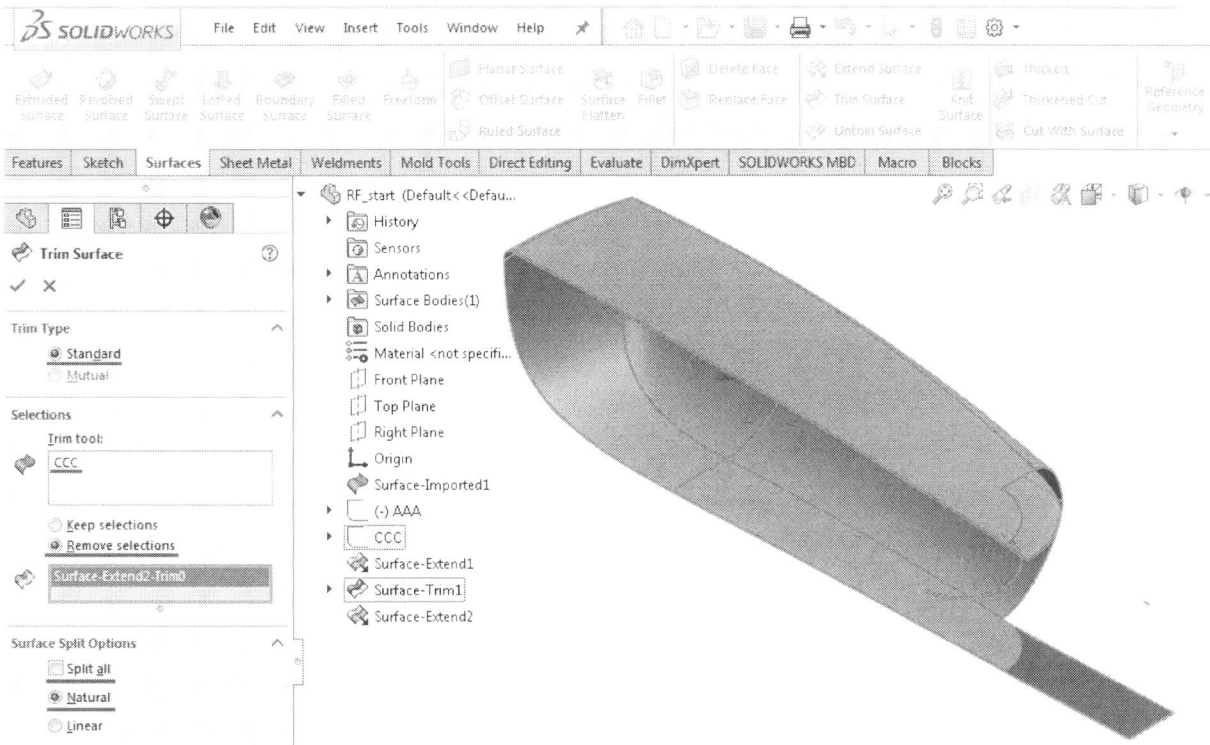

Figure 43 - Trimming a surface using a sketch

Click OK and your part should now appear as shown in the following image.

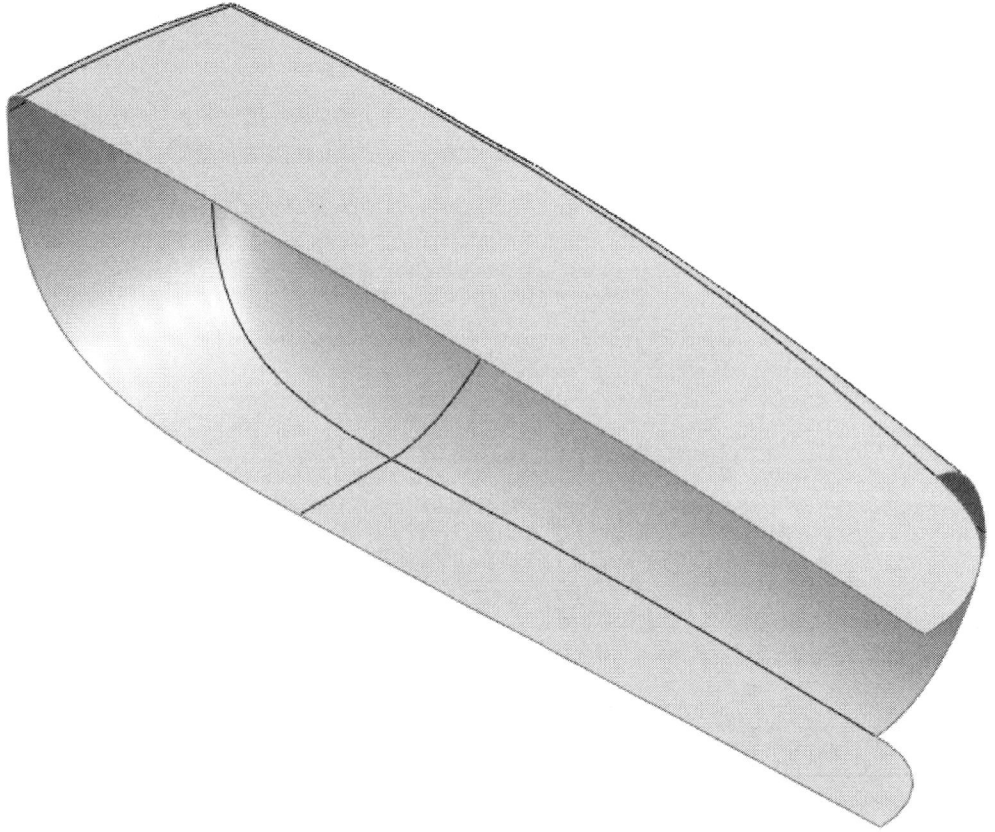

Figure 44 - Surface Model current status

MEASURING THE AREA OF A SURFACE

In the Graphics Area, click on the surface shown in the following image.

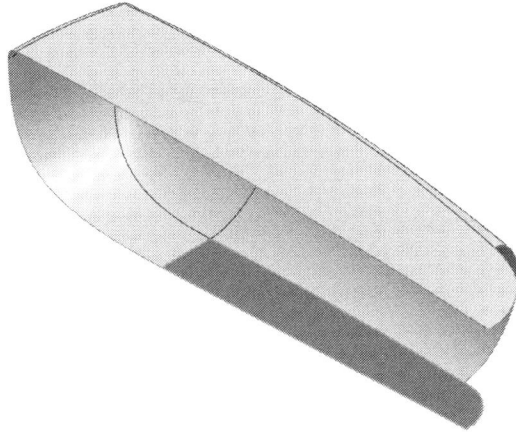

Figure 45 - Select a surface on the Surface Model

While the surface is still highlighted, Click Measure (Tools toolbar) or Tools > Evaluate > Measure. The Measure Dialog Box appears as shown in the following image. The surface area is underlined in the following image and thus the answer is 862.67mm^2. A +/- 1.0mm answer may be acceptable. Save your part.

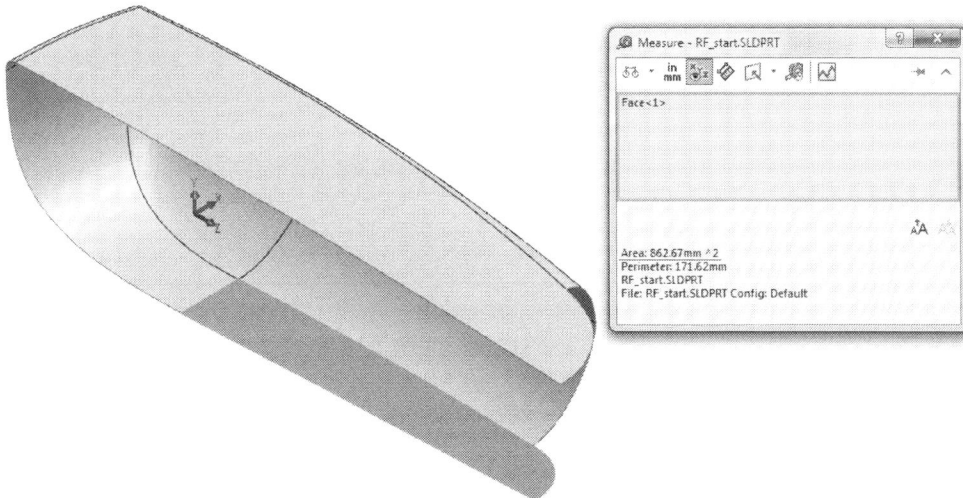

Figure 46 - Chapter Two Answer - Measuring Surface Area

Before we continue to Chapter 3, let us have a closer look at the Trim Surface PropertyManager options since we have already used the Surface Trim control twice already now.

In the Trim Surface PropertyManager Under: -

Trim Type:

1. Standard - Trims surfaces using surfaces, sketch entities, curves, planes, and so on.

2. Mutual - Trims multiple surfaces using the surfaces themselves.

Under Selections:

1. Trim tool (available with Standard selected under Trim Type). Select a surface, sketch entity, curve, or plane in the graphics area as the tool that trims other surfaces.

2. Surfaces (available with Mutual selected under Trim Type). Select multiple surfaces in the graphics area for Trimming Surfaces to use to trim themselves.

Trim action:

1. Keep selections - Retains the surfaces listed under Pieces to Keep. Intersecting surfaces not listed under Pieces to Keep are discarded.

2. Remove selections. Discard the surfaces listed under Pieces to Remove. Intersecting surfaces not listed under Pieces to Remove are retained.

Select surfaces in Pieces to Keep or in Pieces to Remove, based on the trim action.

Under Surface Split Options, select an item:

1. Natural - Boundaries extend tangent from the ends of the Trim tool.

2. Linear - Boundaries extend from the Trim tool endpoints to the nearest edge.

Optionally, select Split All to display all splits in the graphics area.

Open the downloaded part in the Chapter 2 Folder named STANDARD TRIM.sldprt to further explore and understand the standard trim tool. The Part contains two surfaces, three sketch entities, a curve and one non standard plane.

The part looks as shown in the following image.

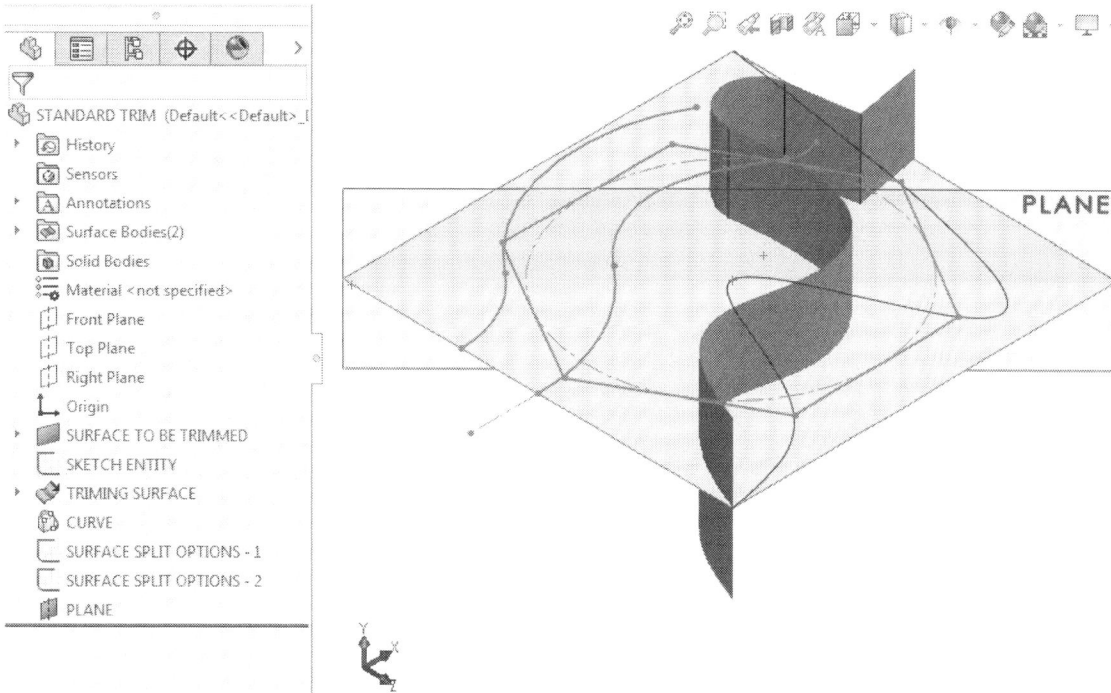

Figure 47 - Understanding the standard trim split options - STANDARD TRIM.sldprt

TRIMMING A SURFACE USING A SKETCH

Click Trim Surface on the Surfaces toolbar, or click Insert > Surface > Trim. In the PropertyManager, under Trim Type, select Standard. Under Selections, choose the sketch named SKETCH ENTITY in the Feature Manager Design Tree or in the Graphics Area. Select the Remove selections Radio Button and in the graphics area, select the area shown in purple in the following image. Under Surface Split Options, uncheck the Split all Checkbox and select the Natural Radio Button.

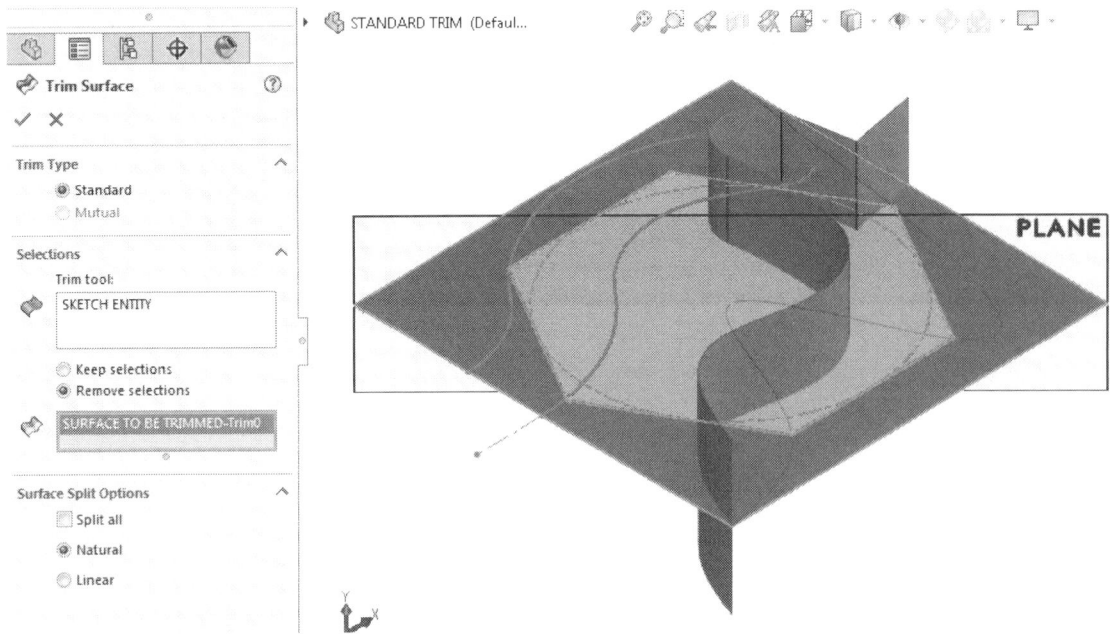

Figure 48 - Understanding the standard trim split options - STANDARD TRIM.sldprt

In this case, however changing the split options have no effect since our trim tool (SKETCH ENTITY) is a closed entity that completely fits on the surface to be cut. You may try changing the split option from Natural to Linear or split all and you will notice there is no effect to the end result. Click OK and your part should now look as shown in the following image.

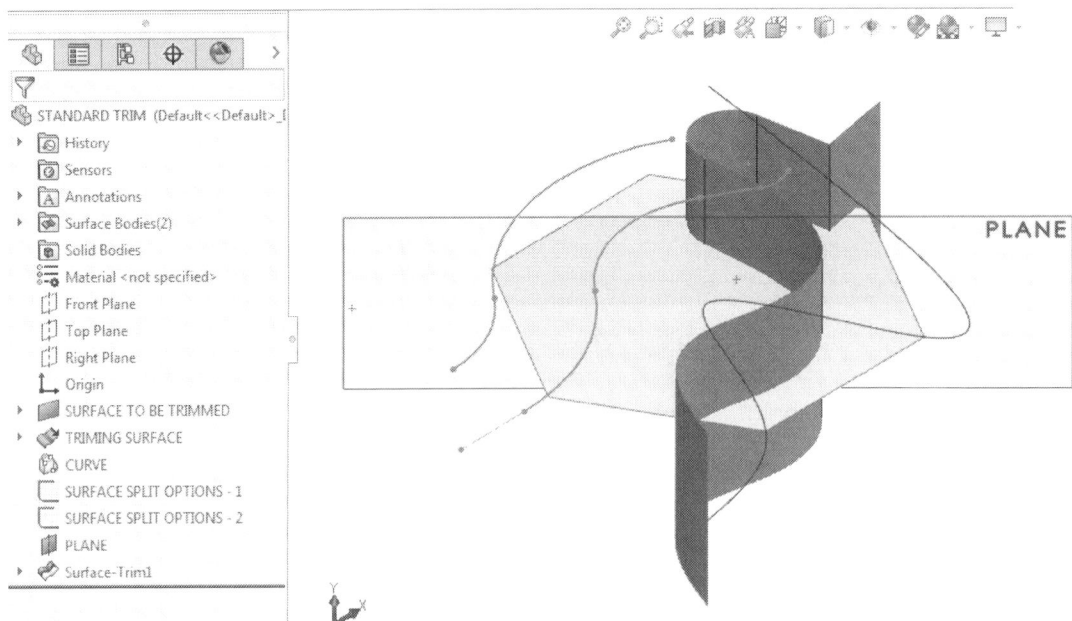

Figure 49 - Understanding the standard trim split options

Now suppress the trim feature we just created and lets run the Trim command again but this time we select the sketch named SURFACE SPLIT OPTIONS SKETCH-1 as our trim tool.

Click Trim Surface on the Surfaces toolbar, or click Insert > Surface > Trim. In the PropertyManager, under Trim Type, select Standard. Under Selections, choose the sketch named SURFACE SPLIT OPTIONS - 1 in the Feature Manager Design Tree or in the Graphics Area. Select the Remove selections Radio Button and in the graphics area, select the area shown in purple in the following image. Under Surface Split Options, uncheck the Split all Checkbox and select the Natural Radio Button. Change the Surface Split Options from Natural to Linear and observe the difference which is summarized in the following two images.

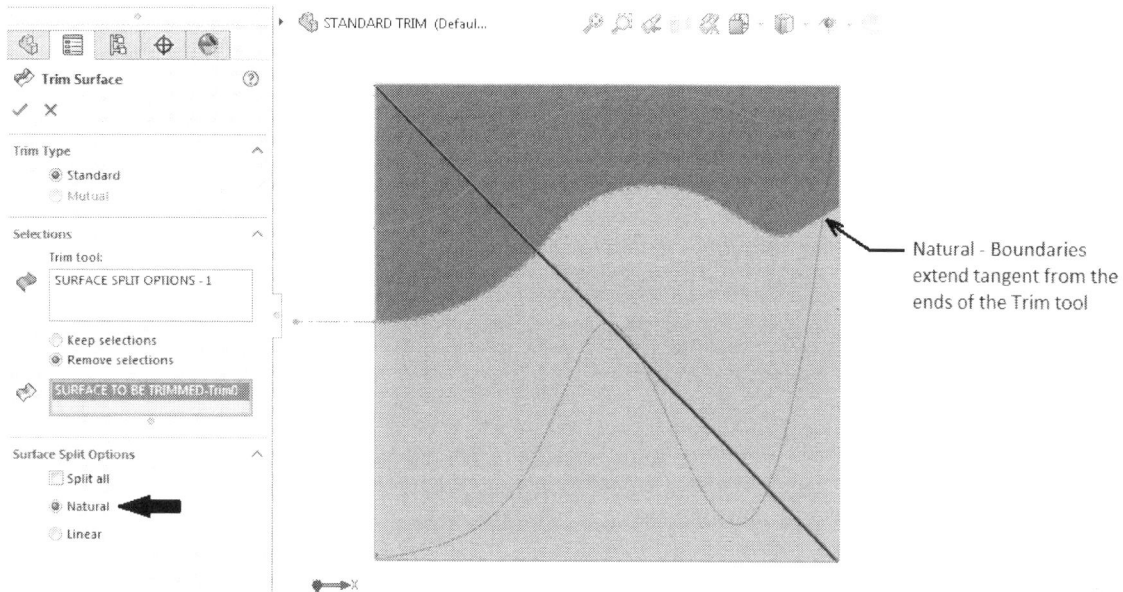

Figure 50 - Understanding the standard trim split options - NATURAL

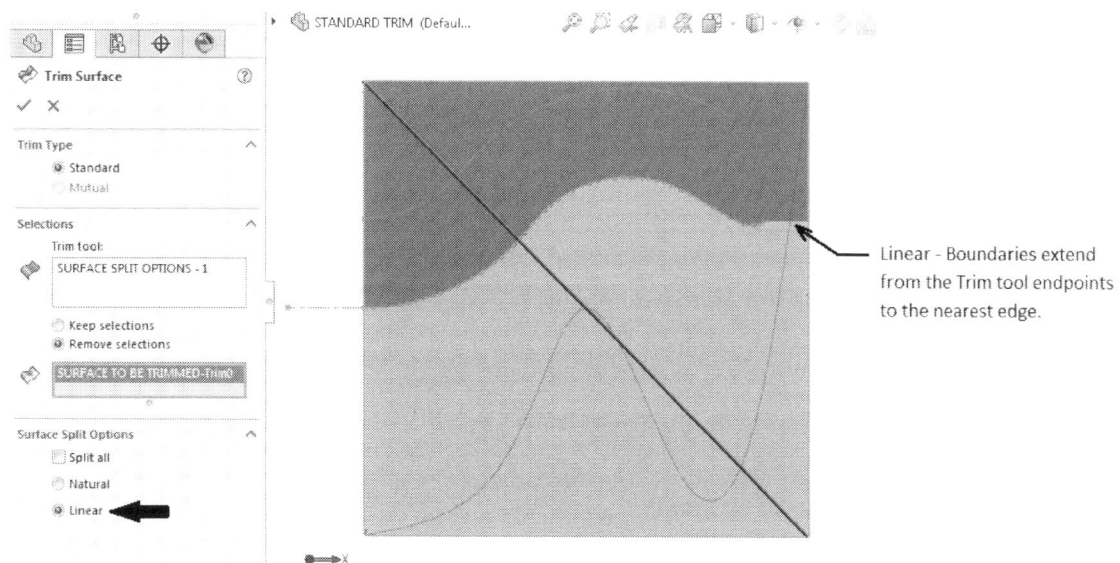

Figure 51 - Understanding the standard trim split options - LINEAR

31

Click Ok and your part should now look as shown in the following image.

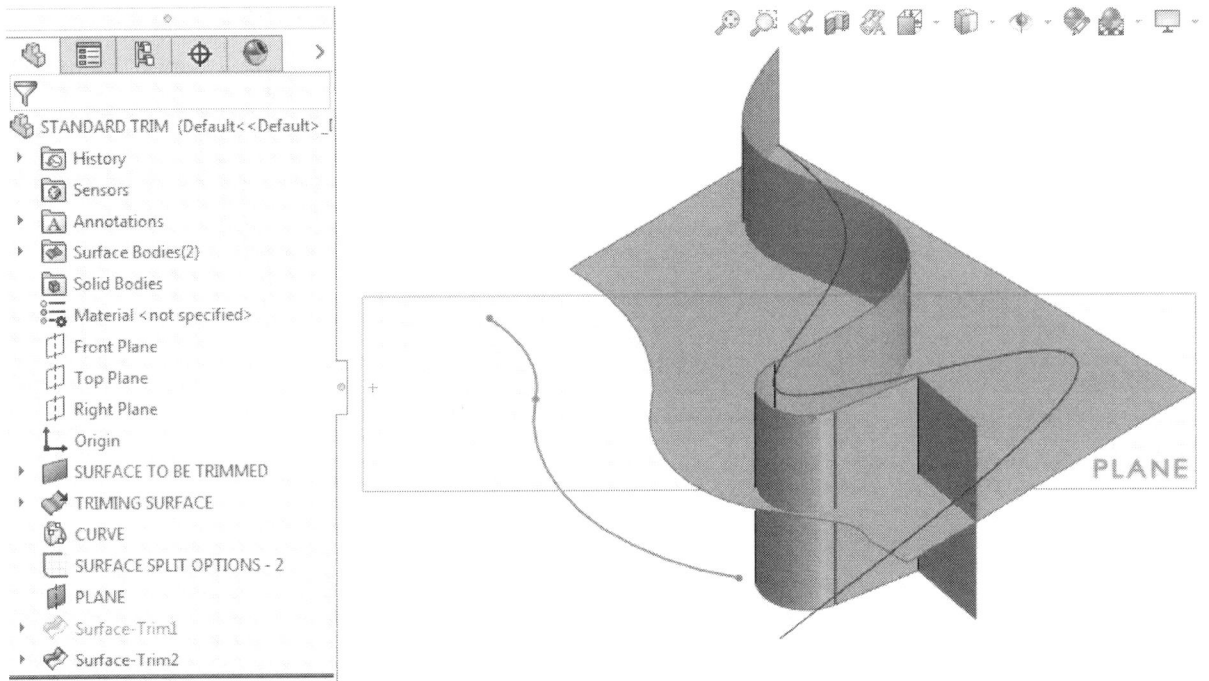

Figure 52 - Understanding the standard trim split options - LINEAR

Now suppress the trim feature we just created and play around with the Trim command again using the SURFACE SPLIT OPTIONS - 2 sketch, CURVE, TRIMING SURFACE AND PLANE respectively as your trim tool and changing the Split Options in each case to observe the different effect these split options have. Split All brings both the natural and linear split options into effect - select and deselect the Split All checkbox as well and observe how it works.

Last but not least, let's have a look at the Mutual Trim Option.

MUTUAL TRIM EXAMPLE ONE

A Mutual Trim trims multiple surfaces using the surfaces themselves. Open the downloaded part MUTUAL TRIM.sldprt or download it from this Google drive location - *http://bit.ly/CSWPA-SU* or Scan the QR Code shown below:

If you experience any problems with downloading any files you may send an email to *cswpasmebook@gmail.com* with the title of the book indicated in your email subject. Open and save the downloaded part to your PC.

The part looks as shown in the following image:

Figure 53 - MUTUAL TRIM.sldprt downloaded part

Click Trim Surface on the Surfaces toolbar, or click Insert > Surface > Trim.

In the PropertyManager, under Trim Type, select Mutual.

Under Selections, Select two surfaces - Surface-Extrude1 and Surface-Extrude4 in the graphics area or in the Feature Manager Design Tree for Trimming Surfaces to use to trim themselves as shown in the following image.

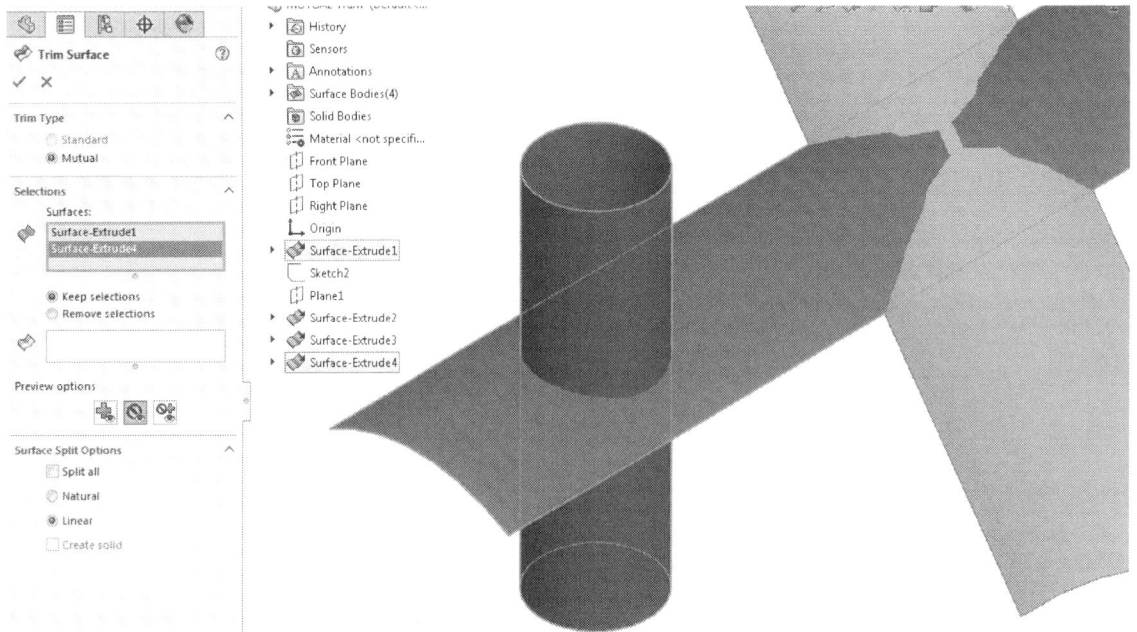

Figure 54 - Mutual Trim

Select a trim action:

Keep selections - retains the surfaces listed under Pieces to Keep. Intersecting surfaces not listed under Pieces to Keep are discarded.

Remove selections - discards the surfaces listed under Pieces to Remove. Intersecting surfaces not listed under Pieces to Remove are retained.

In this example, we will select Keep selections and Select surfaces in Pieces to Keep as shown in the following image.

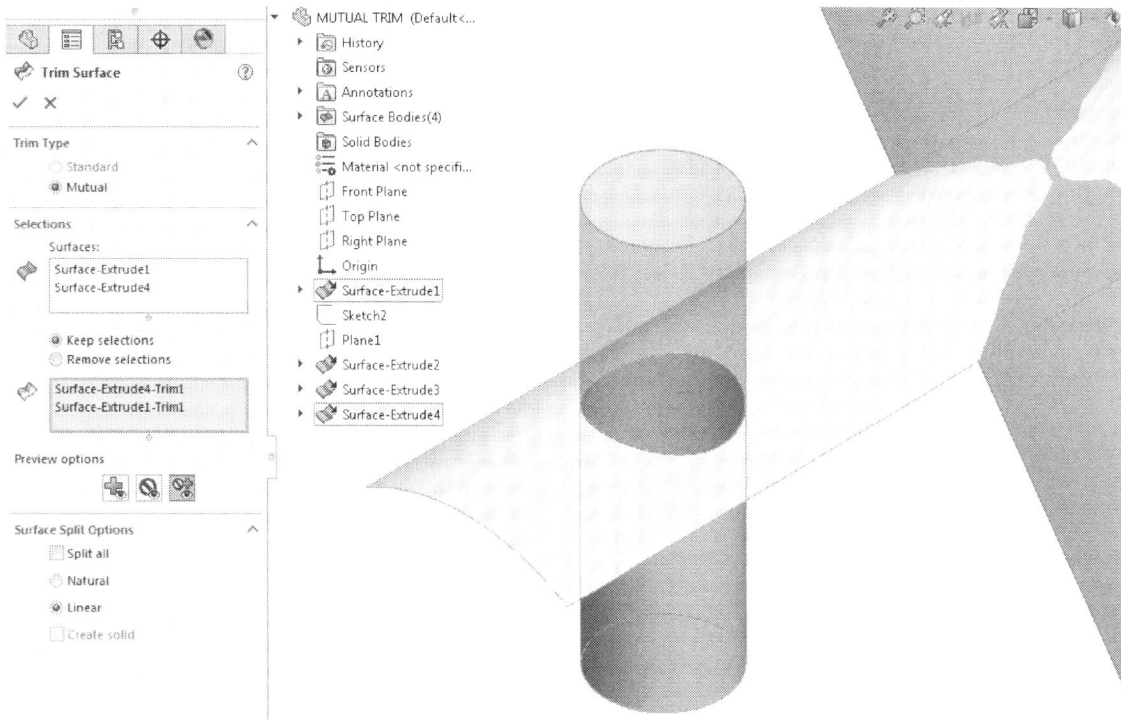

Figure 55 - Mutual Trim - Trim Action

MUTUAL TRIM PREVIEW OPTIONS

Preview options are available when using the Mutual Trim Surface Control. Under Preview Options, Select one of the following and observe the result in the Graphics Area as listed below and shown in the following images:

1. Show included surfaces - Displays the regions that you are including as surfaces. All other surfaces are hidden.

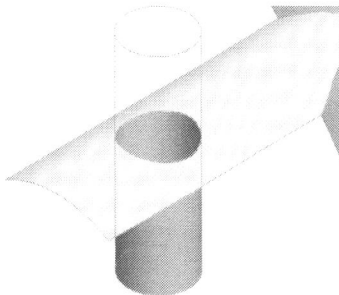

Figure 56 - Mutual Trim Preview Options - Show Included Surfaces

2. Show excluded surfaces - Displays the surfaces that you are excluding as transparent. All other surfaces are hidden.

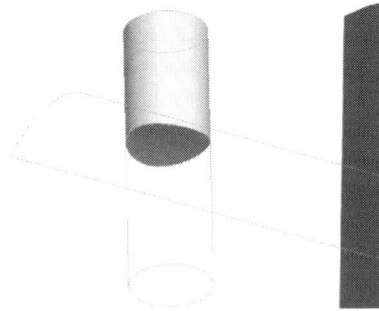

Figure 57 - Mutual Trim Preview Options - Show Excluded Surfaces

3. Show both included and excluded surfaces - Displays both included and excluded surfaces. Excluded surfaces appear transparent.

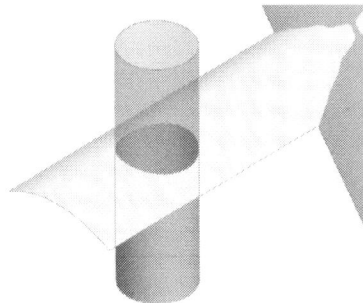

Figure 58 - Mutual Trim Preview Options - Show both Included and Excluded Surfaces

Under Surface Split Options, select any option - Natural or Linear since there is really no difference in this particular example. Optionally, select Split All. Notice that when you change the Surface Split Option you have to reselect the pieces to keep - you may also have to change the preview option if the selections become hidden before you can proceed to select the pieces to keep.

Click Ok.

Save your part. Your part should now look as shown in the following image.

Figure 59 - Mutual Trim Result

As you might have noticed, the number of Surface Bodies in the Surface Bodies Folder in the Feature Manager Design Tree have changed from 4 Off to 3 Off because the Mutual Trim Control automatically knits or combine the kept surfaces into one body.

MUTUAL TRIM EXAMPLE TWO

Click Trim Surface on the Surfaces toolbar, or click Insert > Surface > Trim.

In the PropertyManager, under Trim Type, select Mutual.

Under Selections, Select two surfaces - Surface-Trim1 and Surface-Extrude2 in the graphics area or in the Feature Manager Design Tree for Trimming Surfaces to use to trim themselves as shown in the following image.

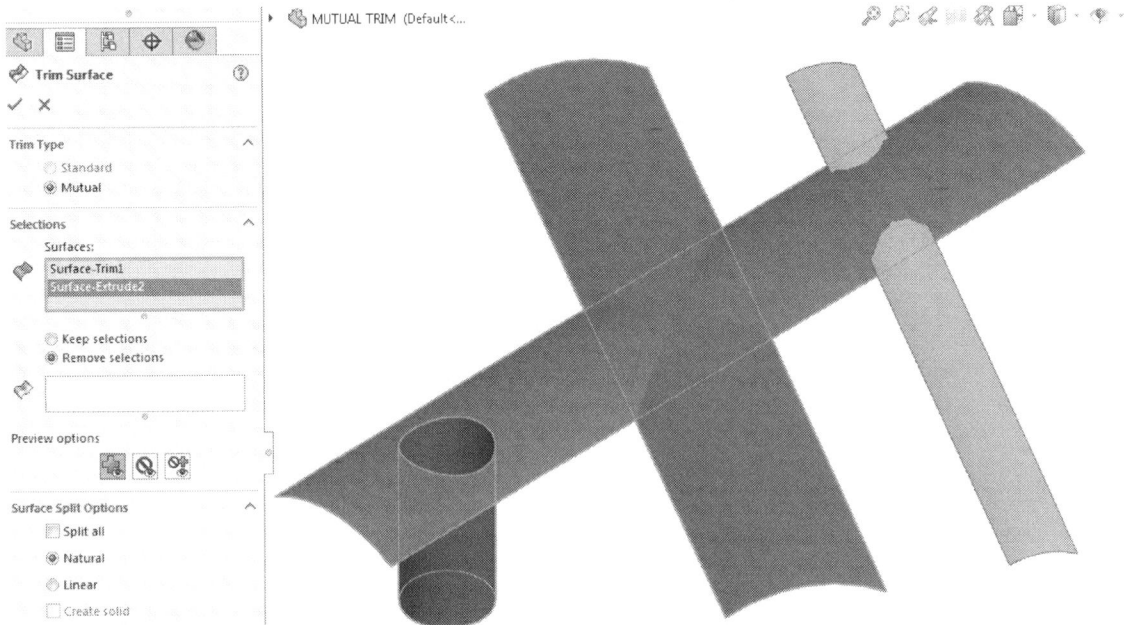

Figure 60 - Mutual Trim - Selections

Select a trim action:

In this example, we will select Remove selections and Show Excluded Surfaces [icon] under Preview Options then select surfaces in Pieces to Remove as shown in the following image.

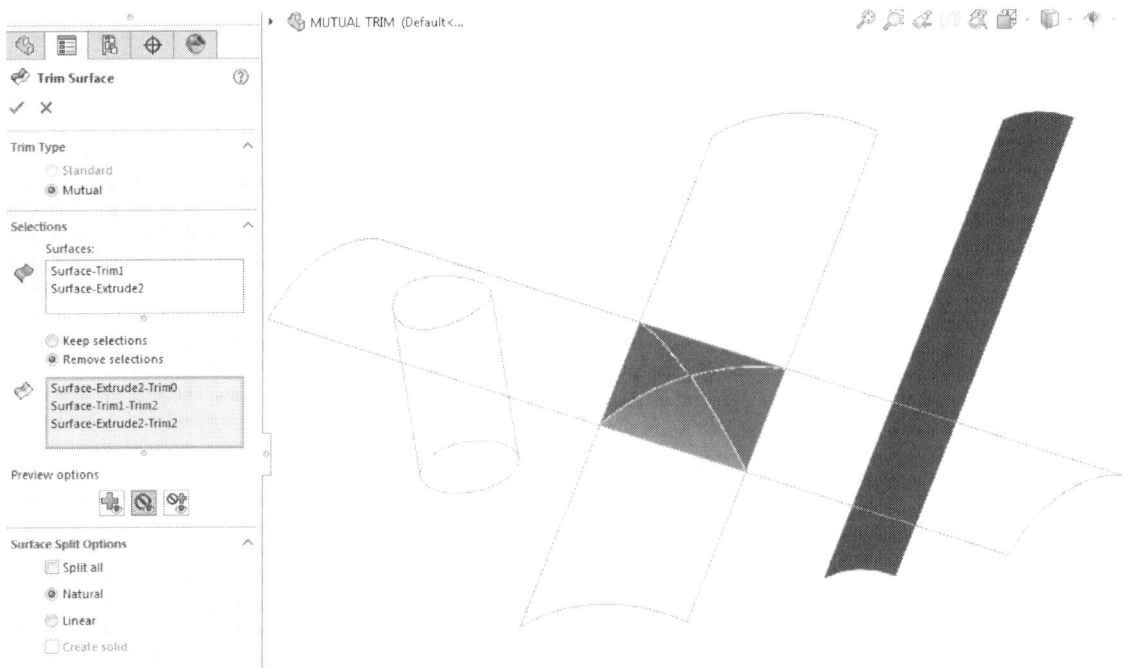

Figure 61 - Mutual Trim - Remove Selections - Show Excluded Surfaces

Click Ok and your part should now look as shown in the following image. Save your part.

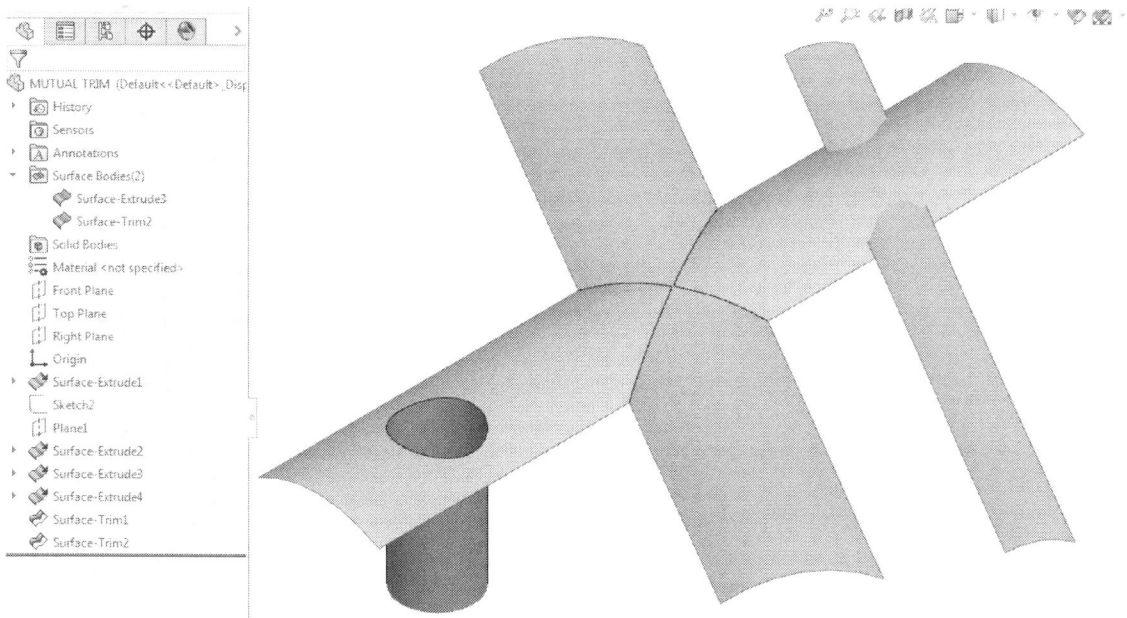

Figure 62 - Mutual Trim Result

MUTUAL TRIM EXAMPLE THREE - SURFACE SPLIT OPTIONS

Click Trim Surface on the Surfaces toolbar, or click Insert > Surface > Trim. In the PropertyManager, under Trim Type, select Mutual. Under Selections, Select two surfaces - Surface-Extrude3 and Surface-Trim2 in the graphics area or in the Feature Manager Design Tree for Trimming Surfaces to use to trim themselves as shown in the following image.

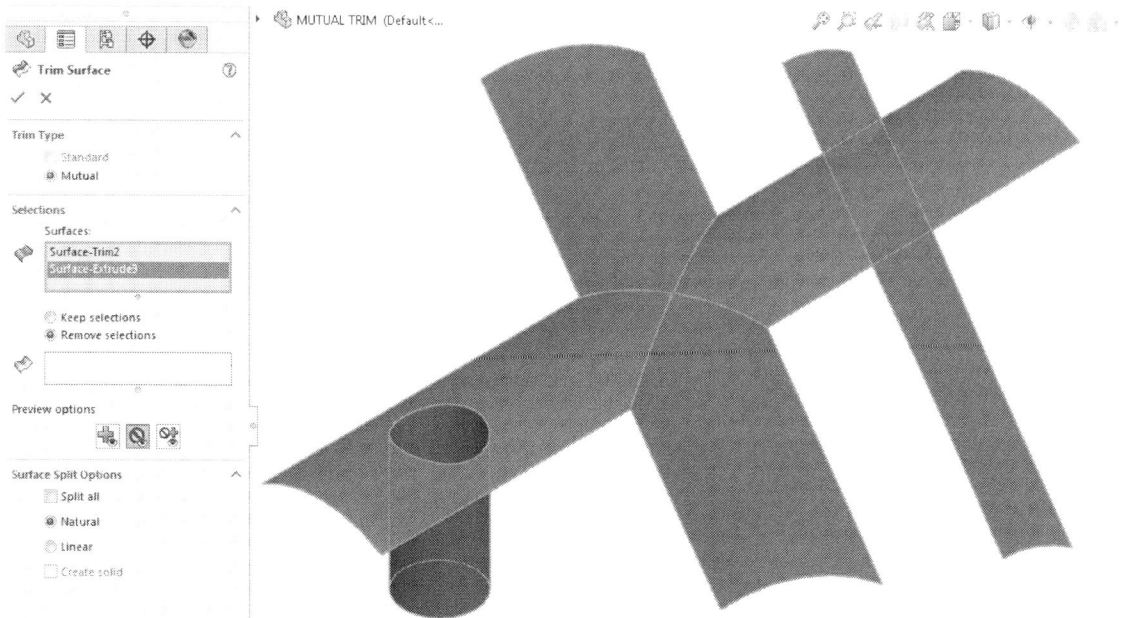

Figure 63 - Mutual Trim - Selections

Select a trim action:

In this example, we will select Remove selections and Show Excluded Surfaces ![icon] under Preview Options. Select Natural under Surface Split Options then select surfaces in Pieces to Remove as shown in the following image.

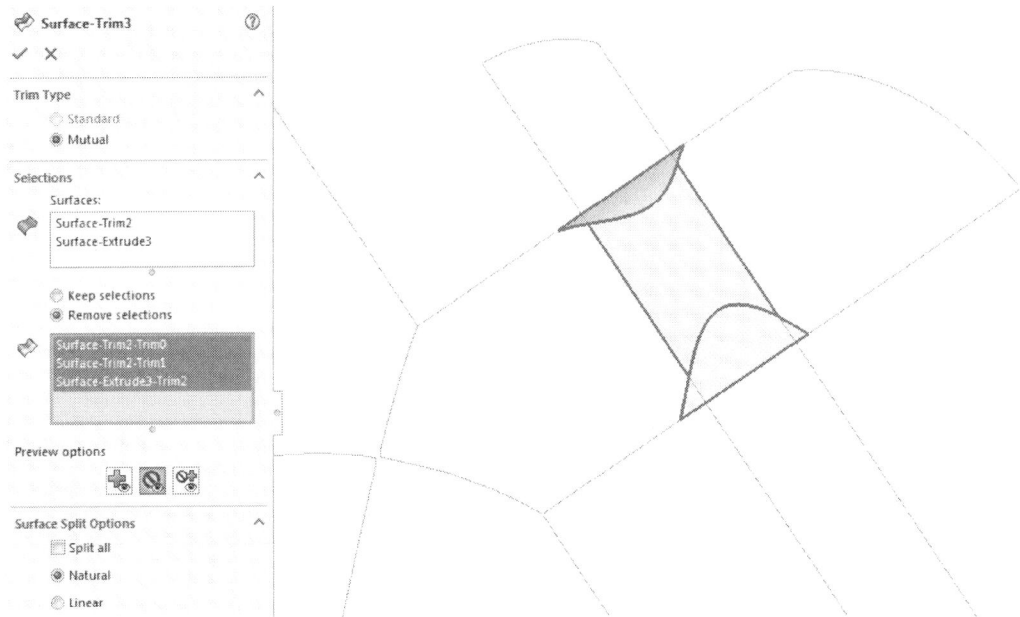

Figure 64 - Mutual Trim - Surface Split Options - Natural

Notice how the Natural Split Option's boundaries extend tangent from the ends of the surface Surface-Extrude3 Trim tool.

Click OK. Save your part. Your part should now look as shown in the following image.

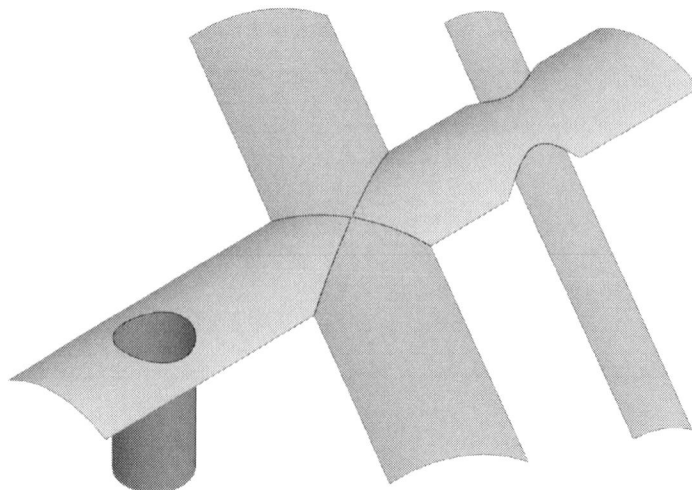

Figure 65 - Mutual Trim - Surface Split Options - Natural - Result

Suppress the Surface Trim Command we executed above - Surface-Trim3 in the Feature Manager Design Tree.

Click Trim Surface on the Surfaces toolbar, or click Insert > Surface > Trim. In the PropertyManager, under Trim Type, select Mutual. Under Selections, Select two surfaces - Surface-Extrude3 and Surface-Trim2 in the graphics area or in the Feature Manager Design Tree as shown in the following image.

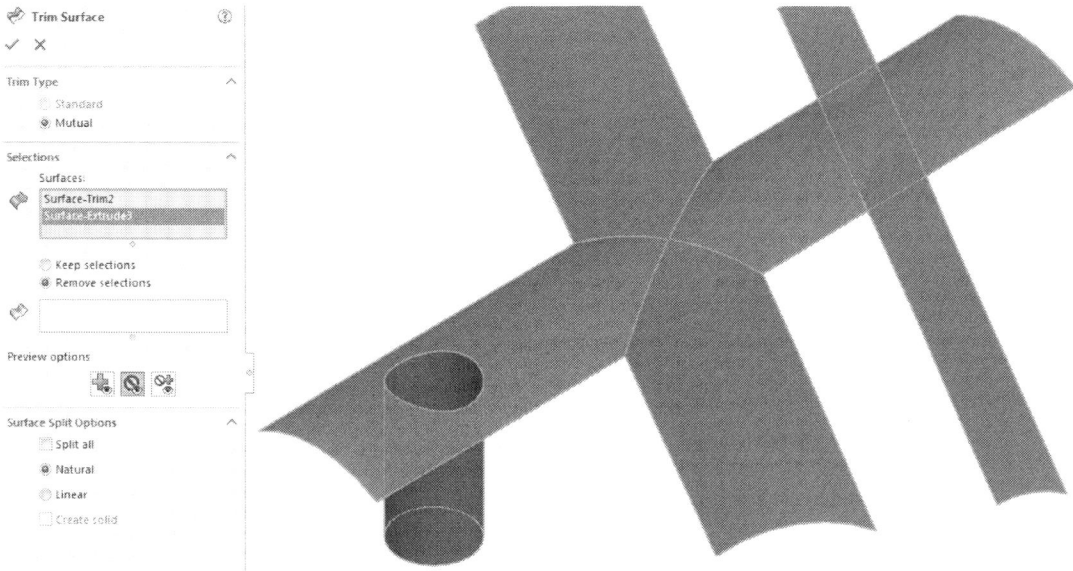

Figure 66 - Mutual Trim - Selections

Select a trim action: In this example, we will select Remove selections and Show Excluded Surfaces 🚫 under Preview Options. Select Linear under Surface Split Options then select surfaces in Pieces to Remove as shown in the following image.

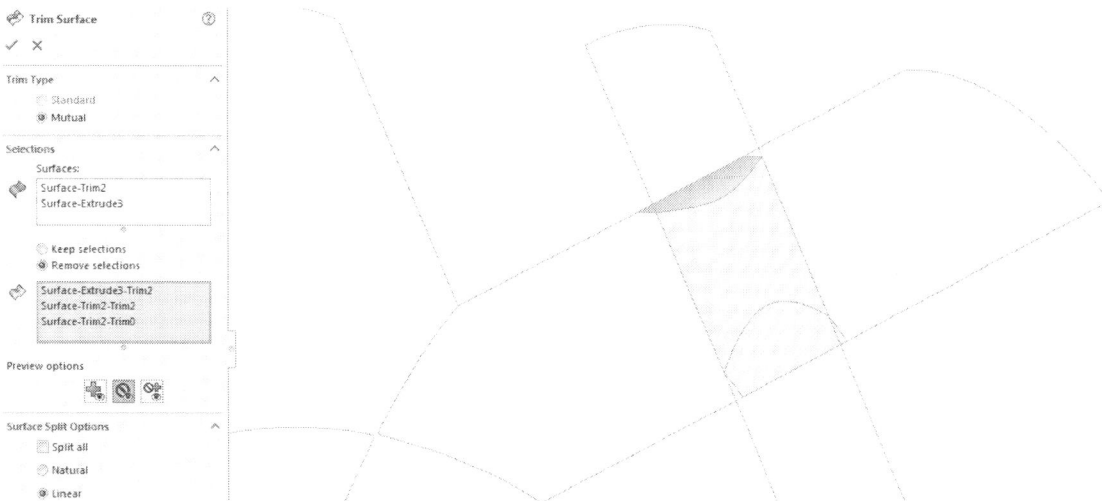

Figure 67 - Mutual Trim - Selections

Notice how the Linear Split Option's boundaries extend from the Trim tool Surface-Extrude3 endpoints to the nearest edge.

Click OK. Save your part. Your part should now look as shown in the following image.

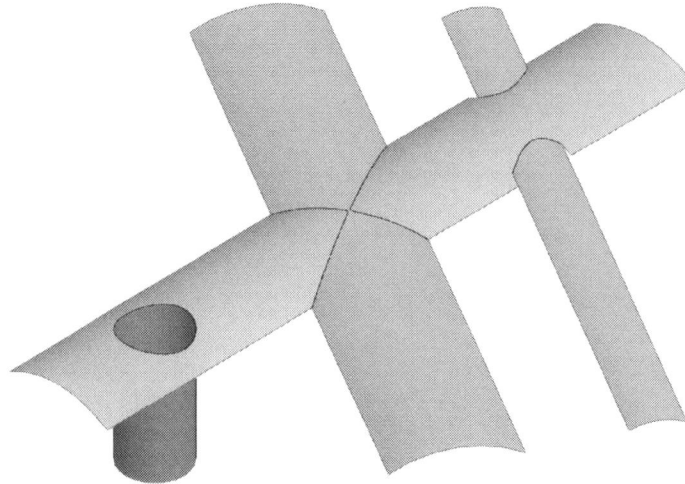

Figure 68 - Mutual Trim - Surface Split Options - Linear - Result

Suppress the Surface Trim Command we executed above - Surface-Trim4 in the Feature Manager Design Tree.

Click Trim Surface on the Surfaces toolbar, or click Insert > Surface > Trim. In the PropertyManager, under Trim Type, select Mutual. Under Selections, Select two surfaces - Surface-Extrude3 and Surface-Trim2 in the graphics area or in the Feature Manager Design Tree as shown in the following image.

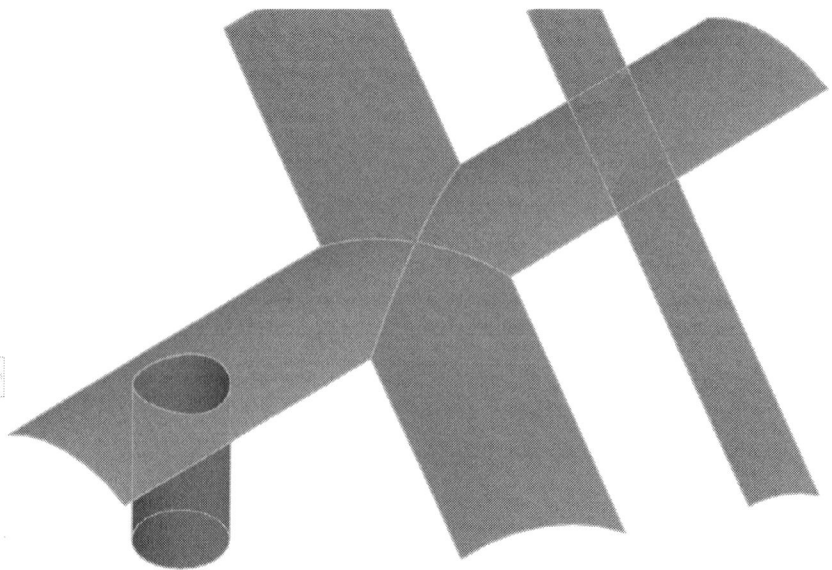

Figure 69 - Mutual Trim - Selections

42

Select a trim action: In this example`, we will select Remove selections and Show Excluded
Surfaces ![icon] under Preview Options. Select Split All under Surface Split Options and you will
notice that the surfaces are split using both Linear and Natural Options as well as creating other
segments unique to the Split All Option as shown in the following image.

Figure 70 - Mutual Trim - Split All

You may play around with this option to see how it's splitting the surfaces. In practice you
usually have to start with either Natural or Linear then change the options depending on the
result you are looking for. Press the escape key on your keyboard. Save and close the part
MUTUAL TRIM.sldprt then move on to Chapter 3.

In this Chapter, we will continue working with the same part - *RF_Start.sldprt* from Chapter Two.

You are required to recreate the missing portion of the parting line shown in the following image using 2D Sketches, 3D Sketches or 3D curves as well as the included sketch pictures. You are then required to measure the length of this recreated portion of the parting line and select a range that contains the value of this measured length from four off options which are: -

1. 39.45 TO 41.35

2. 41.50 TO 43.35

3. 37.21 TO 39.11

4. 34.95 TO 36.98

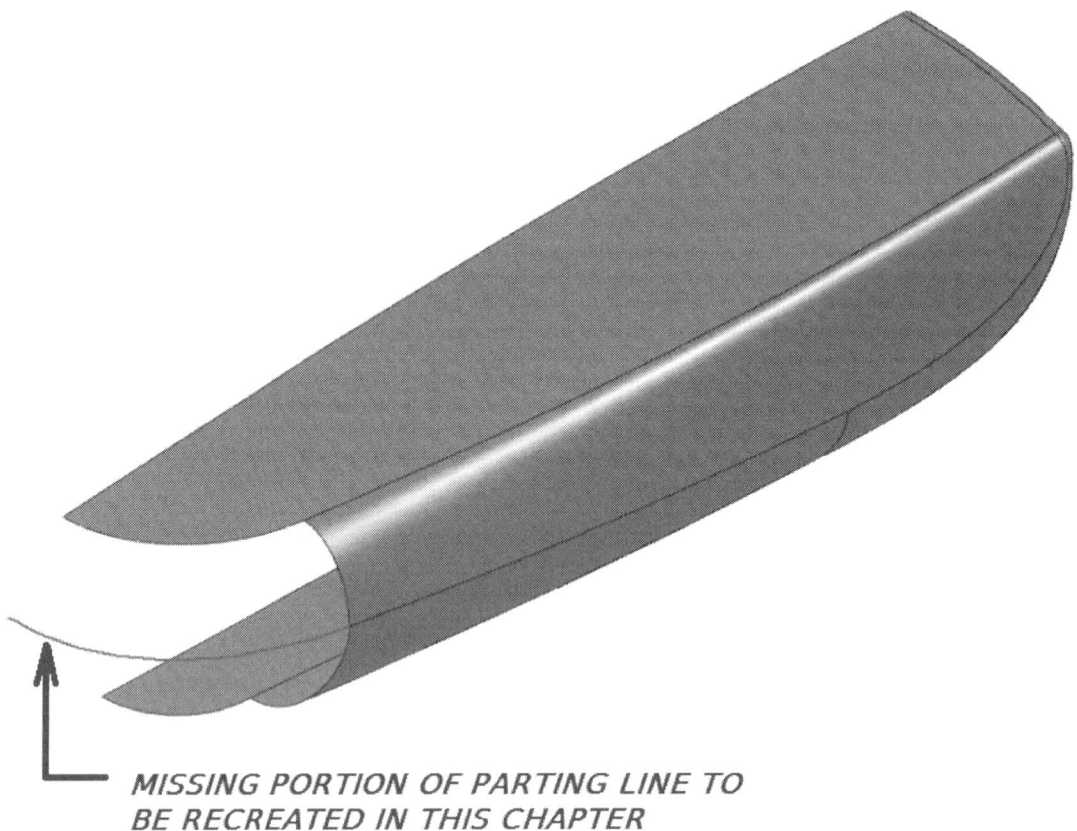

MISSING PORTION OF PARTING LINE TO
BE RECREATED IN THIS CHAPTER

Figure 71 - Recreating the missing portion of a parting line

CREATING SKETCH ENTITIES USING A SKETCH PICTURE

Click or Right Click on sketch AAA in the Feature Manager Design Tree and select Edit Sketch as shown in the following image.

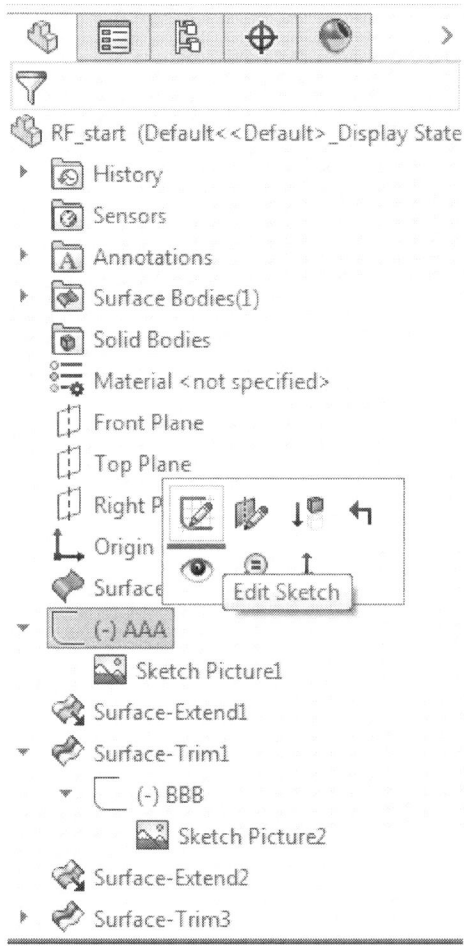

Figure 72 - Editing an Existing Sketch

In the Sketch Mode, zoom to the area shown in the following image.

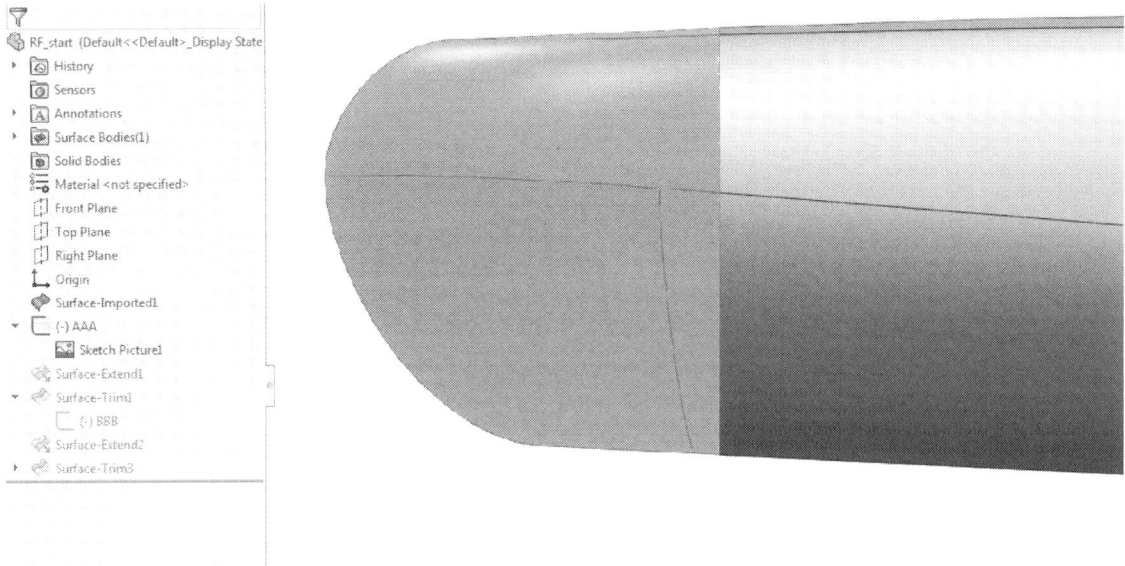

Figure 73 - Creating a sketch using a sketch picture

SKETCHING A STYLE SPLINE

Click Style Spline (Sketch toolbar) or Tools > Sketch Entities > Style Spline.

In the graphics area, zoom to the end point as shown in the following image and click to create the first point of the spline as shown in the following image. Press the Ctrl button on your keyboard before you click on the point to make sure no Automatic Relation is added as you create this first point.

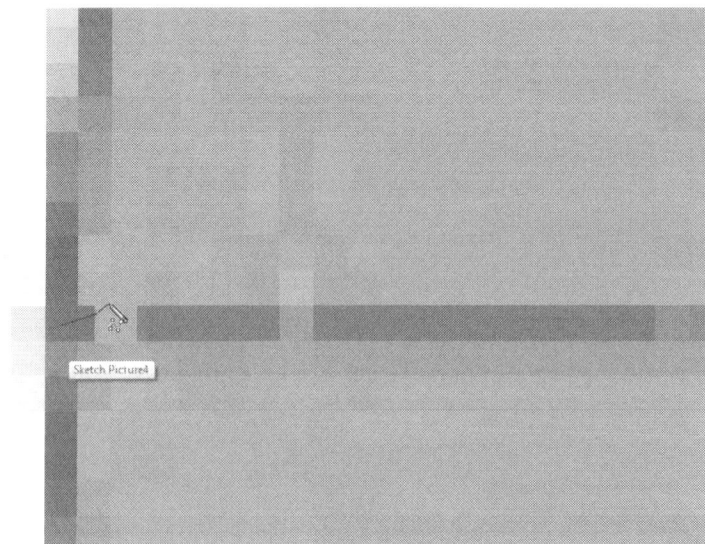

Figure 74 - Sketching a Style Spline

46

Use your mouse scroll wheel or button to zoom out and click to add a second control vertex point as shown in the following image.

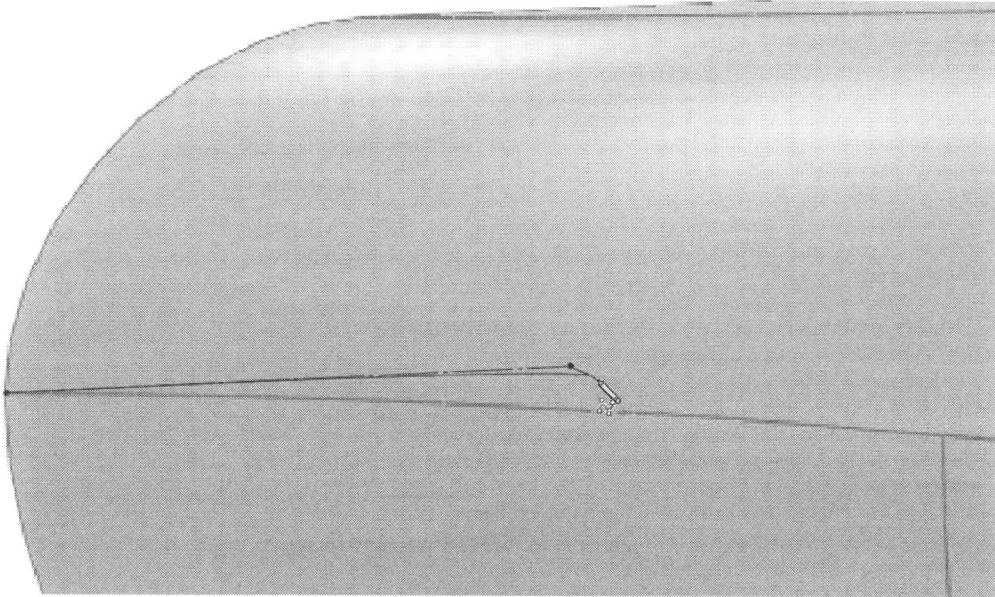

Figure 75 - Sketching a Style Spline

Use your mouse scroll wheel or button to zoom out and zoom in and click to add the end point of the Style Spline making sure a coincident relation is automatically added as shown in the following image. Press the Escape key on your keyboard to complete the style spline.

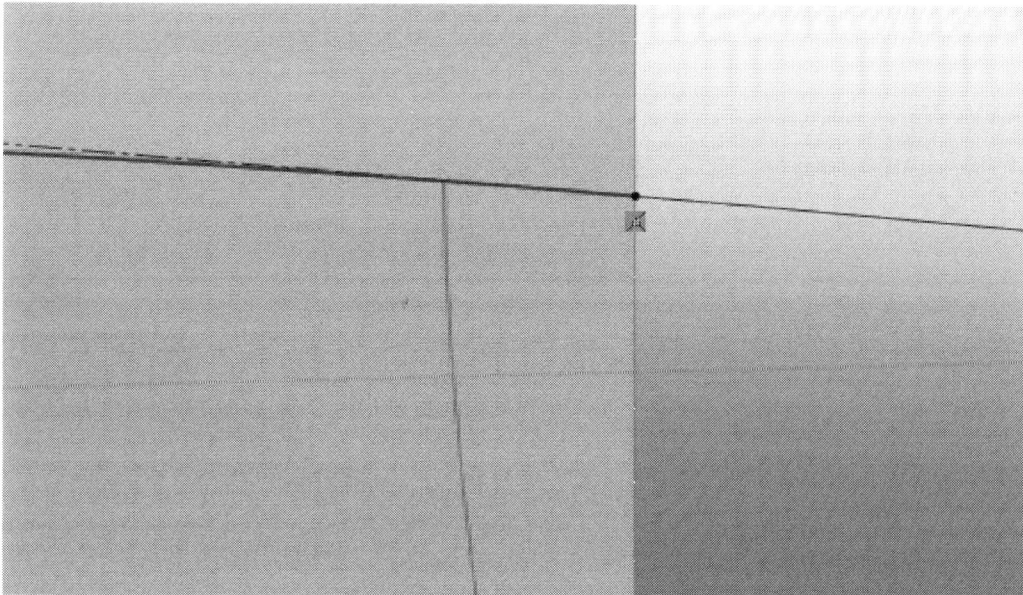

Figure 76 - Sketching a Style Spline

Add a fixed relation first then add a tangent relation between the parting line and the newly created style spline *(see the following image)*. Last but not least, add 10.45mm dimension as shown in the following image to Fully Define the Style Spline.

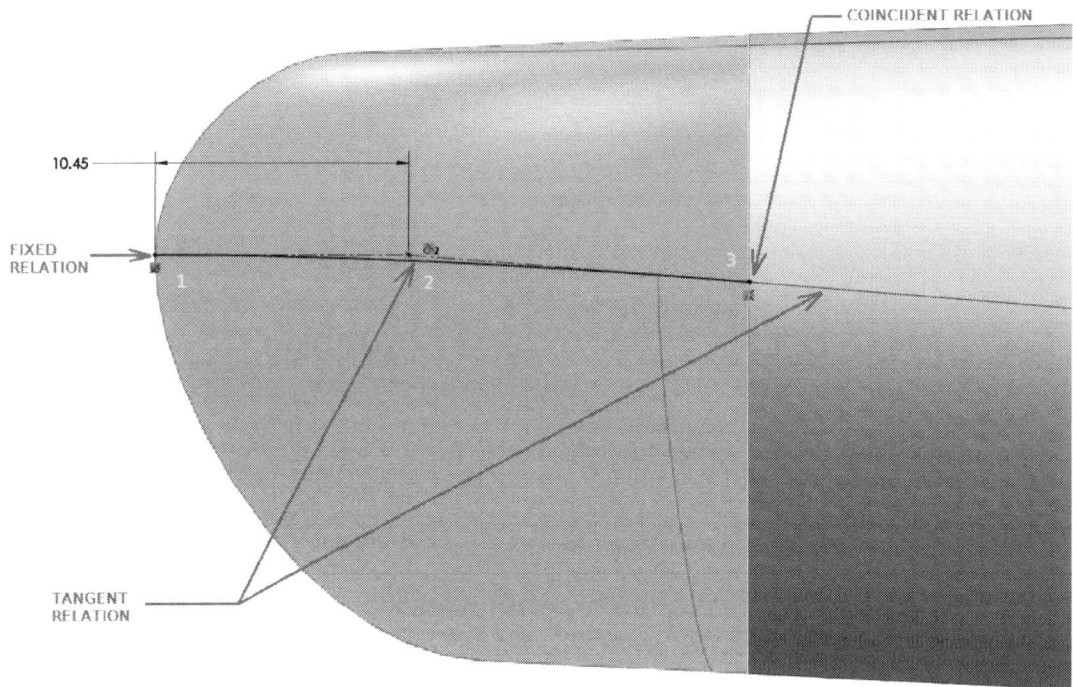

Figure 77 - Fully Defined Style Spline

You may be wondering how we got the 10.45mm dimension. The 10.45mm dimension was arrived at by first manually dragging the control vertex number 2 while zoomed in as close as possible to the sketch picture and then using the Thumbwheel on the Modify Dialog box with a spin increment value of 0.05mm.

USING THE MODIFY DIALOG BOX

If you double click a dimension, the Modify Dialog Box appears as shown in the following image.

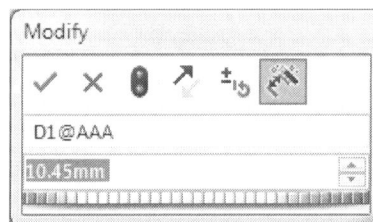

Figure 78 - Modify Dialog Box

On the Modify Dialog Box, click ±'↺ (Reset spin increment value) underlined in the following image.

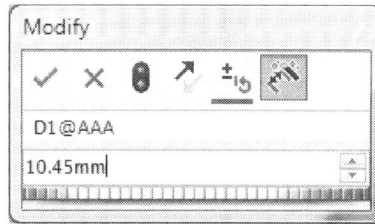

Figure 79 - Modify Dialog Box - Resetting the Spin Increment Value

The Increment Dialog Box appears. Change the increment used by the spin box arrows by entering a value of 0.05mm then press Enter to accept the value. Select Make Default to make your new specification the default spin box increments in System Options as shown in the following image.

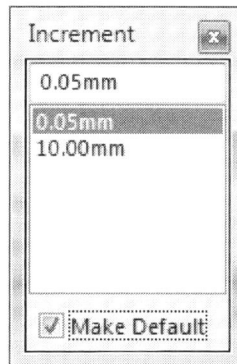

Figure 80 - Increment Dialog Box - Setting an Increment Value

Close the Increment Dialog Box. Click and drag the Thumbwheel to the left or right and you will notice that the dimension is now increasing or decreasing by 0.05mm - the following image shows the thumbwheel.

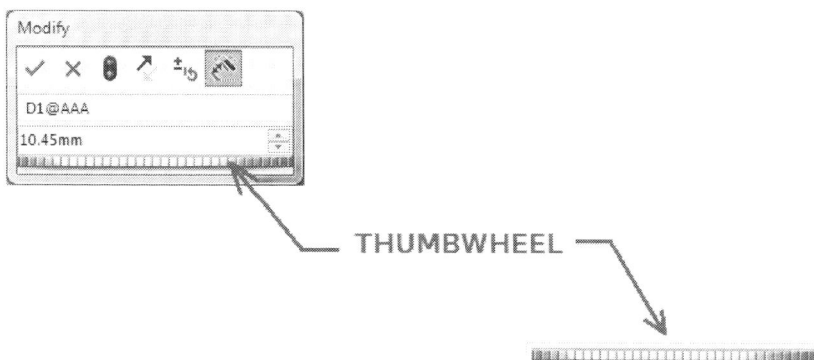

Figure 81 - Modify Dialog Box - Using the thumbwheel to fine tune a dimension

49

Your dimension doesn't have to be 10.45mm but it can be any dimension that brings the Style Spline as close as possible to the line we are tracing on the sketch picture AAA.

If you don't like using the Thumbwheel you may also use the Up / Down Arrows to increase or decrease the dimension by the length increment set in the Spin Box Increments - see the following image.

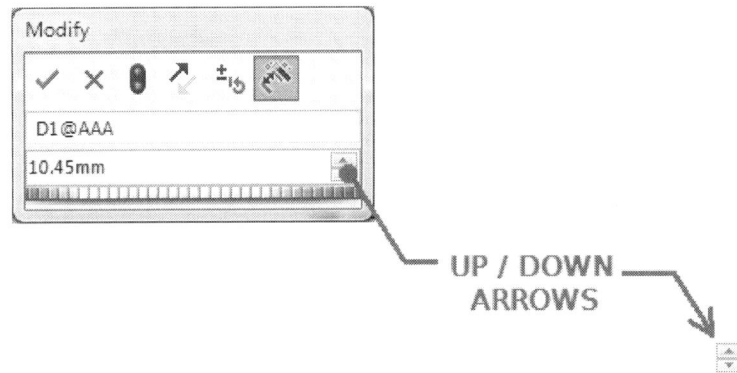

Figure 82 - Modify Dialog Box - Using the UP/DOWN Arrows to fine tune a dimension

NOTE: The Modify dialog box displays a slider as shown in the following image instead of a thumbwheel when the numeric value is bounded, such as an angle that must be greater than 0 and less than 90 degrees.

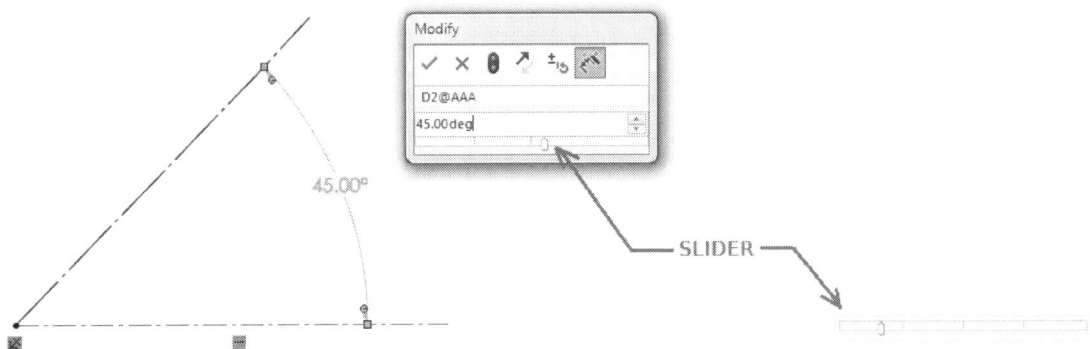

Figure 83 - Modify Dialog Box Slider

Once you are happy with the position of the Style Spline, you may exit the Sketch Mode. Save your part.

CHANGING A SKETCH'S COLOUR

Let's change the colour of the sketch we just created in the section above to make it easier to distinguish this sketch for the purpose of referencing or adding relations to it e.t.c as we create the next sketch in the following section.

Right Click on Sketch AAA in the Feature Manager Design Tree and select Sketch Color as shown in the following image or Click Edit > Appearance > Sketch/Curve Color then select sketch AAA in the Feature Manager Design Tree or in the Graphics Area.

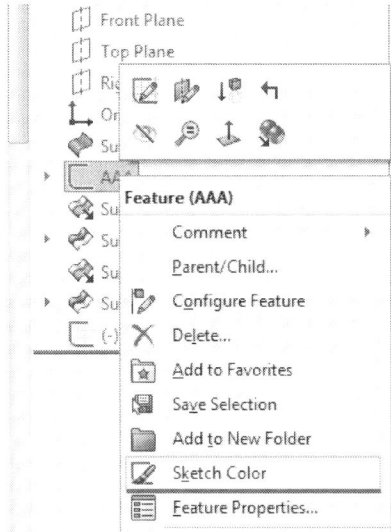

Figure 84 - Changing a sketch colour

The Appearances Property Manager appears and you may pick a color from the Standard Colour Swatch as shown in the following image.

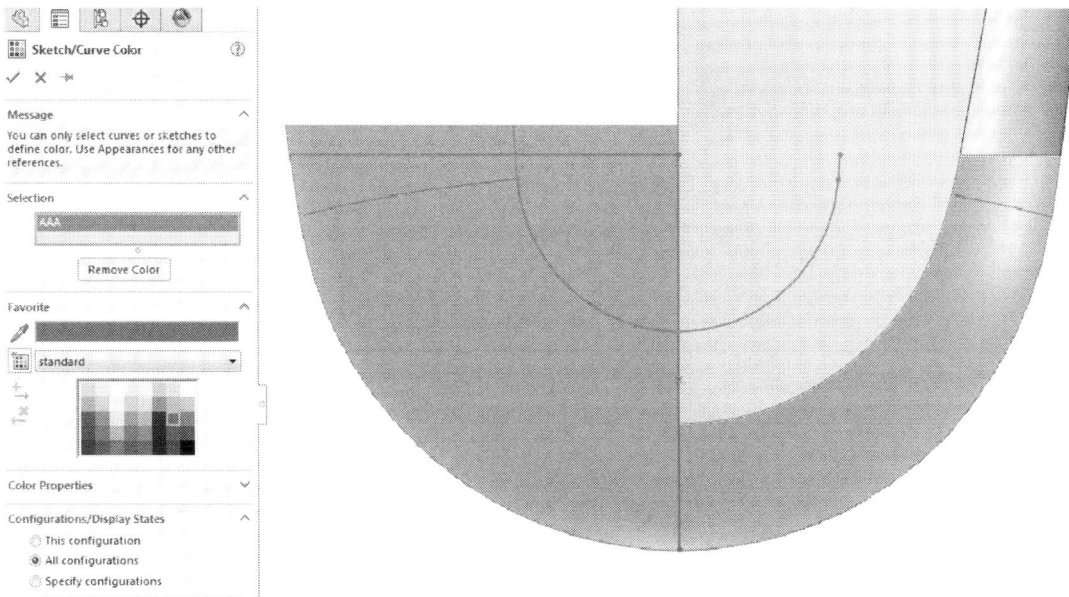

Figure 85 - Changing a sketch colour - Appearances PropertyManager

Click OK to apply the selected colour and close the Appearances PropertyManager. Note that the colour displays only when the sketch is inactive. If the colour does not change, check the Color Display Mode button and make sure it is not selected. To access the Color Display Mode button, Click View > Toolbars > Line Format and the Line Format Toolbar appears as shown in the following image. Selecting and unselecting this Color Display Mode button toggles the colour of edges and sketch entities between their applied line or layer colour and the system status colours. Select and unselect the button by clicking on it and observe the effect on the colour we applied to sketch AAA. Leave the button unselected so that the colour we applied on sketch AAA remains visible. Dock or close the Line Format Toolbar.

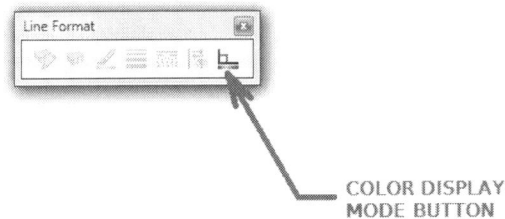

COLOR DISPLAY
MODE BUTTON

Figure 86 - Line Format Toolbar

STARTING A SKETCH ON THE TOP PLANE

Click the Top Plane in the FeatureManager Design Tree and select Sketch as shown in the following image or Click Sketch on the Sketch toolbar or Click Insert > Sketch then select the Top Plane.

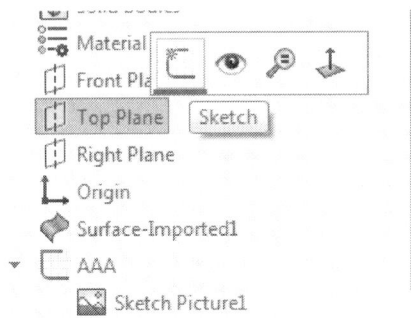

Figure 87 - Starting a sketch on the Top Plane

SKETCHING A 3 POINT ARC

Click 3 Point Arc on the Sketch Toolbar or Click Tools > Sketch Entities > 3 Point Arc. Click to set a start point on position 1 *(see the following image)* in the Graphics Area. Drag the pointer, then click to set an end point at position 2 *(see the following image)*. Zoom in and Drag to set the radius. Click to set the arc radius to as close as possible to the line on the sketch picture which we are tracing - see the following image.

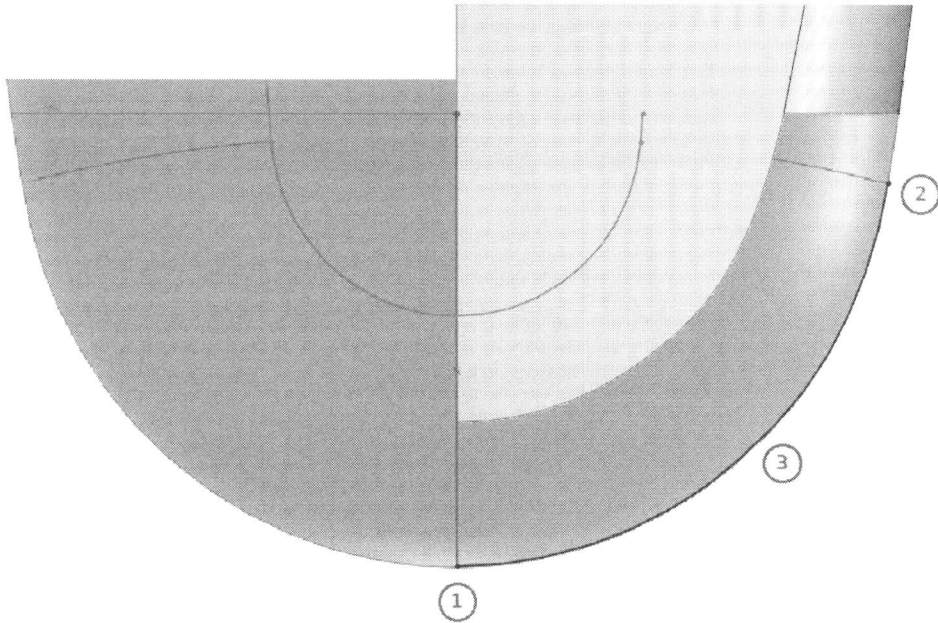

Figure 88 - Sketching a 3 Point Arc - Tracing a Sketch Picture

APPLYING SKETCH RELATIONS

Change your view to Isometric View as shown in the following image.

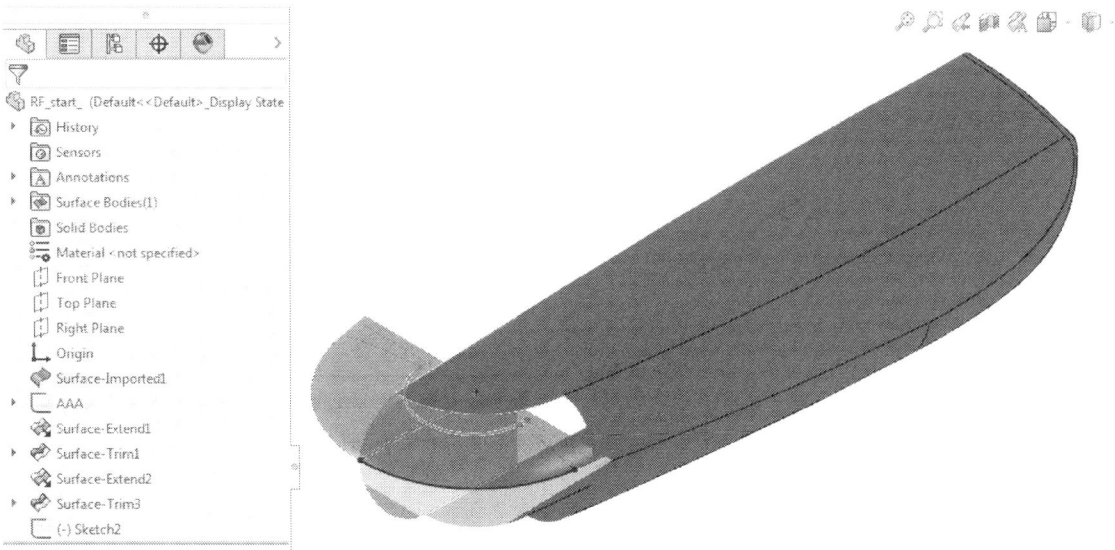

Figure 89 - Sketching - Isometric View

Zoom in to area shown in the following image and add a coincident relation between the end point of the 3 point Arc and Sketch AAA.

53

Figure 90 - Adding a coincident relation

Once you have added the coincident relation, change the view orientation to a Top View as shown in the following image.

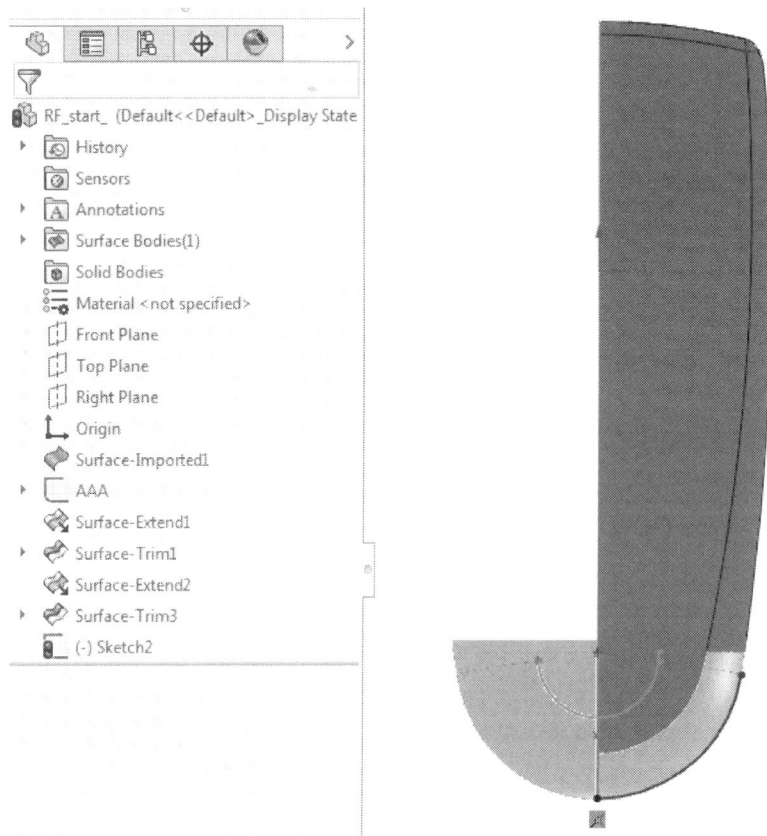

Figure 91 - Top View - Coincident Relation

54

Zoom in to the area shown in the following image.

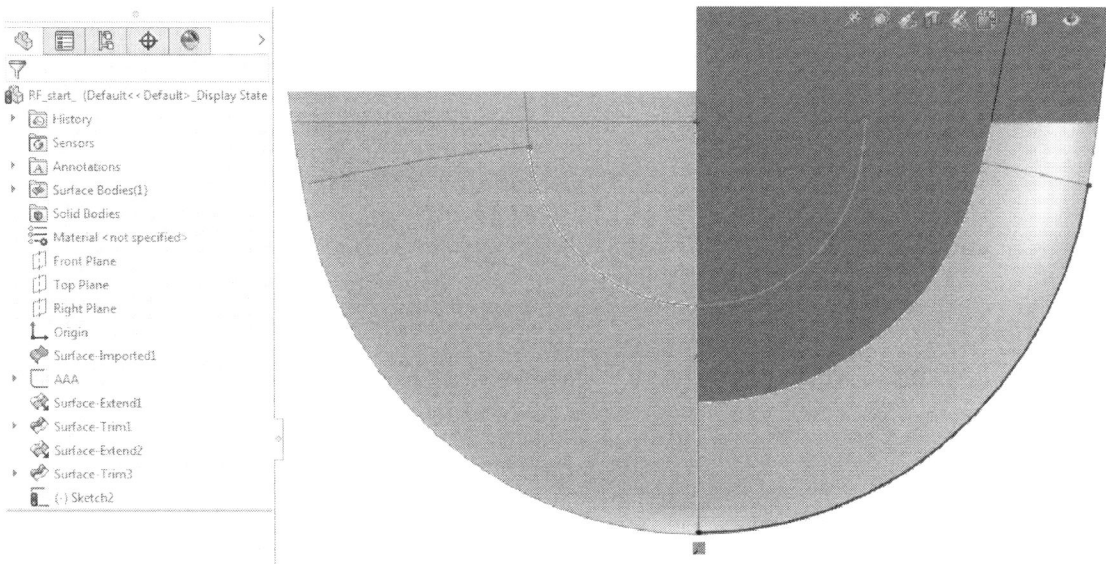

Figure 92 - Adding relations

Click Centerline (Sketch toolbar) or click Tools > Sketch Entities > Centerline. Click to start the centerline at the end point or point of coincidence of the Arc and sketch AAA. Drag, or move the pointer and click, to set the end of the centerline. Apply a Horizontal relation if this is not automatically applied as shown in the following image.

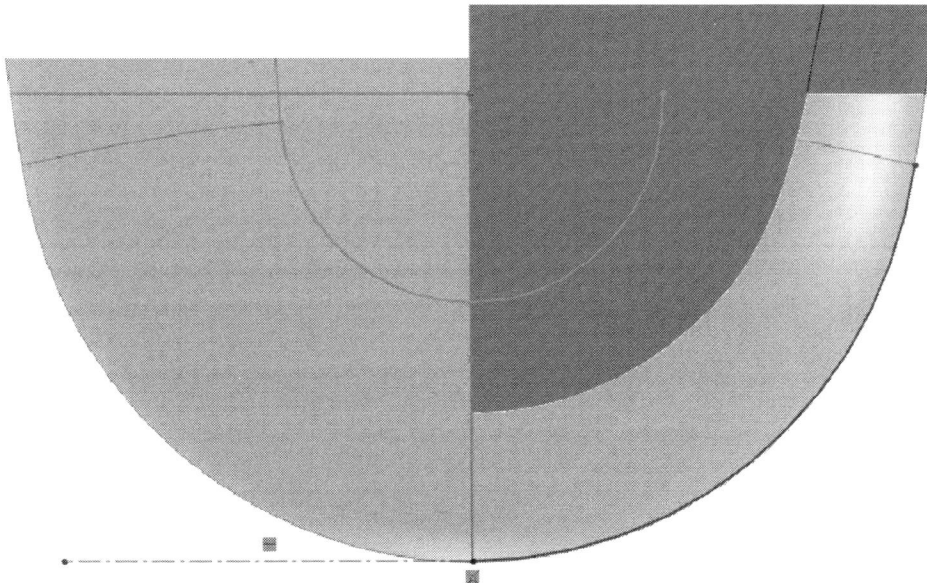

Figure 93 - Adding relations - Coincident and Horizontal Relation

Press and hold the Shift Key on your keyboard then select *(click)* the Arc and the Construction Line in the Graphics Area. Release the Shift Key and in the Property Manager, click on Tangent under Add Relations and then click OK to close the Property Manager. A Tangent Relation is applied between the horizontal construction line and the Arc.

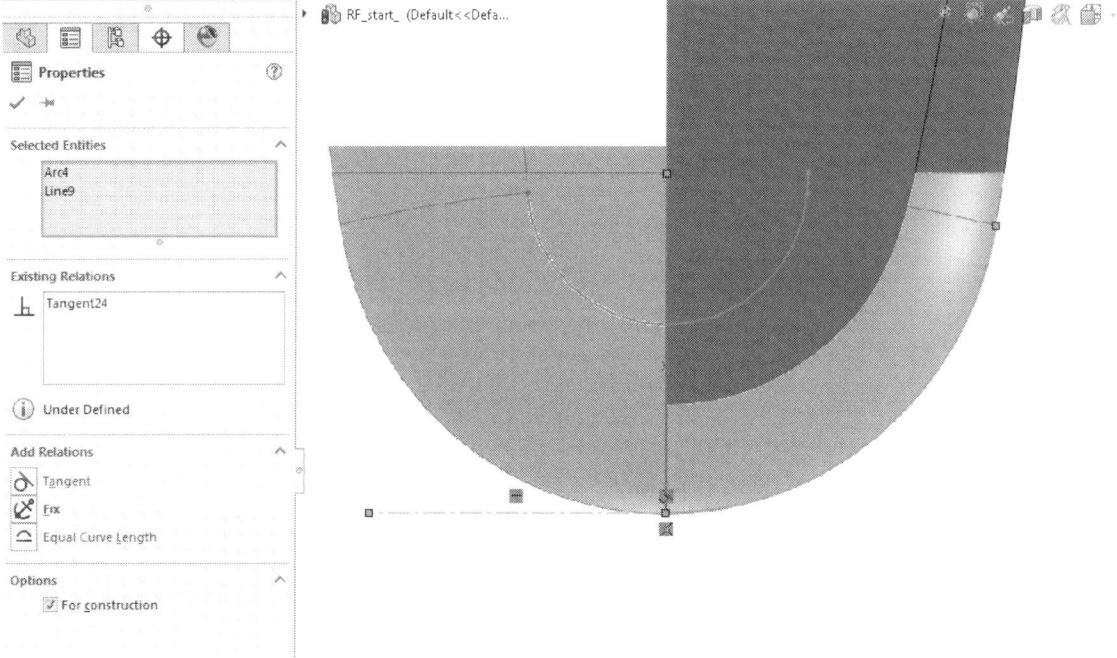

Figure 94 - Adding relations - Tangent Relation

Add a 25mm radius dimension to the arc as shown in the following image. Drag the 25mm dimension so that it is as close as possible to the area shown in the following image.

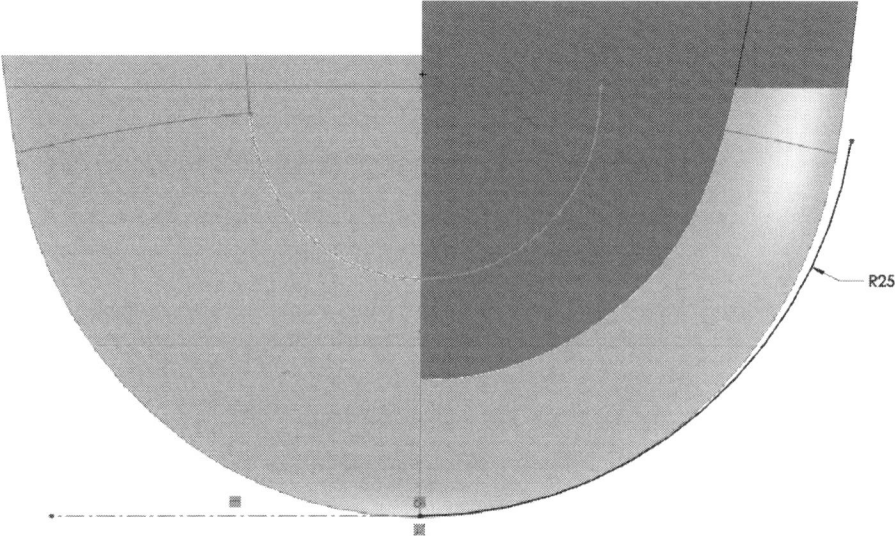

Figure 95 - Adding a Dimension

Double click the 25mm Dimension and zoom in to the area close to the 25mm dimension in the previous image. In the Modify Dialog Box, change the increment value to 0.05mm using the same procedure described in the previous sections. Use the Down Arrow to reduce the radius dimension as shown in the following images.

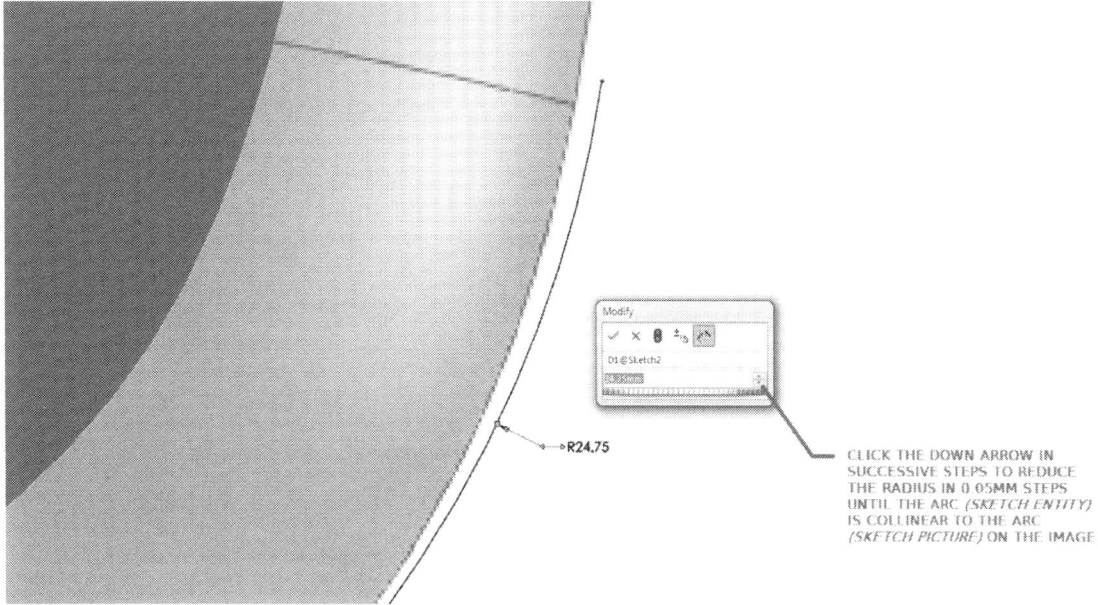

R24.75

CLICK THE DOWN ARROW IN
SUCCESSIVE STEPS TO REDUCE
THE RADIUS IN 0.05MM STEPS
UNTIL THE ARC *(SKETCH ENTITY)*
IS COLLINEAR TO THE ARC
(SKETCH PICTURE) ON THE IMAGE

Figure 96 - Modify Dialog Box - Using the UP / DOWN Arrow to fine tune a dimension

You can see in the following image that our Arc is moving closer to the image.

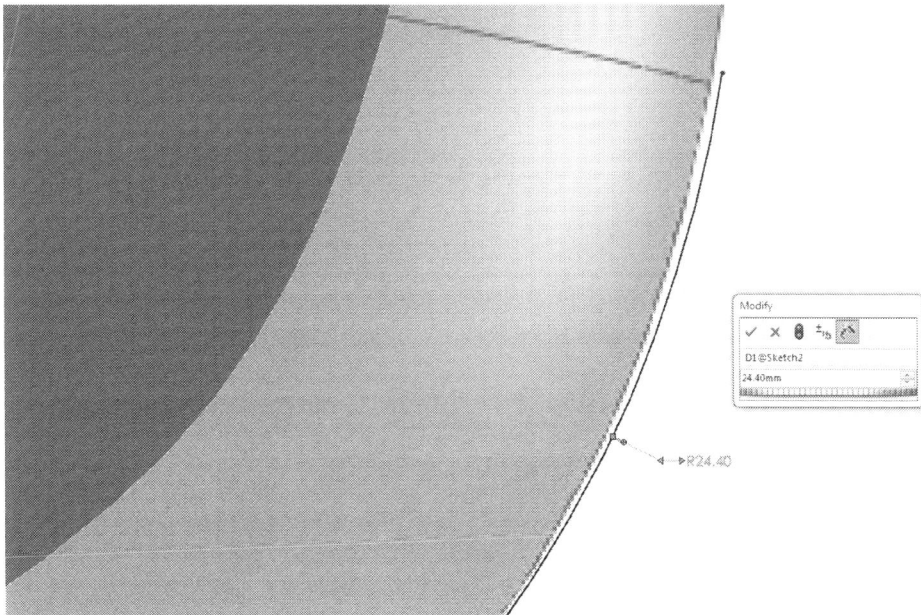

R24.40

Figure 97 - Modify Dialog Box - Using the UP / DOWN Arrow to fine tune a dimension

Zoom in and zoom out and zoom in again as you reduce the radius until you are happy with the position of the Sketch Entity Arc relative to the Sketch Picture Arc - I am happy with a radius of 24.15mm as shown in the following image. Click OK to close the Modify Dialog Box.

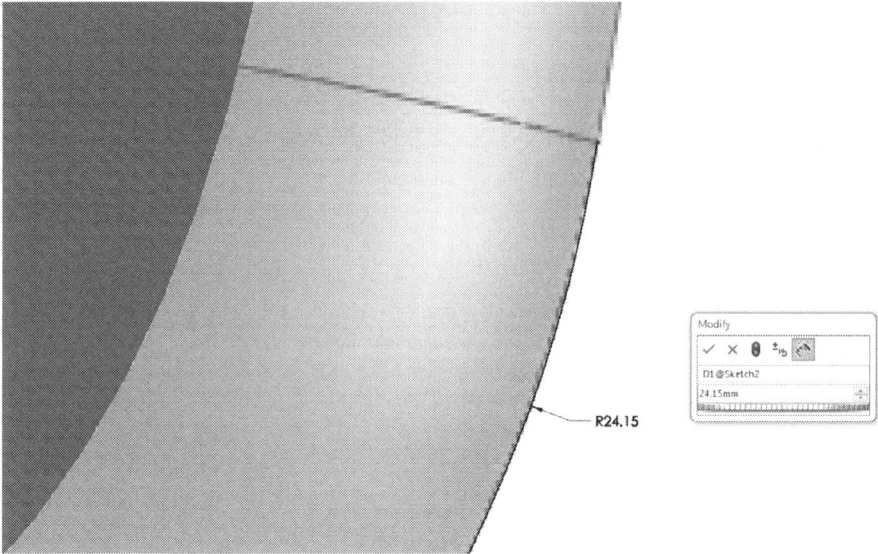

Figure 98 - Modify Dialog Box - Using the UP / DOWN Arrow to fine tune a dimension

You part should now look as shown in the following image.

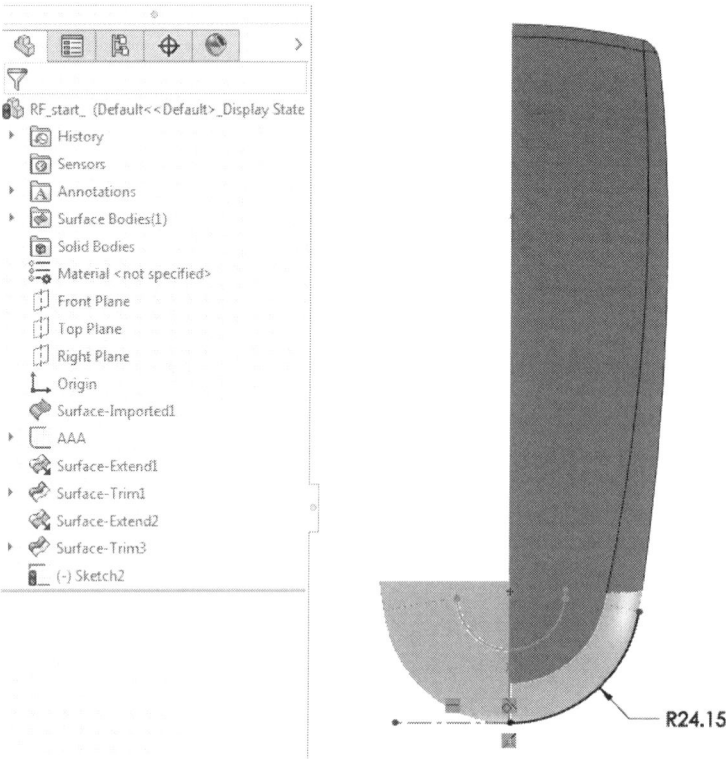

Figure 99 - Surface Model Current Status

58

USING CONVERT ENTITIES

Convert Entities is used to create one or more curves in a sketch by projecting an edge, loop, face, curve or external sketch contour, set of edges, or set of sketch curves onto the sketch plane.

Change your view to Isometric view and then in the Graphics Area select the existing split line on the part as shown in the following image then Click Convert Entities (Sketch toolbar) or Tools > Sketch Tools > Convert Entities.

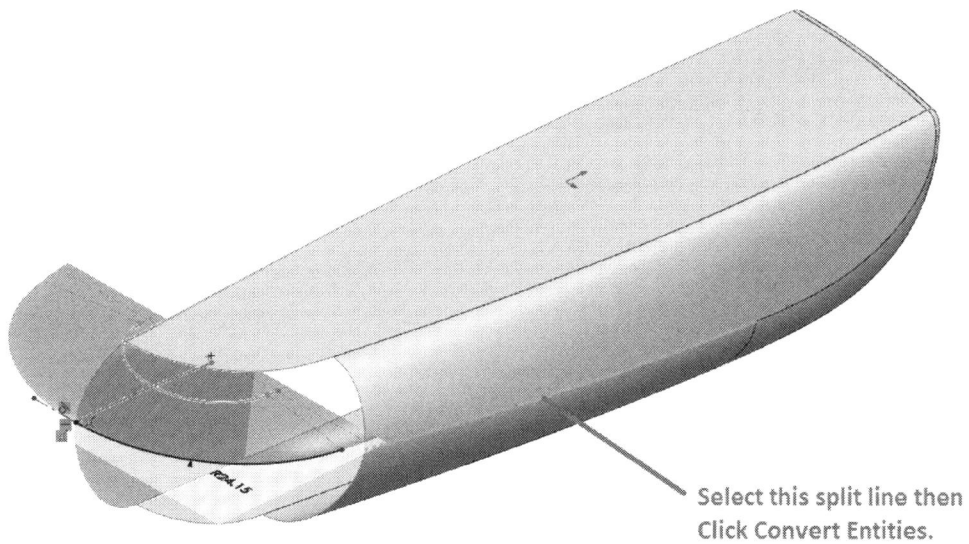

Select this split line then
Click Convert Entities.

Figure 100 - Using Convert Entities

A new curve is created and your part should now look as shown in the following image.

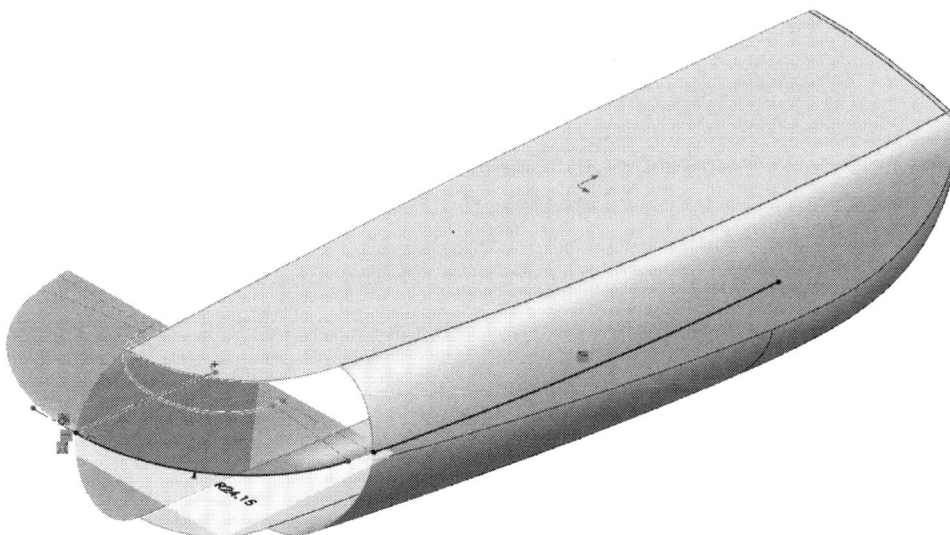

Figure 101 - Using Convert Entities

59

CONVERTING SKETCH ENTITIES TO CONSTRUCTION GEOMETRY

Sketch entities in a sketch or drawing can be converted to construction geometry. Construction geometry is used only to assist in creating the sketch entities and geometry that are ultimately incorporated into the part. Construction geometry is ignored when the sketch is used to create a feature. Construction geometry uses the same line style as centerlines.

Any sketch entity can be specified for construction. Points and centerlines are always construction entities.

To convert the curve we just created using Convert Entities to Construction Geometry do one of the following: -

1. Select the Curve the Graphics Area and Select For Construction in the PropertyManager then Click OK.

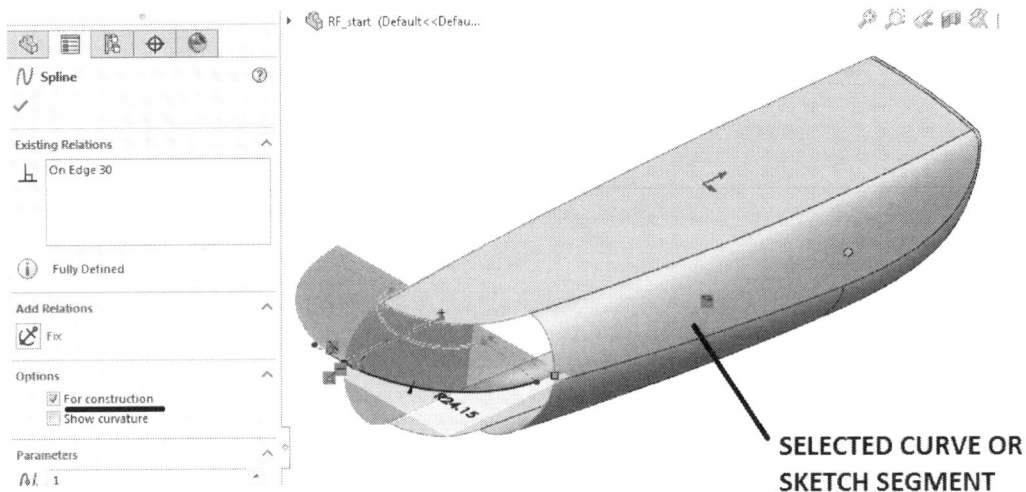

Figure 102 - Converting Sketch Entities to Construction Geometry

2. Click Tools > Sketch Tools > Construction Geometry then select the sketch entity to convert to Construction Geometry.

    Click Close in the Confirmation Corner - Top Right Corner of the Graphics Area. Or alternatively, press the letter "D" on your keyboard and the confirmation corner will appear next to your mouse pointer where you may Click ✓.

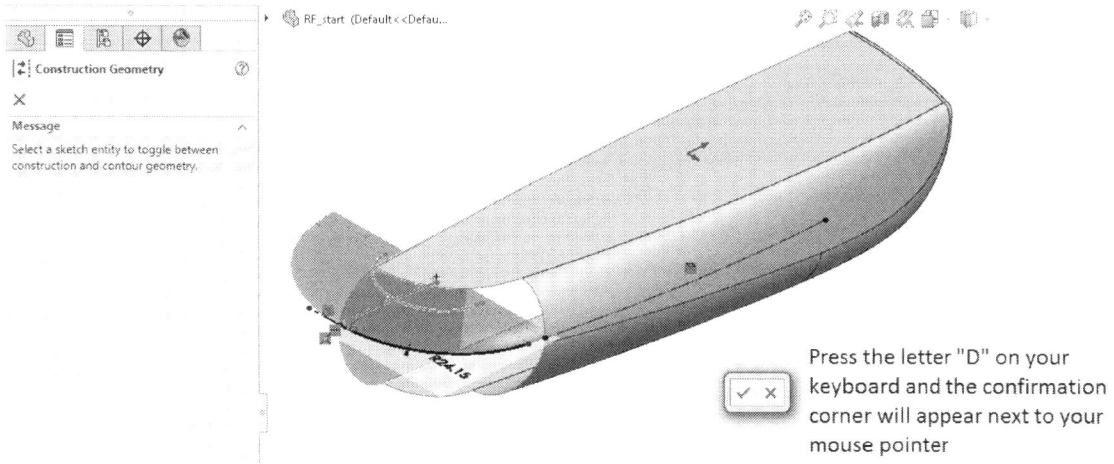

Press the letter "D" on your keyboard and the confirmation corner will appear next to your mouse pointer

Figure 103 - Converting Sketch Entities to Construction Geometry

3. Right-click the curve or sketch segment we would like to convert to construction geometry and select Construction Geometry as shown in the following image.

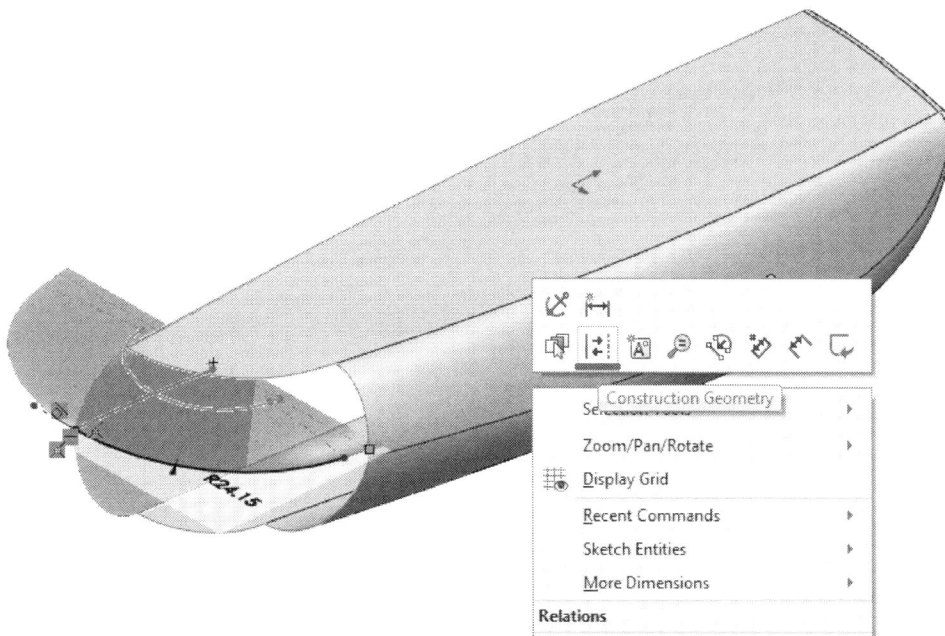

Figure 104 - Converting Sketch Entities to Construction Geometry

Your part should now look as shown in the following image:

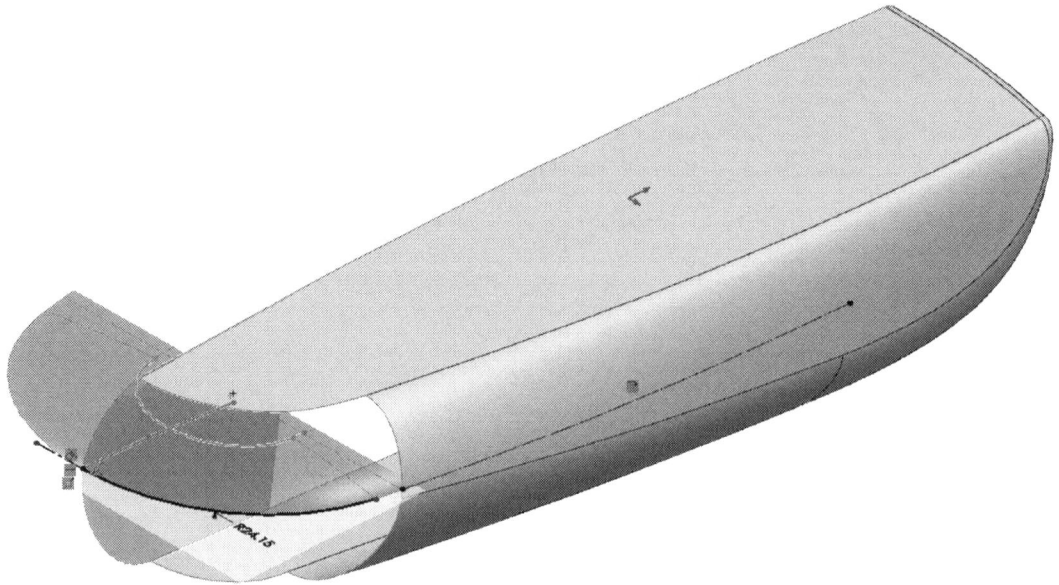

Figure 105 - Construction Geometry - Part Current Status

Zoom in to the area shown in the following image and sketch a spline with a relations as shown in the following two images - thus a tangent relation and a coincident relation to the end point of the sketch segment or curve we created using the convert entities command and we have just converted to construction geometry in the section above as well as to the arc which we gave a 24.15mm radius.

R24.15

Figure 106 - Sketching a Spline - coincident and tangent relation

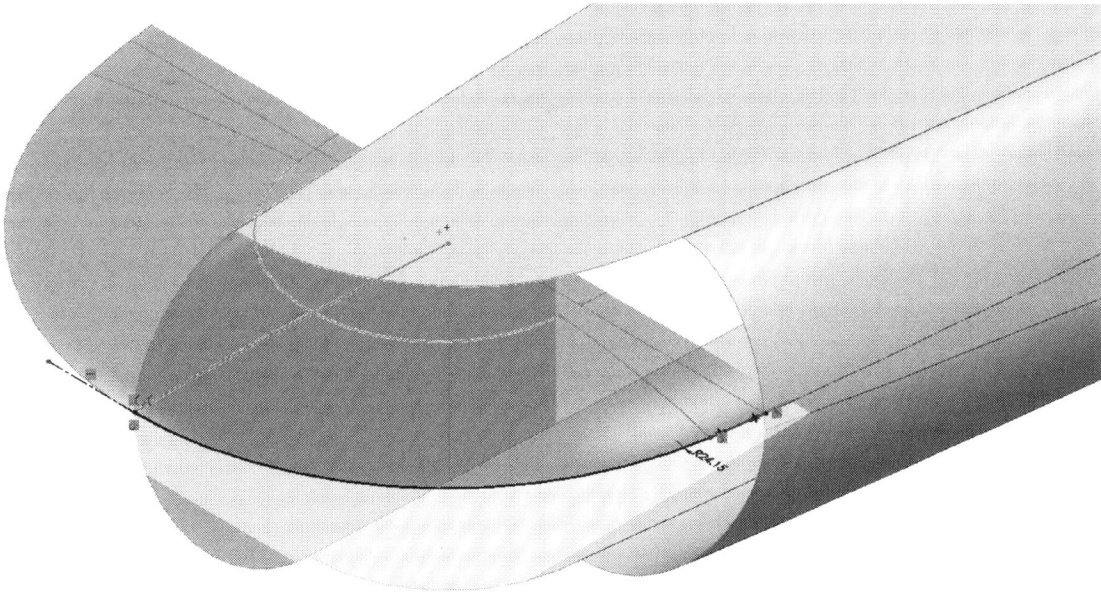

Figure 107 - Sketching a Spline and adding relations

Exit the sketch and save your part.

Change the colour of the Sketch we just created using the same procedure described in sections above.

RENAMING A SKETCH

To rename a Sketch or any Feature in the Feature Manager Design Tree do one of the following for the feature whose name you want to change: -

1. Click-pause-click the feature name.

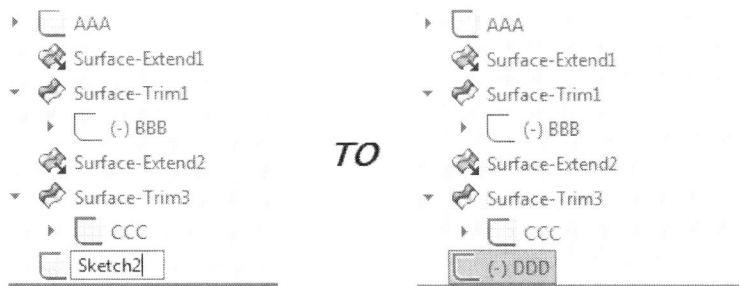

Figure 108 - Renaming a Sketch

2. Right-click the feature name and select Feature Properties. The Feature Properties Dialog Box opens. Change the Name in the name text area from Sketch2 to DDD then click OK.

63

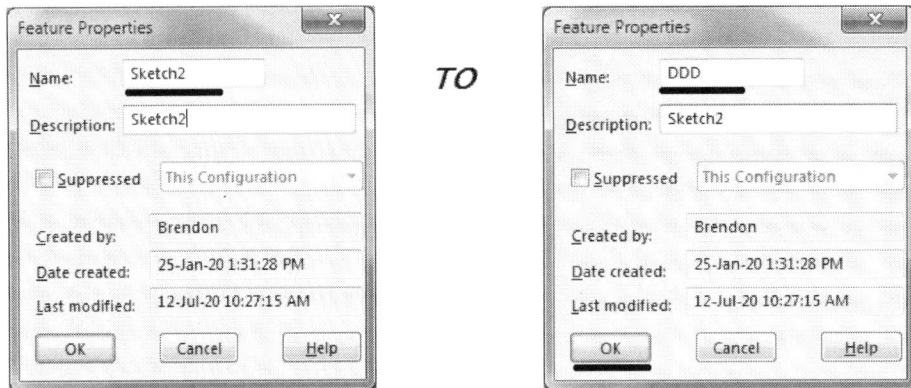

Figure 109 - Renaming a Sketch

3. Select the Feature Name (Sketch2) and press F2. Enter a new name and press Enter.

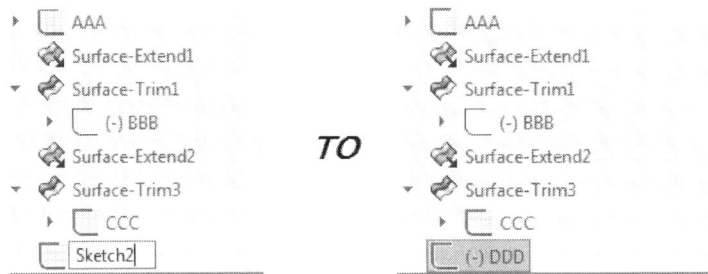

Figure 110 - Renaming a Sketch

CREATING A PROJECTED CURVE

Project Curve is a tool in Solidworks used to form 3D Curves from two 2D Curves from two different planes.

Click Project Curve on the Curves toolbar, or Insert > Curve > Projected.

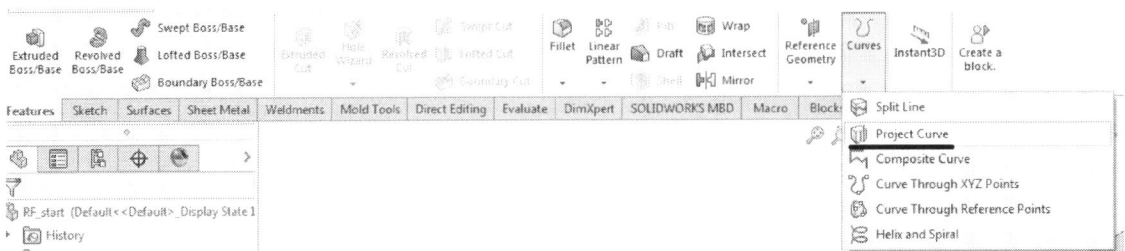

Figure 111 - Project Curve

In the PropertyManager:-

Under Selections, set Projection type to Sketch on Sketch.

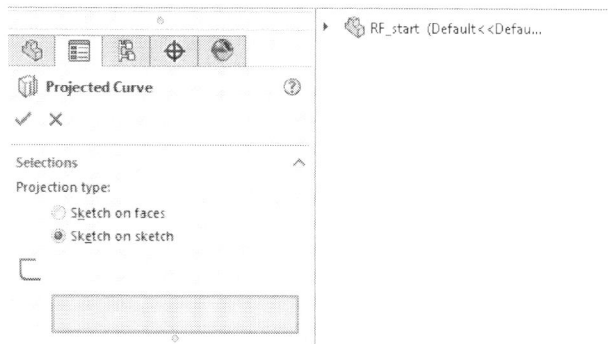

Figure 112 - Projected Curve - Sketch On Sketch

The <u>Sketch on Sketch</u> selection is used to create a curve that represents the intersection of sketches from two intersecting planes. The sketch profiles must be aligned such that when they are projected normal to their sketch plane; the implied surfaces will intersect, creating the desired result.

Under Sketches to Project, select AAA and DDD sketches in the flyout FeatureManager Design Tree or the graphics area as shown in the following image.

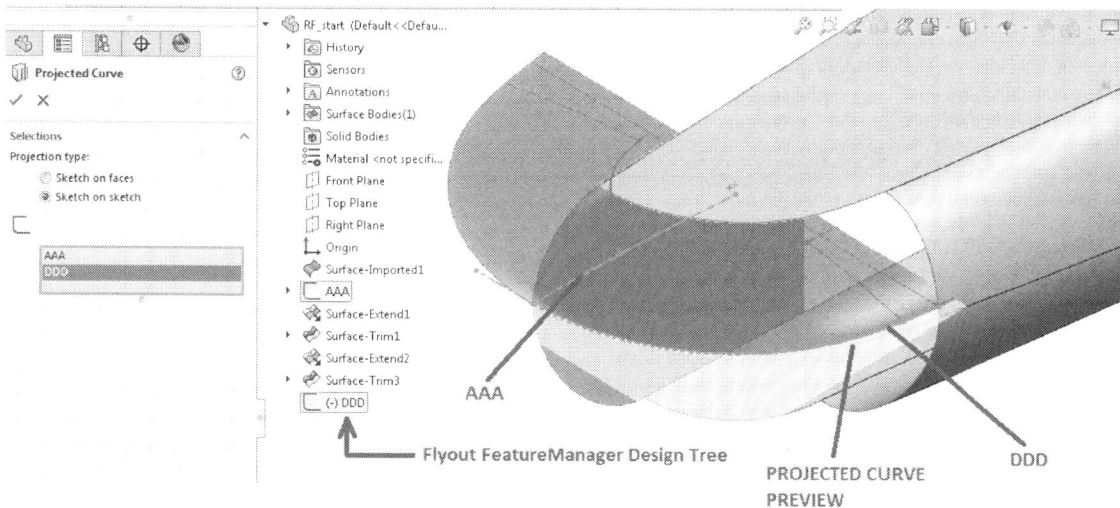

Figure 113 - Projected Curve - Sketch On Sketch - Sketches to Project

The two sketches (AAA & DDD) project onto each other to form a 3D curve . A preview of the projected curve appears.

Click OK.

You will notice that a new feature - Curve1 is created in your Feature Manager Design Tree and a 3D Curve is also created in the Graphics Area as shown in the following image.

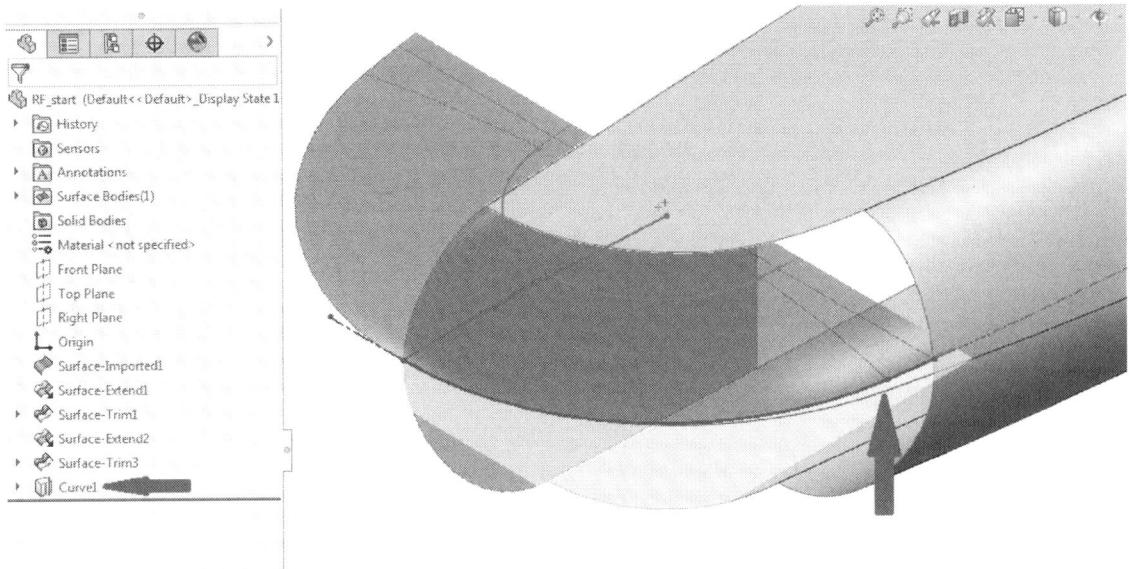

Figure 114 - Projected Curve

HIDING OR SHOWING SKETCHES

You can turn the display of sketches on or off by doing one of the following:

1. To hide or show individual sketches, Right Click the sketch you want to hide in the Feature Manager Design Tree or Graphics Area and select Hide or Show.

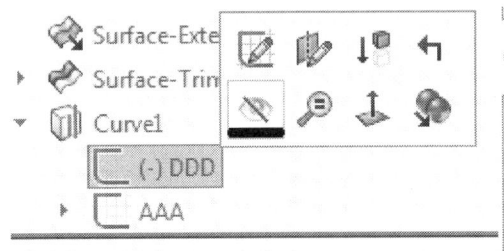

Figure 115 - Hide or Show an Individual Sketch

2. To hide or show all sketches in the Graphics Area do one of the following:

   • Click View > Hide/Show > Sketches

   • In the Heads-up View toolbar, click Hide/Show Items ⚇ > View Sketches

66

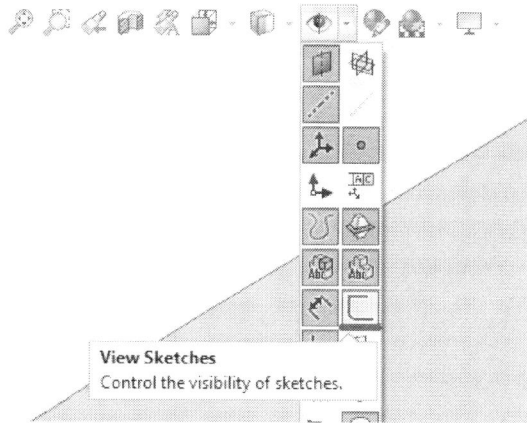

Figure 116 - Hide or Show All Sketches using the Heads Up View Toolbar

Your part should now look as shown in the following image with all the sketches hidden.

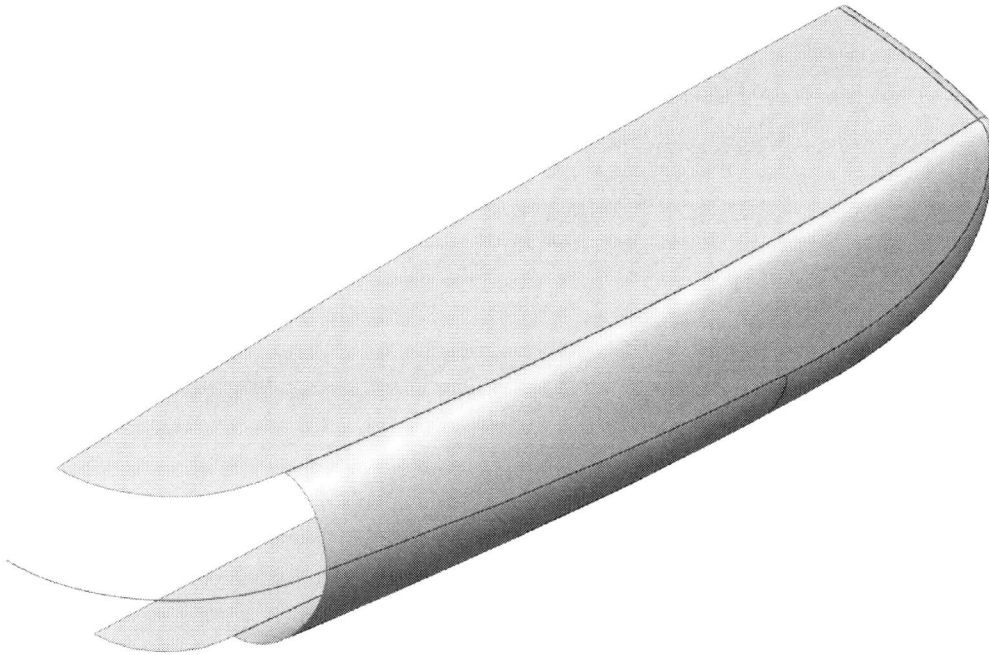

Figure 117 - Recreated Portion of the Parting Line

To measure the length of the recreated portion of the parting line, do one of the following: -

Click Measure ![measure icon] (Tools toolbar) or Tools > Evaluate > Measure.

In the Graphics Area, select the two segments that form Curve1 and the two appear in the Measure Tool Input Box as shown in the following image with the Total Length indicated as 38.12mm.

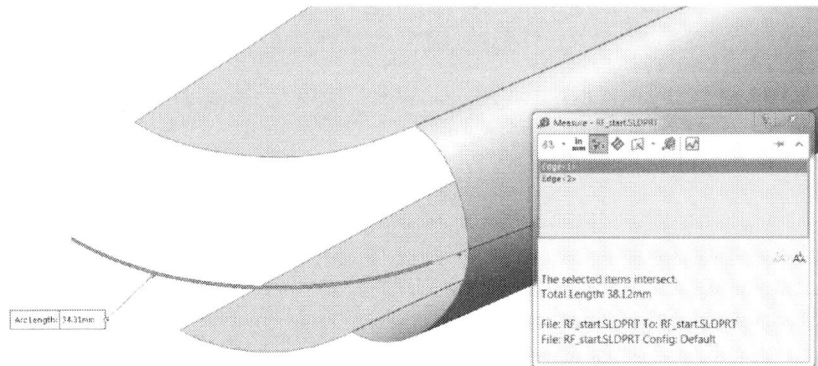

Figure 118 - Measure Tool

MEASURE TOOL QUICK COPY

When you hover over a numeric value, the numeric value gets highlighted in the dialog box and

Copy ![copy icon] is displayed. Click ![copy icon] to copy the value to the clipboard. You can paste the value at the required location or in the exam you can paste the answer in the answer text area where applicable.

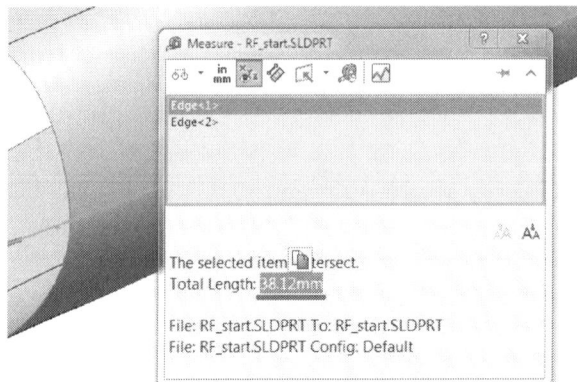

Figure 119 - Measure Tool Quick Copy Functionality

At the beginning of this chapter you were required to measure the length of the recreated portion of the parting line and select a range that contains the value of this measured length from four off options which are: -

1. 39.45 TO 41.35

2. 41.50 TO 43.35

3. 37.21 TO 39.11

4. 34.95 TO 36.98

That means our answer is option 3 - or the 37.21 TO 39.11 range since our measured length is 38.12mm.

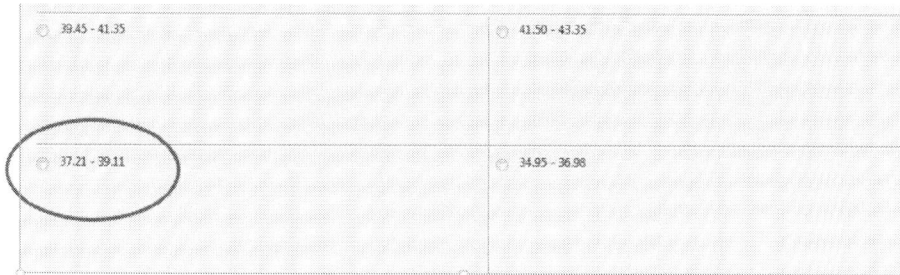

Figure 120 - Chapter Three Answer

You may move on to Chapter Four or go through the three following examples to learn more about Projected Curves including the Sketch On Face Projected Curve.

SKETCH ON SKETCH CURVE

Open the downloaded part SKETCH_ON_SKECTH 01.SLDPRT or download the part from this Google drive location - *http://bit.ly/CSWPA-SU* or Scan the QR Code shown below:

If you experience any problems with downloading any files you may send an email to *cswpasmebook@gmail.com* with the title of the book indicated in your email subject. Open and save the downloaded part to your PC.

The part should look as shown in the following image - it consists of two solid bodies and two sketches - PATH SIDE VIEW which is on the Right Plane and PATH TOP VIEW which is on the Top Plane. Change the view in the Graphics Area to Top View and then to Right View observing the relationship between the two sketches and the solid bodies in this part from either of these views.

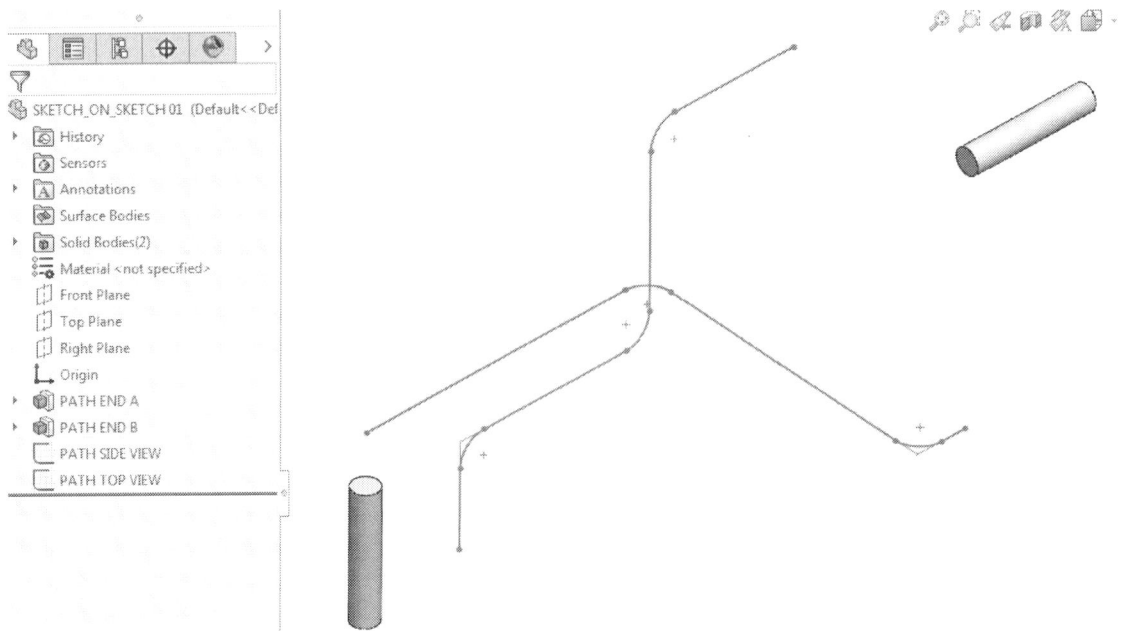

Figure 121 - Sketch On Sketch Curve

Click Project Curve on the Curves toolbar, or Insert > Curve > Projected.

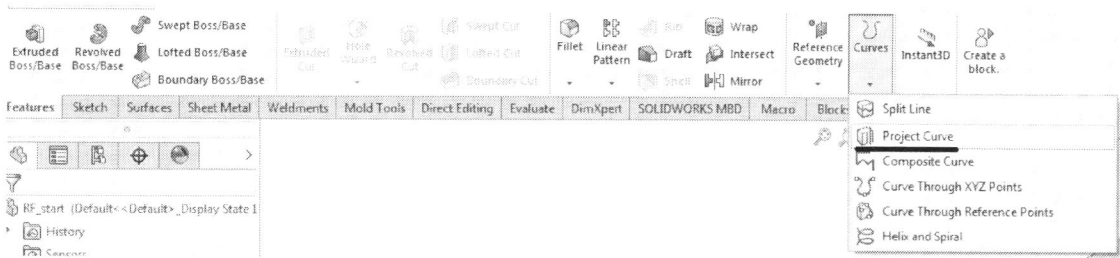

Figure 122 - Project Curve

In the PropertyManager:-

Under Selections, set Projection type to Sketch on Sketch.

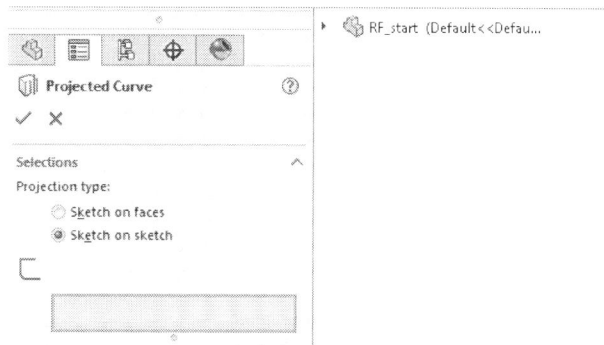

Figure 123 - Projected Curve - Sketch On Sketch

Under Sketches to Project, select PATH SIDE VIEW and PATH TOP VIEW sketches in the flyout FeatureManager Design Tree or in the graphics area as shown in the following image.

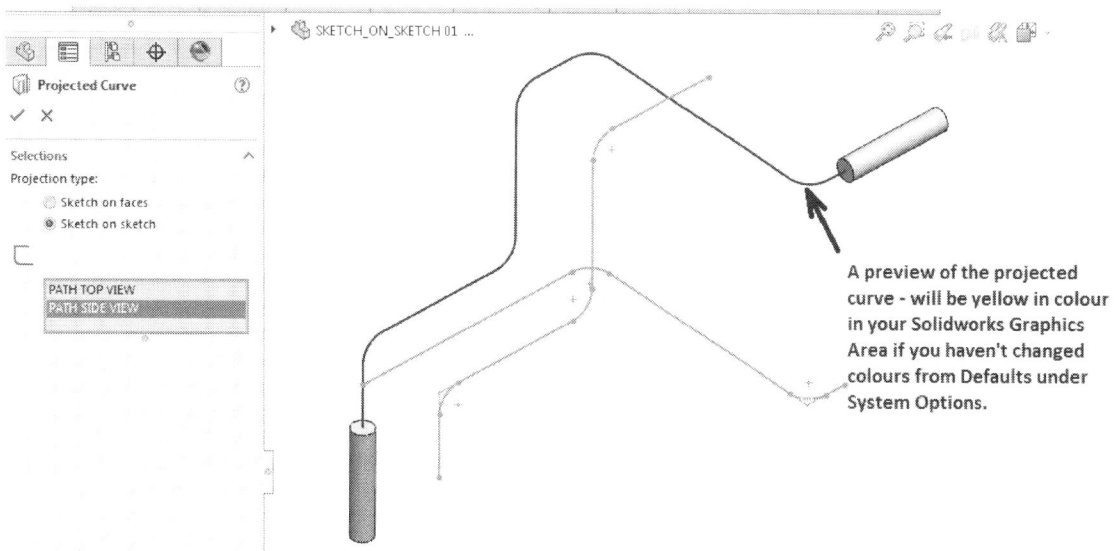

Figure 124 - Projected Curve Preview - Colour Changed for legibility purposes

71

The two sketches (PATH SIDE VIEW & PATH TOP VIEW) project onto each other to form a 3D curve . A preview of the projected curve appears.

Click OK.

You will notice that a new feature - Curve1 is created in your Feature Manager Design Tree and a 3D Curve is also created in the Graphics Area as shown in the following image.

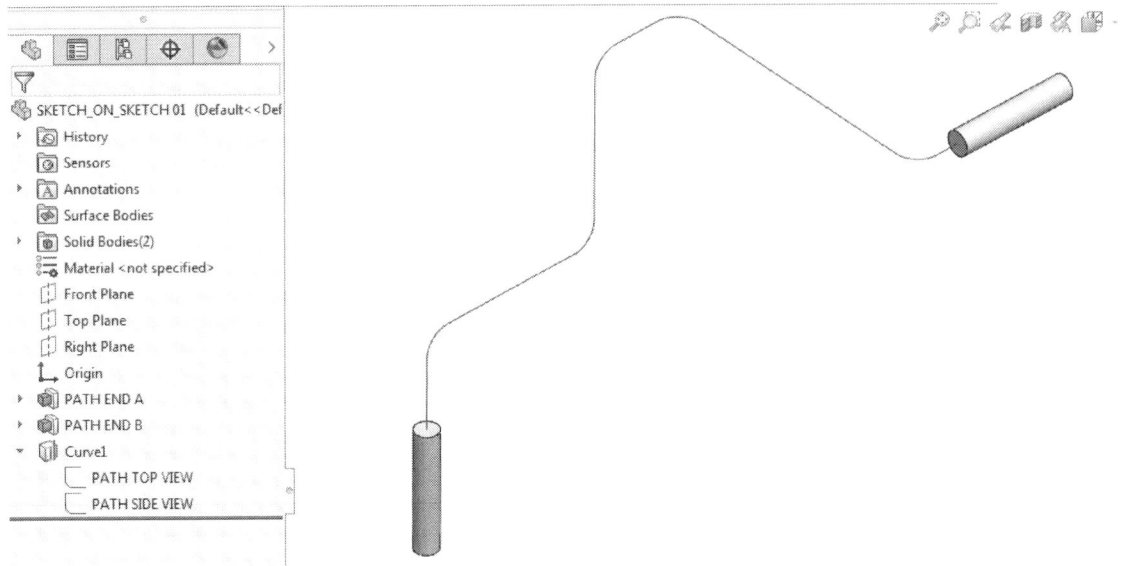

Figure 125 - Projected Curve

CREATING A SWEPT FEATURE USING A PROJECTED CURVE AS THE PATH

Click Swept Boss/Base (Features toolbar).

In the Property Manager - Select the Sketch Profile radio button and under Profile, select the top edge or face of the solid body PATH END A as shown in the following image.

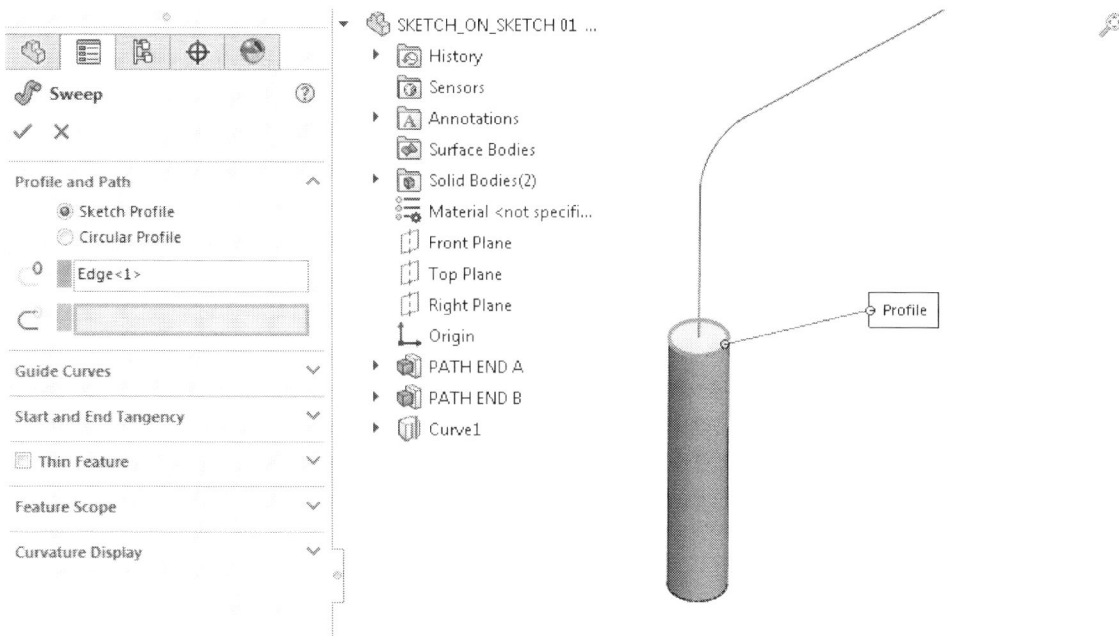

Figure 126 - Using a Projected Curve as a Sweep Profile

Under Path, select Curve1 in the Graphics Area or in the Flyout Feature Manager Design Tree as shown in the following image - A preview of the swept feature appears.

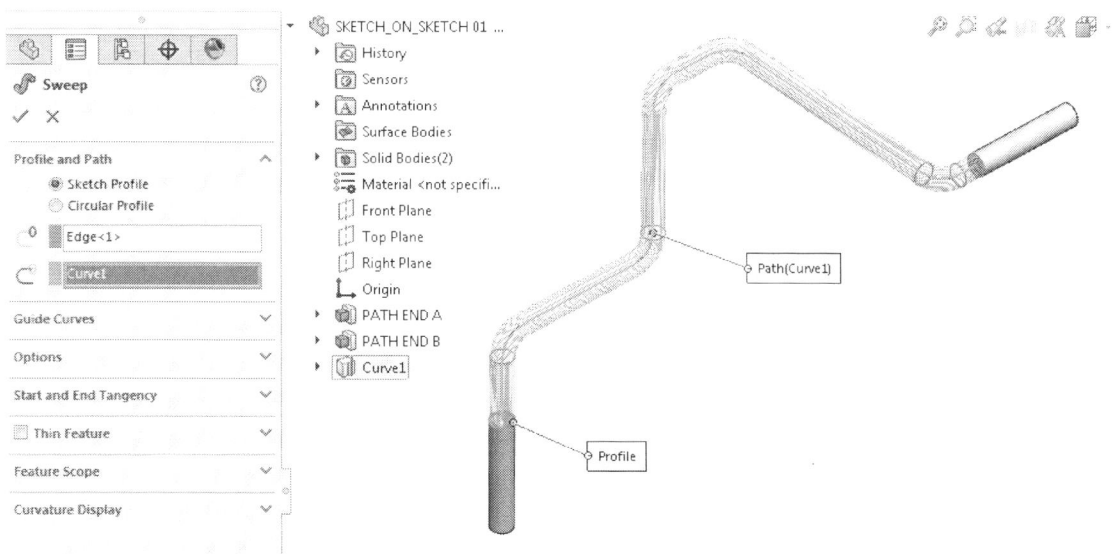

Figure 127 - Sweep Path

Click OK. Your part should now look as shown in the following image with the curve hidden provided you selected Merge Result under Options in the Sweep Property Manager.

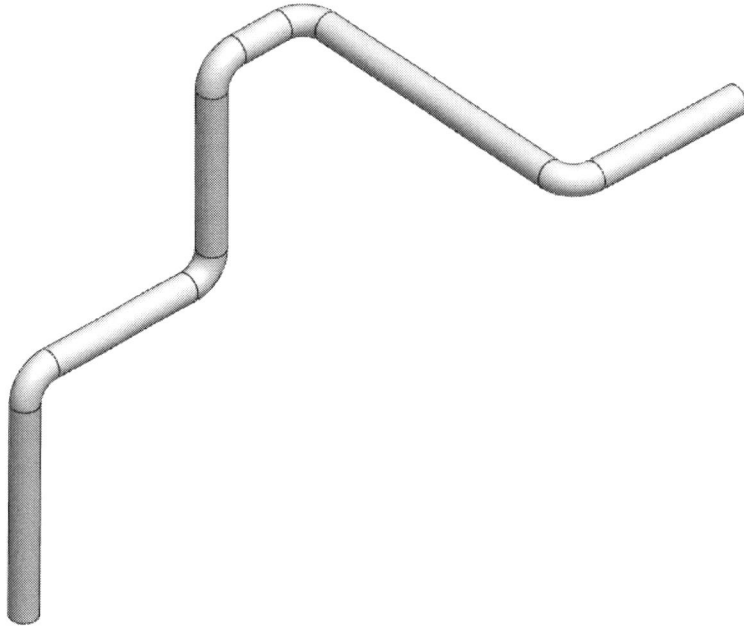

Figure 128 - Sweep Feature

Save your part. Close the Part and move on to the next example.

SKETCH ON FACES CURVE

Open the downloaded part SKETCH_ON_FACE 02.SLDPRT or download the part from this Google drive location - *http://bit.ly/CSWPA-SU* or Scan the QR Code shown below:

If you experience any problems with downloading any files you may send an email to *cswpasmebook@gmail.com* with the title of the book indicated in your email subject. Open and save the downloaded part to your PC.

The part should look as shown in the following image - it consists of one surface body (Surface-Extrude1) and one sketch (EEE) as shown in the following image.

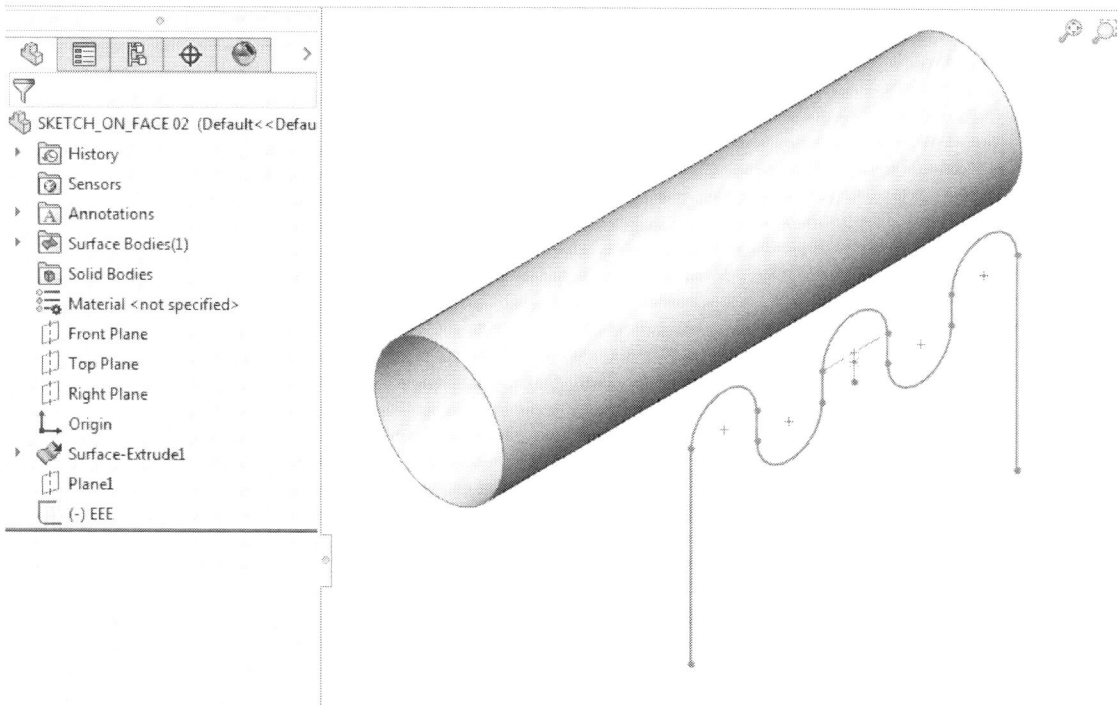

Figure 129 - Sketch on Face 02.sldprt Part

Click Project Curve on the Curves toolbar, or Insert > Curve > Projected.

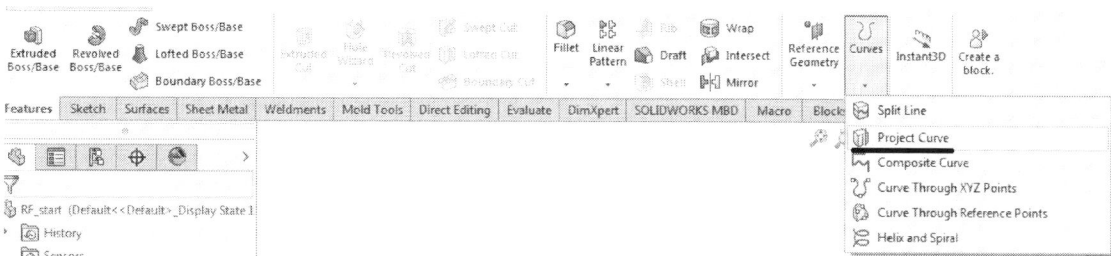

Figure 130 - Project Curve

In the PropertyManager:-

Under Selections, set Projection type to Sketch on Faces.

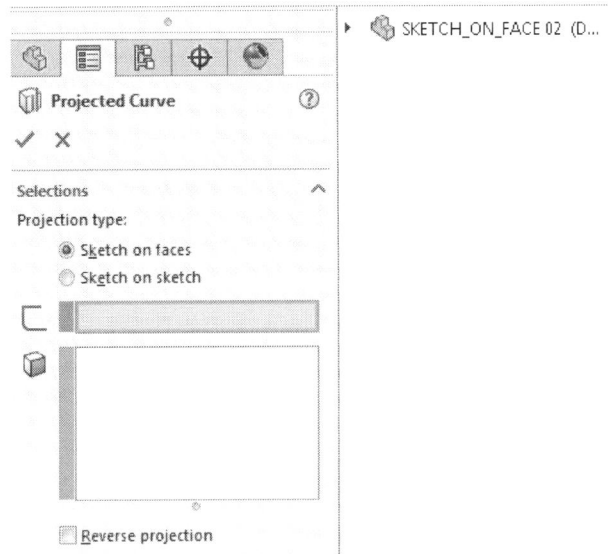

Figure 131 - Projected Curve - Sketch On Faces

Under Sketch to Project, select the Sketch EEE in the flyout FeatureManager Design Tree or in the graphics area as shown in the following image.

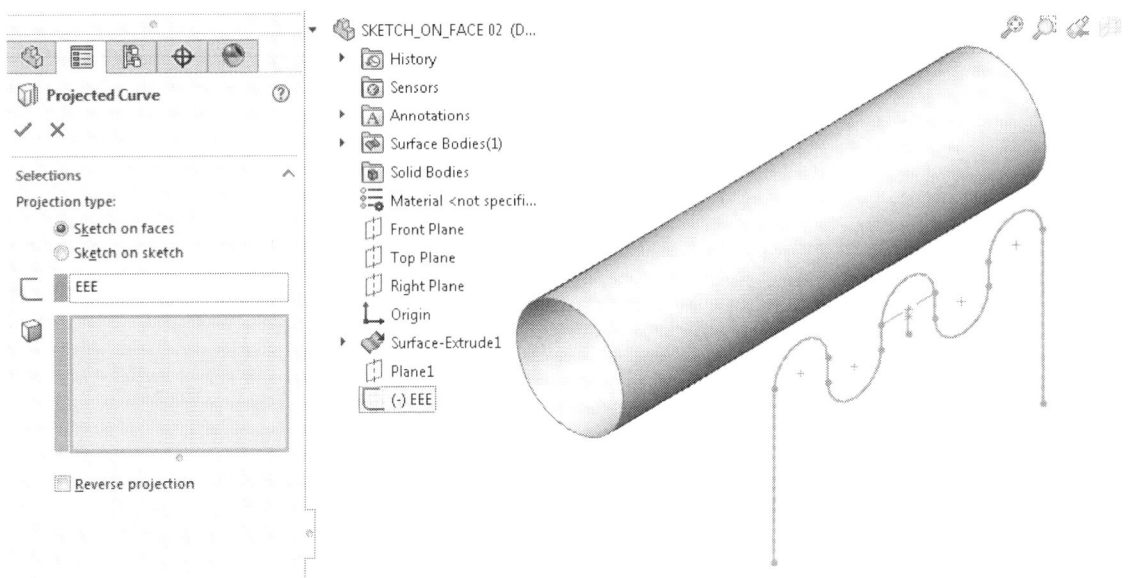

Figure 132 - Projected Curve - Sketch On Faces

Under Projection Faces, select the cylindrical face on the model where we want to project the sketch. Select the Reverse projection check box, or click the handle in the graphics area, if necessary or to get a preview of the projected curve correctly as shown in the following image.

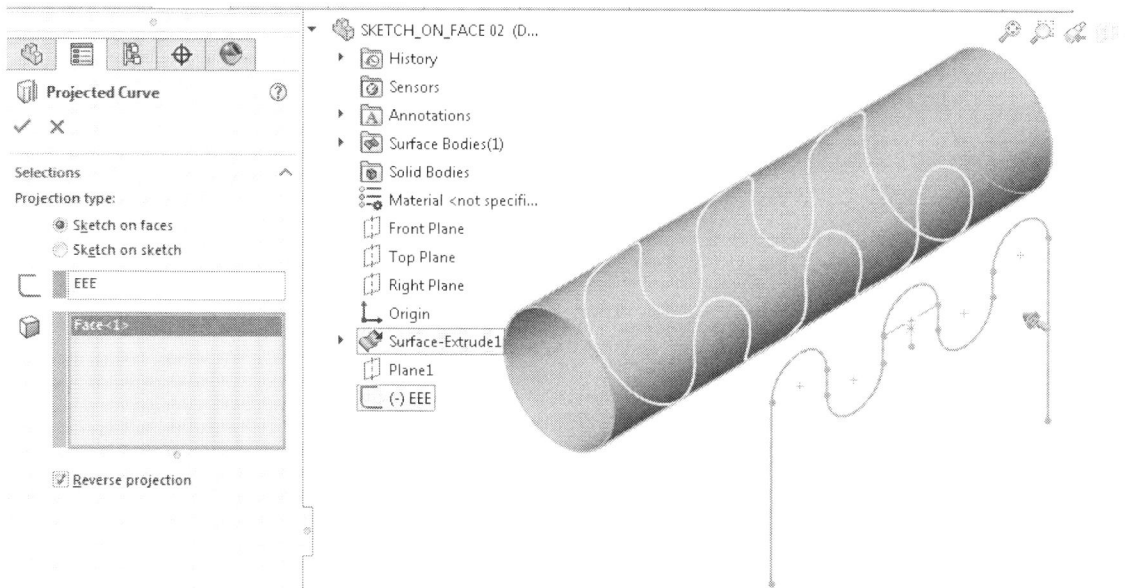

Figure 133 - Projected Curve - Sketch On Face

Click OK and your part should now look as shown in the following image.

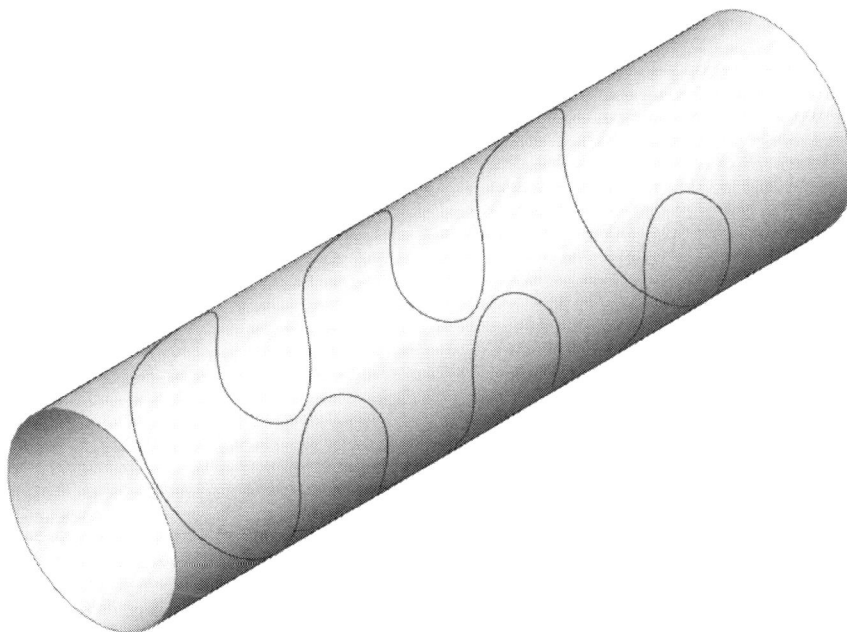

Figure 134 - Sketch On Faces Projected Curve

USING THE SWEPT SURFACE FEATURE

Click Swept Surface (Surfaces toolbar) or Insert > Surface > Sweep.

In the Property Manager - Select the Circular Profile radio button and under Profile, select the Projected Curve we just created in the section above and enter a 6.35mm Dimension under Diameter as shown in the following image. A preview of the Swept Surface appears.

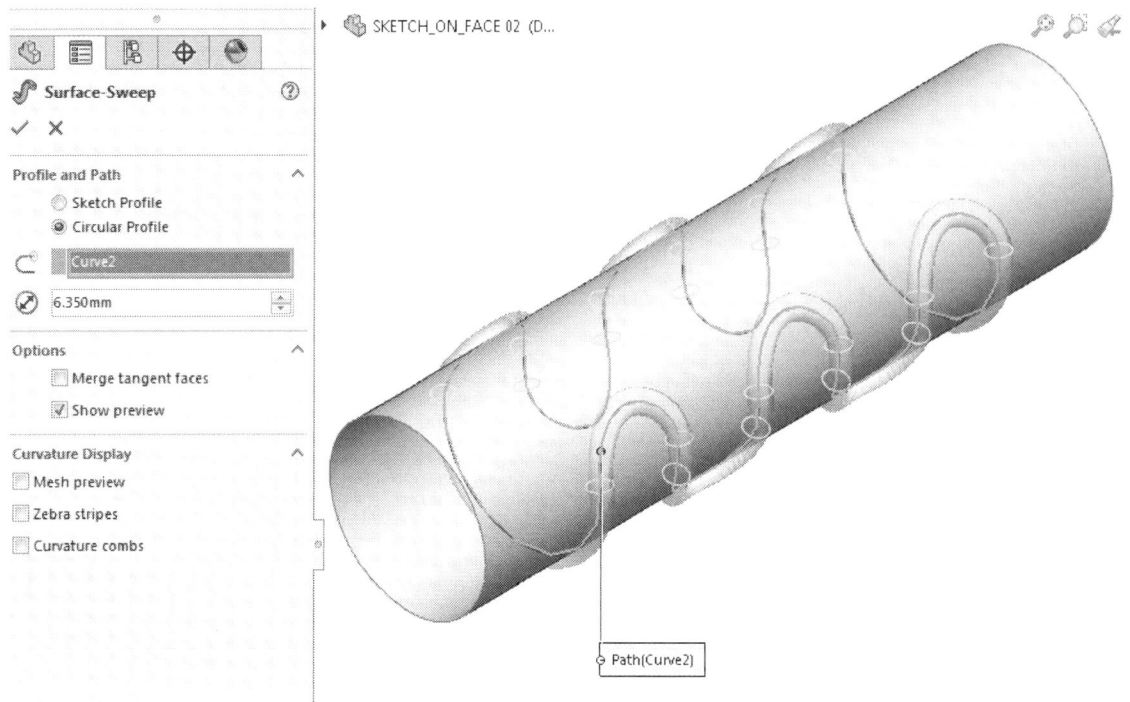

Figure 135 - Swept Surface Preview

Click Ok and your part should now look as shown in the following image *(with the Curve Hidden)*.

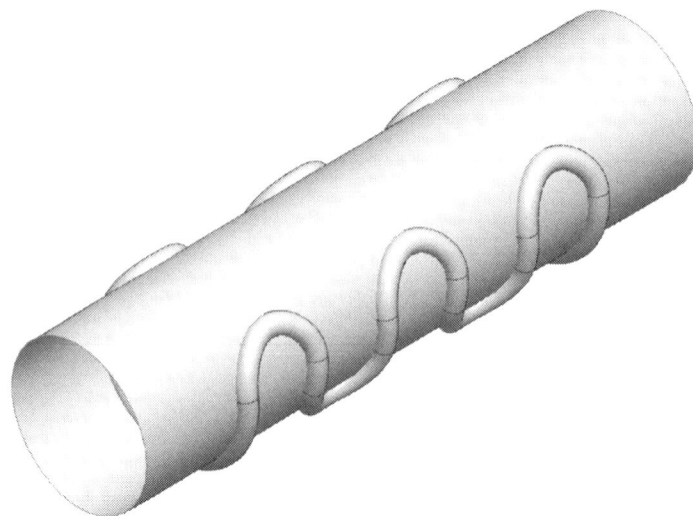

Figure 136 - Swept Surface created using a projected curve as the Sweep Path

MUTUAL TRIM

Click Trim Surface on the Surfaces toolbar, or click Insert > Surface > Trim.

In the PropertyManager, under Trim Type, select Mutual.

Under Selections, Select two surfaces - Surface-Extrude1 and Surface-Sweep1 in the graphics area or in the Flyout Feature Manager Design Tree for Trimming Surfaces to use to trim themselves as shown in the following image.

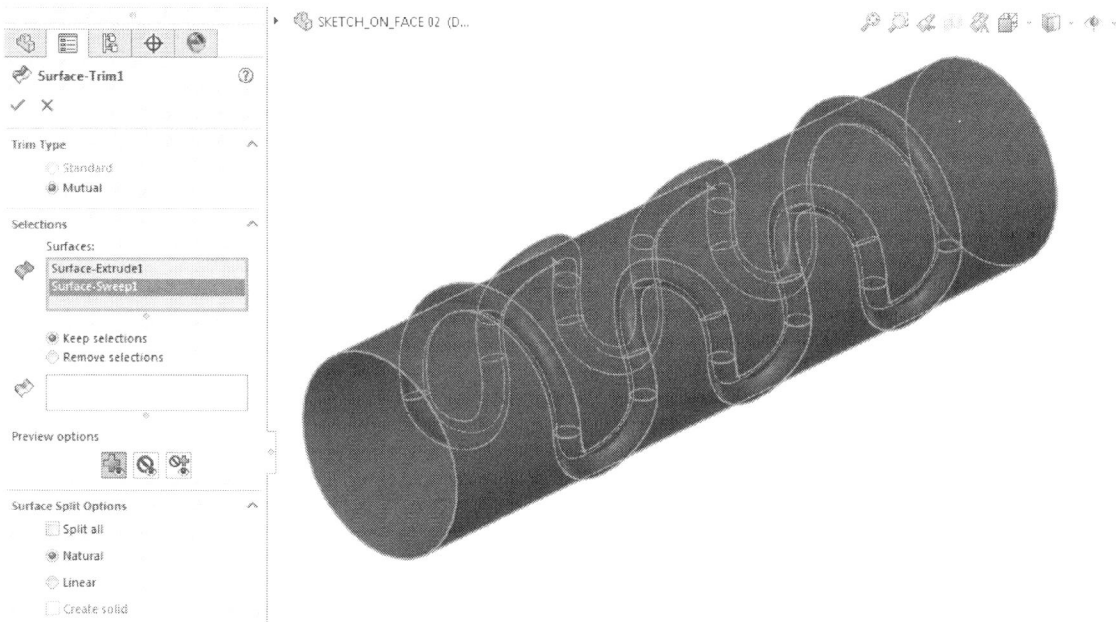

Figure 137 - Mutual Trim - Selections

Select a trim action:

In this example, we will select Keep selections and Show Included Surfaces under Preview Options then select surfaces in Pieces to Keep as shown in the following image.

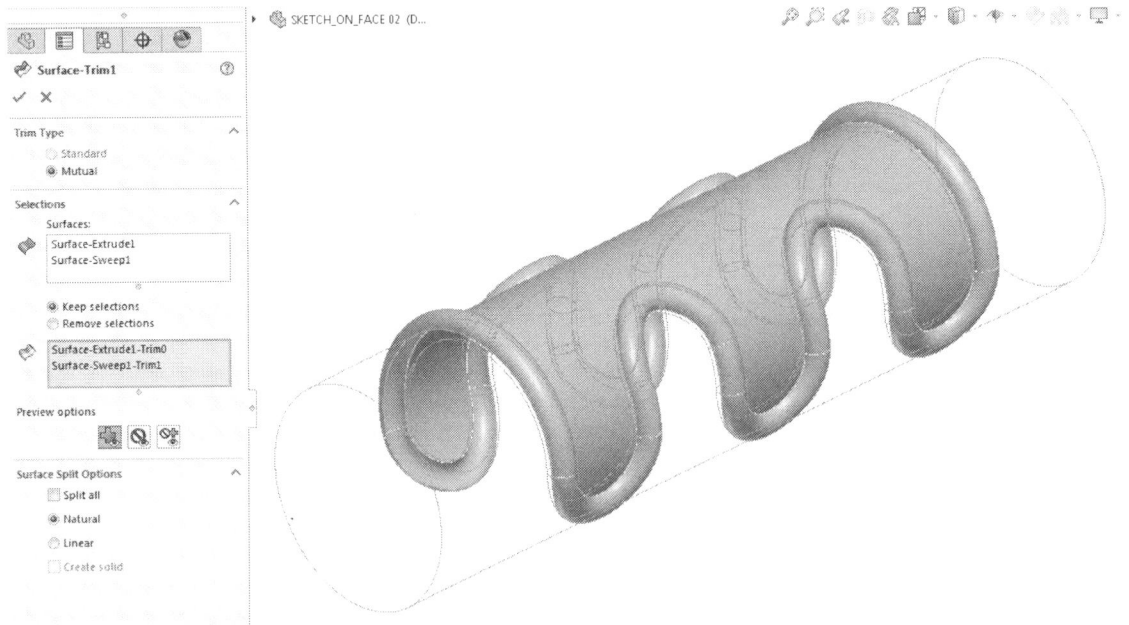

Figure 137 - Mutual Trim - Keep Selections - Show Included Surfaces

Click Ok and your part should now look as shown in the following image. Save your part.

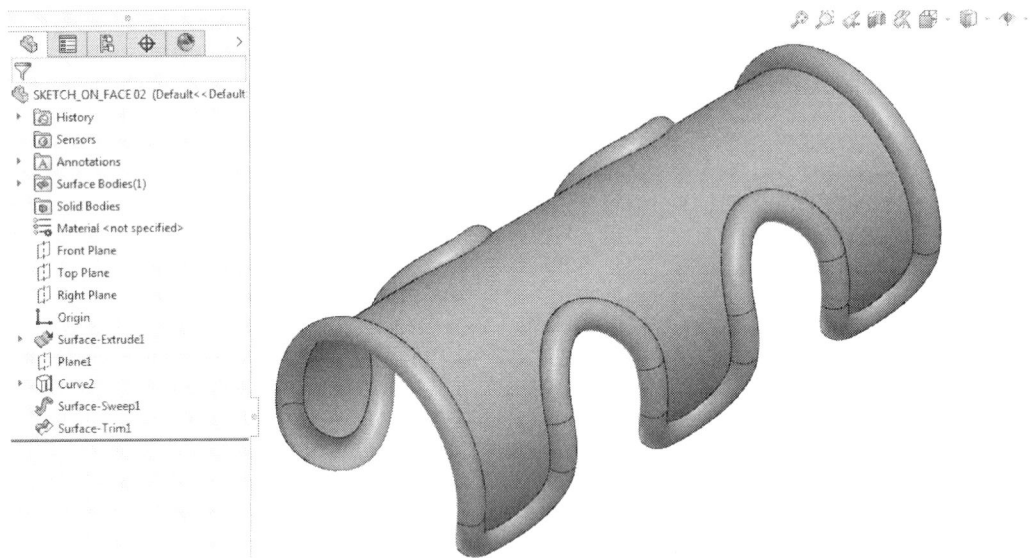

Figure 139 - Mutual Trim Result

FILLET SURFACE

Click Fillet (Features toolbar) or Insert > Surface > Fillet/RoundInsert.

In the Property Manager: -

- Select the Tangent propagation checkbox

- Select the Full preview radio button

- Under Fillet Parameters, Select Symmetric and enter a radius value of 3mm

- Under Profile, select Circular

- Then Select an edge to fillet in the Graphics area as shown in the following image

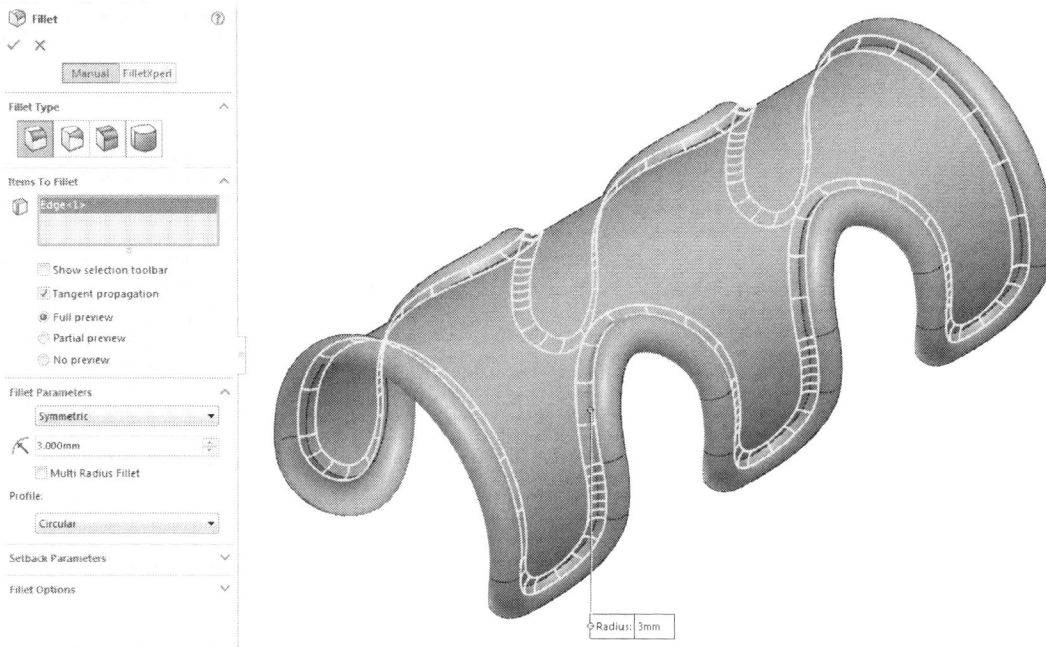

Figure 140 - Fillet Surface

MOVING THE CONFIRMATION CORNER OPTIONS TO THE POINTER

Click in the Graphics Area then Press and Release the letter "**D**" on your keyboard. Click Ok to Accept and Finish the current command on the Pop Up Confirmation Corner near your mouse pointer - see the following image.

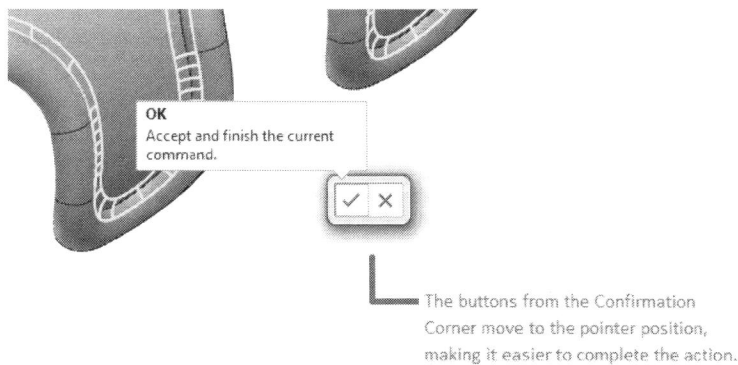

OK
Accept and finish the current command.

The buttons from the Confirmation Corner move to the pointer position, making it easier to complete the action.

Figure 141 - Moving the Confirmation Corner Options to the Pointer

Your part should now look as shown in the following image.

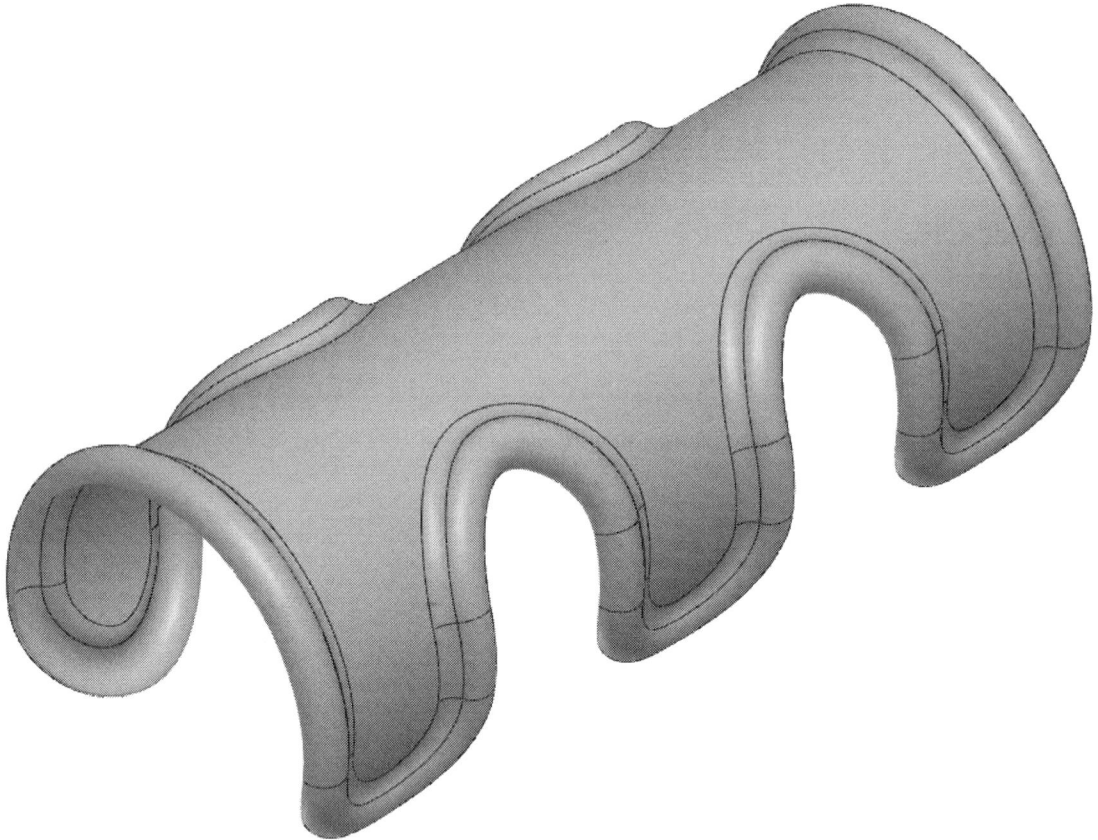

Figure 142 - SKETCH ON FACE 02.SLDPRT Part Current Status

THICKENING A SURFACE

Click Thicken on the Features toolbar, or Click Insert > Boss/Base > Thicken.

In the PropertyManager under Thicken Parameters, select the Surface body in the Graphics Area as the Surface to Thicken.

Examine the preview, and select the side of the surface you want to thicken - in this example we select Thicken Side 2 - thus thickening in one direction to the inside.

Enter a Thickness Value of 2.00mm as shown in the following image. Note that when you select Thicken Both Sides, it adds the Thickness you specify to both sides.

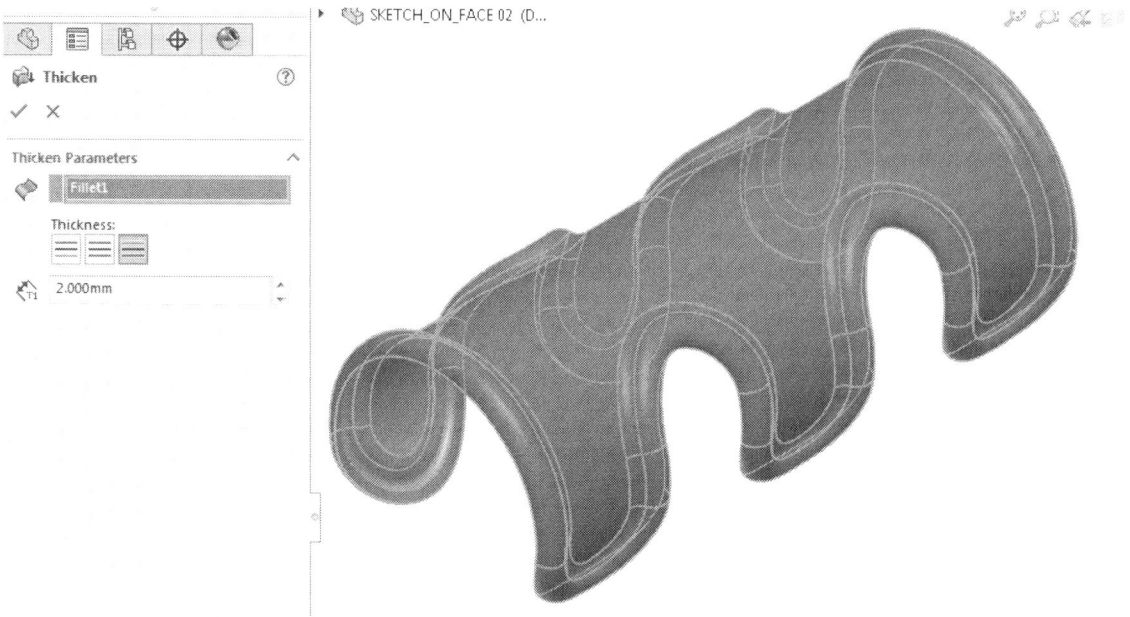

Figure 143 - Thickening a Surface

Click OK and your part should now look as shown in the following image.

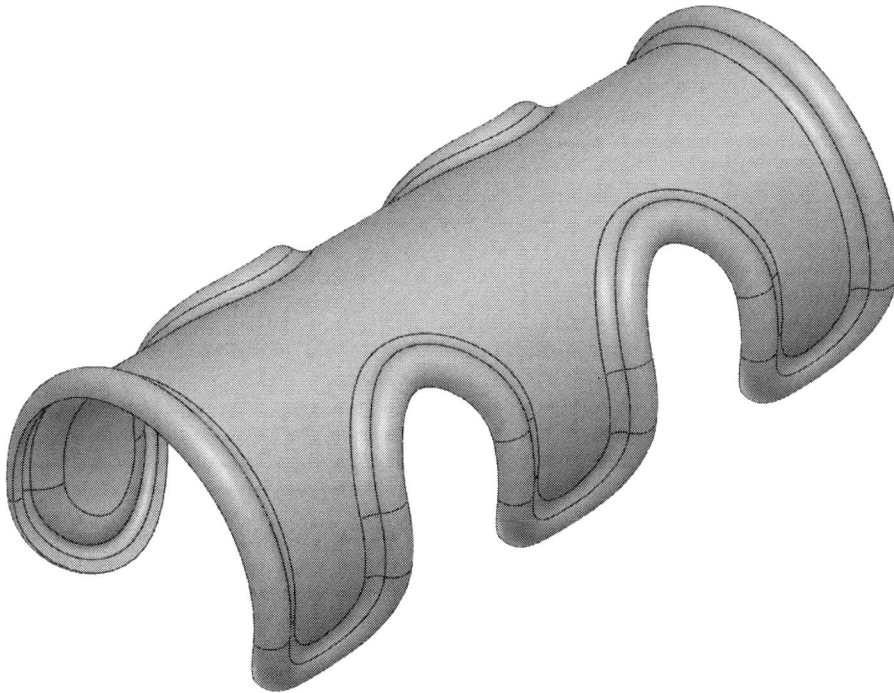

Figure 144 - Thickened Surface

Save and close your part and move on to Chapter Four.

CHAPTER 4

In this Chapter we will continue working with the RF_start.sldprt Part from where we left in Chapter 3.

In this Chapter, you are required to create the final missing surfaces using the Boundary Surface Feature - see the following image.

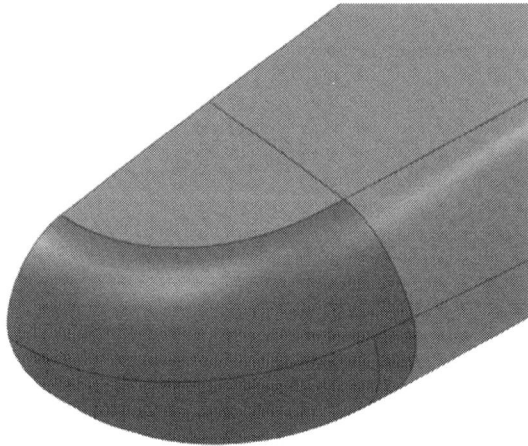

Figure 145 - Recreated Surfaces

The recreated surfaces can either be homogenous or made up of multiple surfaces. Both results are acceptable.

The recreated surface(s) above the parting line should be tangent to the other surfaces above the parting line.

The recreated surface(s) below the parting line should be tangent to the other surfaces below the parting line.

You are then required to measure and provide the total surface area of the recreated surfaces in square millimeters to two decimal places.

CREATING TWO-POINT SPLINES WITH TANGENCY

Hide All Sketches except for Sketch AAA.

Start a new sketch on the Right Plane.

Click Convert Entities (Sketch toolbar) or Tools > Sketch Tools > Convert Entities.

In the PropertyManager, under Entities to Concert, select the two edges in the Graphics Area shown in the following image.

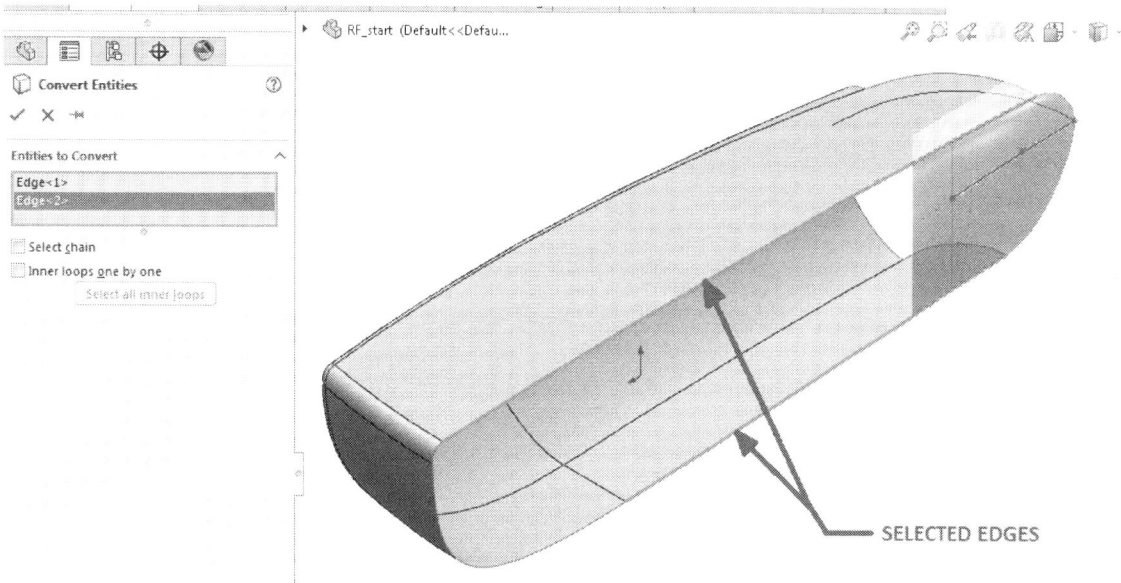

Figure 146 - Selected Edges - Convert Entities

Click OK.

Press and Hold the Ctrl Key on your keyboard and then select the two sketch entities in the Graphics Area. Select the For Construction checkbox under Options in the PropertyManager as shown in the following image.

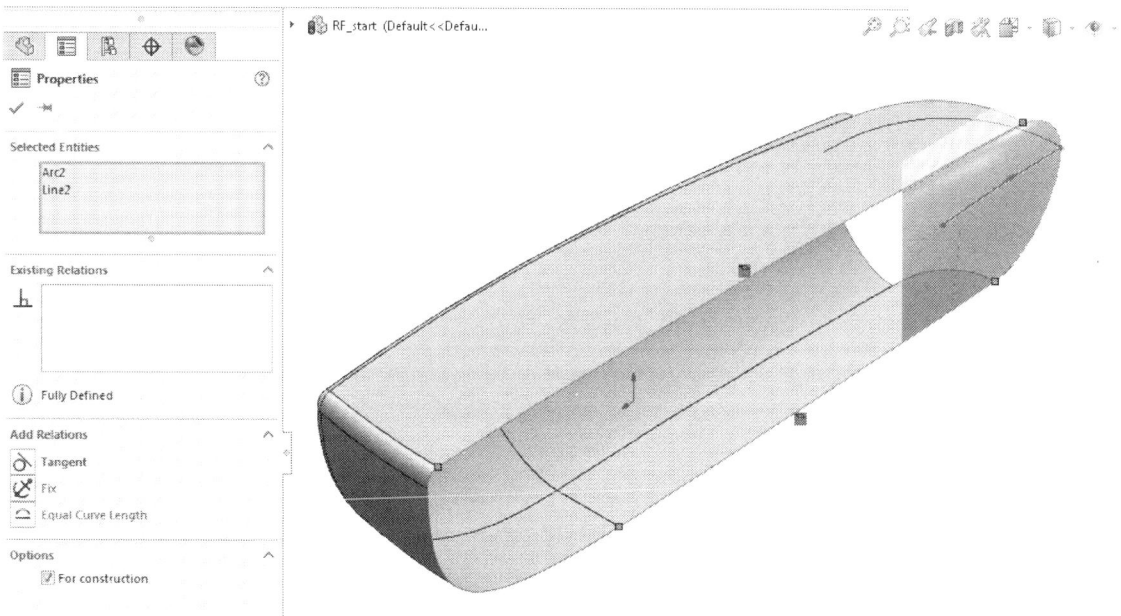

Figure 147 - Covert Sketch Entities to Construction Lines

Click OK.

Your Part should now look as shown in the following image.

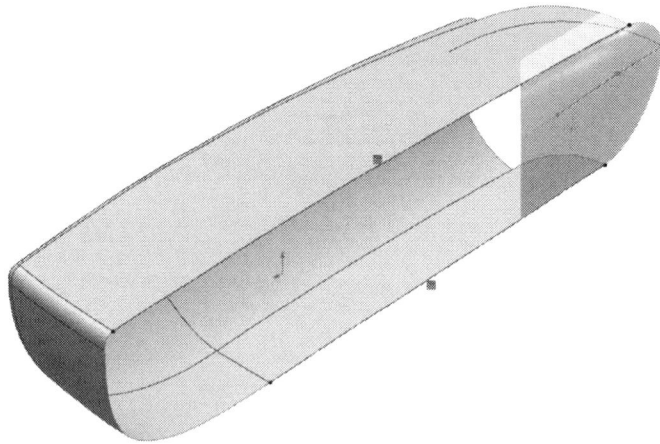

Figure 148 - Part Current Status

Click Normal To (View Orientation flyout – Heads-up View toolbar).

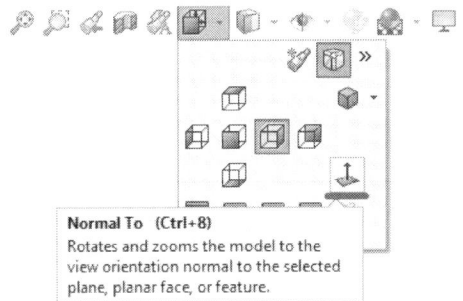

Normal To (Ctrl+8)
Rotates and zooms the model to the view orientation normal to the selected plane, planar face, or feature.

Figure 149 - Rotating view to be normal to the sketch plane

Your part's view orientation should now look as shown in the following image.

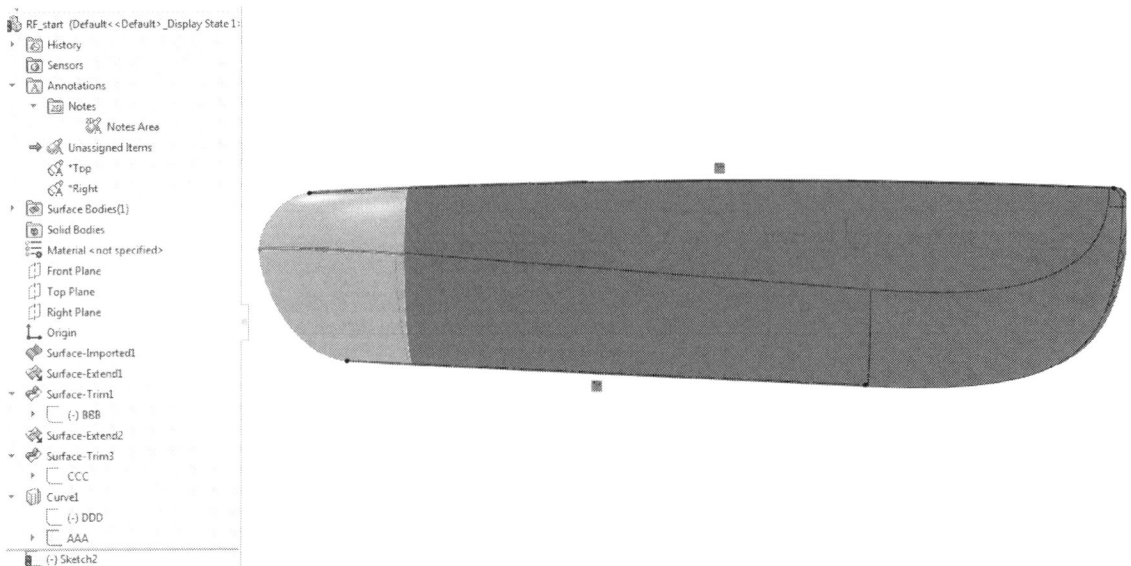

Figure 150 - Part current status - Normal To View

Click Spline (Sketch toolbar) or Tools > Sketch Entities > Spline. The pointer changes to ✎. Click to place the first point coincident with the end point of the curve under sketch AAA (Point 1). Drag out the first segment and click the next point coincident with the top end point of one of the construction lines we created in sections above (Point 2). Hit the Escape Key on your keyboard to complete the spline as shown in the following image.

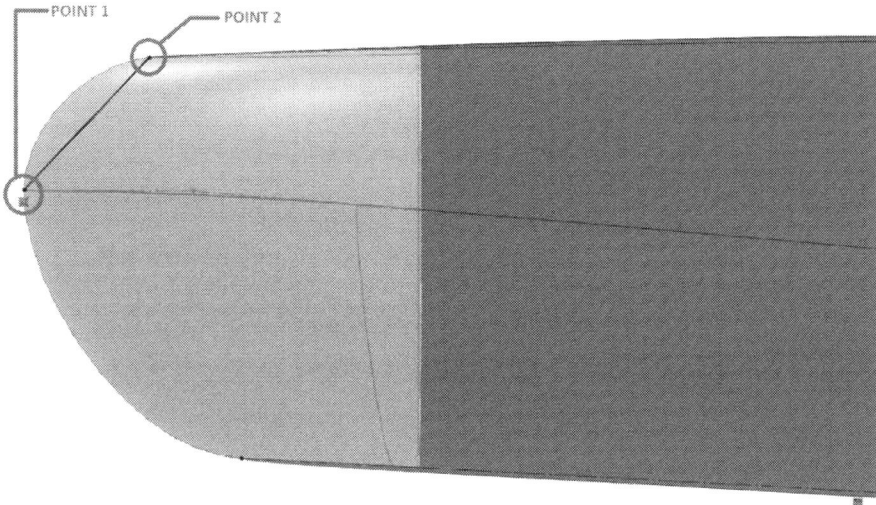

Figure 151 - Two Point Spline

SPLINE HANDLES

Select the Spline we just created above and you will notice what are called Spline handles appear in Grey - **NB:** These may not be visible in the following image due to printing contrast hence you may have to check your Graphics Area in Solidworks.

Figure 152 - Spline Handles appear when you select the spline

Spline handles allow you to control the geometry of the spline about each point you placed e.g. Point 1 or Point 2 in our example. A spline handle includes multiple handle types to control the weight and direction (vector) of the spline at that point.

You can control weight and direction:

- Individually or together

- Asymmetrically: one side only

- Symmetrically: both sides together

There are three parts to the Spline Handle as shown in the following image:

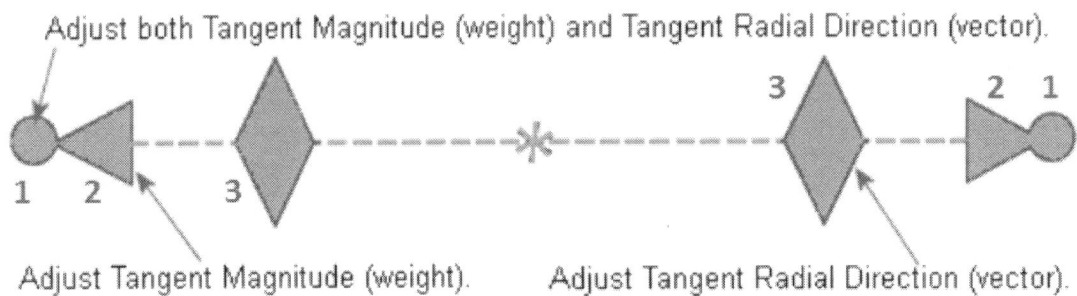

Figure 153 - Structure of a Spline Handle

- The diamond shape (3) allows you to change the angular direction through a point.

- The arrow (2) allows you to modify the tangent magnitude.

- The point on the end (1) allows you to modify both the angular direction and tangent magnitude at the same time.

SHOW / HIDE SPLINE HANDLES

If Spline Handles do not appear, you may Click Tools > Spline Tools > Show Spline Handles.

EDITING SPLINES

Click or Select the Two Point Spline we created to show the Spline Handles. Move your mouse pointer over the diamond shape on the lower Spline Handle and when your mouse changes to

press and hold the mouse left hand side button then drag the spline handle to an angle of +/- 45 degrees as shown in the following image.

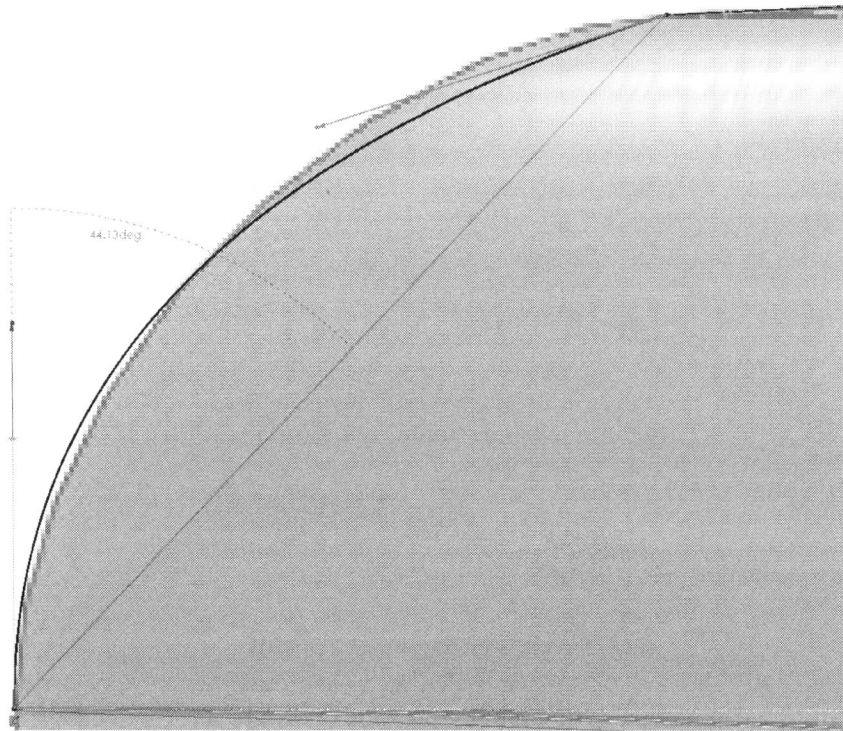

Figure 154 - Editing a Spline using Spline Handles

Release the LHS mouse button and click anywhere in the Graphics Area to finish and accept the action.

You will notice that after you have moved the lower Spline Handle it has gone from grey to blue, indicating it was activated and moved. Deselecting the spline causes the Top Grey Spline Handle to disappear, but the blue one remains visible as shown in the following image - **NB:** Colours will not be visible in the black and white print out version of this book hence you may only observe the aforementioned colours in the Graphics Area in Solidworks.

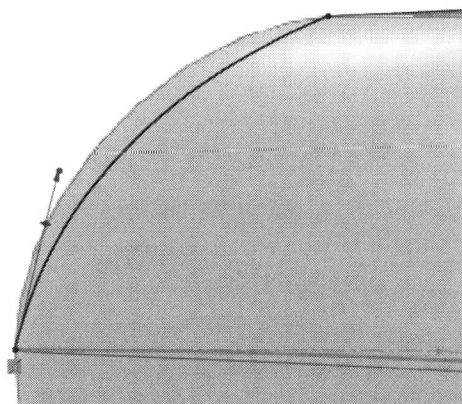

Figure 155 - Editing a Spline using Spline Handles

Select the Spline and move your mouse pointer over the diamond shape on the top Spline Handle and when your mouse changes to ⬚ press and hold the mouse left hand side button then drag the spline handle to an angle of +/- 18 degrees as shown in the following image.

Figure 156 - Editing a Spline using Spline Handles

Release the LHS mouse button and click anywhere in the Graphics Area to finish and accept the action.

Now both Spline Handles remain blue and visible as shown in the following image after deselecting the spline, showing that both have been activated and moved or manipulated.

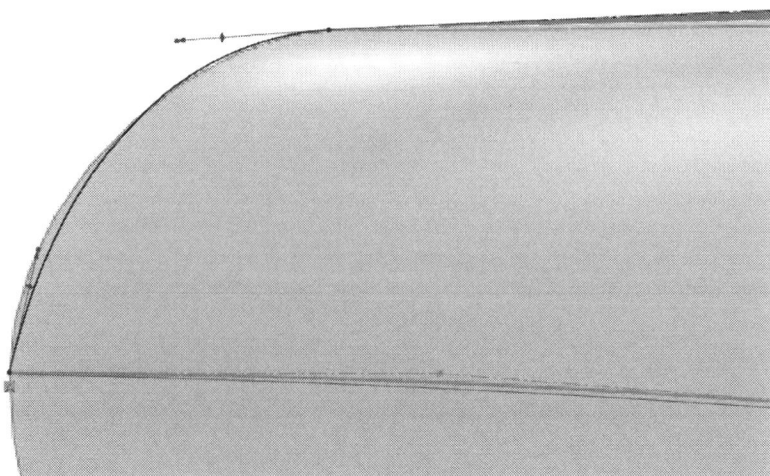

Figure 157 - Editing a Spline using Spline Handles

ADDING RELATIONS

Click Add Relations (Dimensions/Relations toolbar) as shown in the following image or Click Tools > Relations > Add.

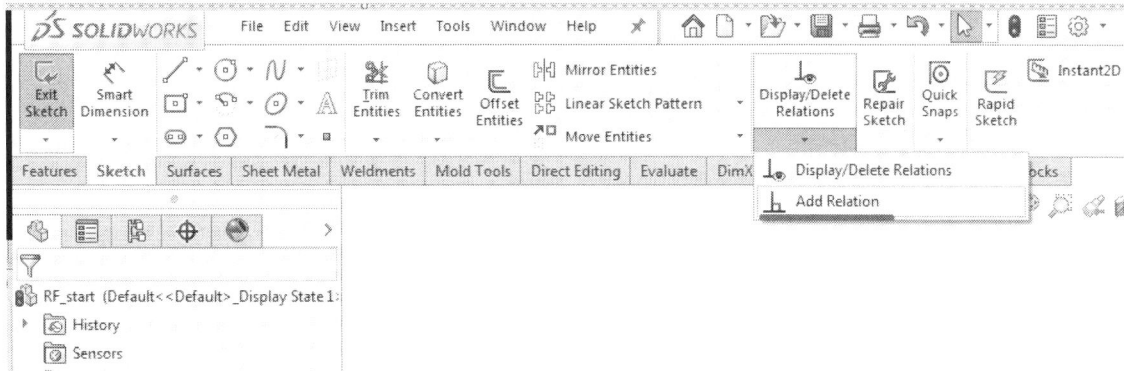

Figure 158 - Add Relations

In the PropertyManager under Selected Entities, select the spline and the top arc (construction line). Under Add Relations, click on Tangent to add the Tangent Relation under Existing Relations as shown in the following image.

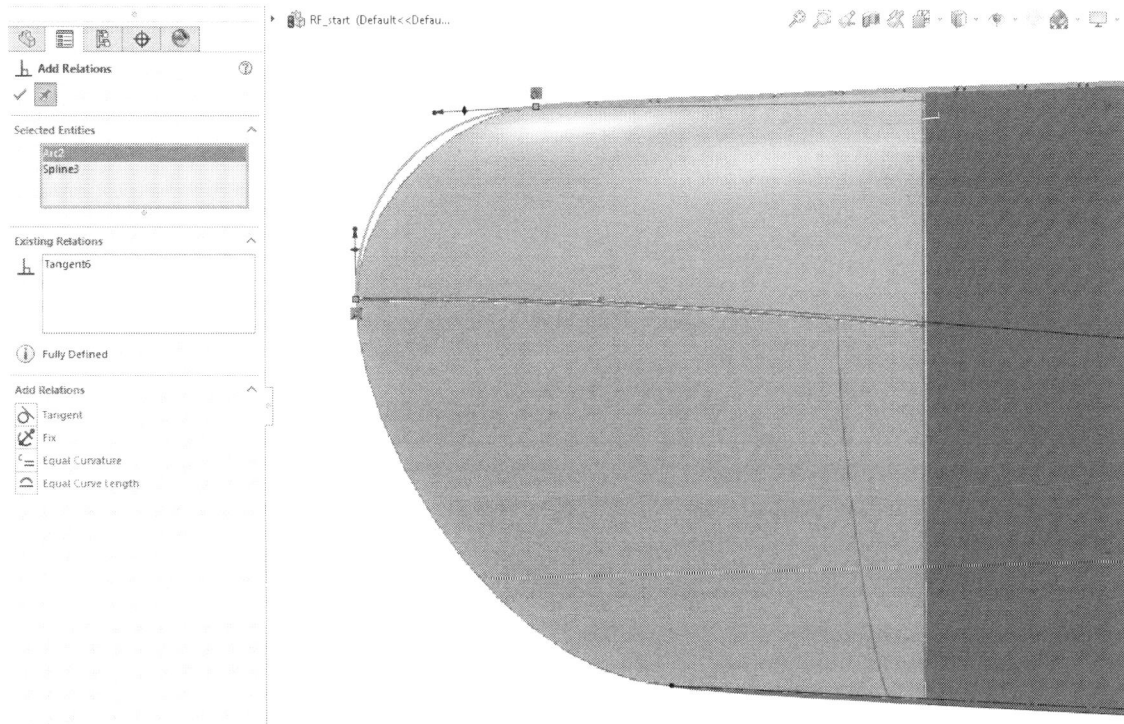

Figure 159 - Adding a Tangent Relation

Click Ok to apply the tangent relation and finish the Add Relations Command. Click the diamond part of the lower Spline Handle and select the Make Vertical Command in the popup menu to add a Vertical Relation as shown in the following image then Click Ok in the Spline PropertyManager (Top LHS Corner).

Figure 160 - Adding a Vertical Relation

HIDE / SHOW SKETCH RELATIONS

Click View > Hide/Show > Sketch Relations or Click View Sketch Relations on the Heads-up View Toolbar to select or clear display of the sketch relation icons in the Graphics Area. If you clear View > Hide/Show > Sketch Relations, but you select a sketch entity in an open sketch, the sketch relation icons appear. With sketch relations set to show, your sketch should now look as shown in the following image.

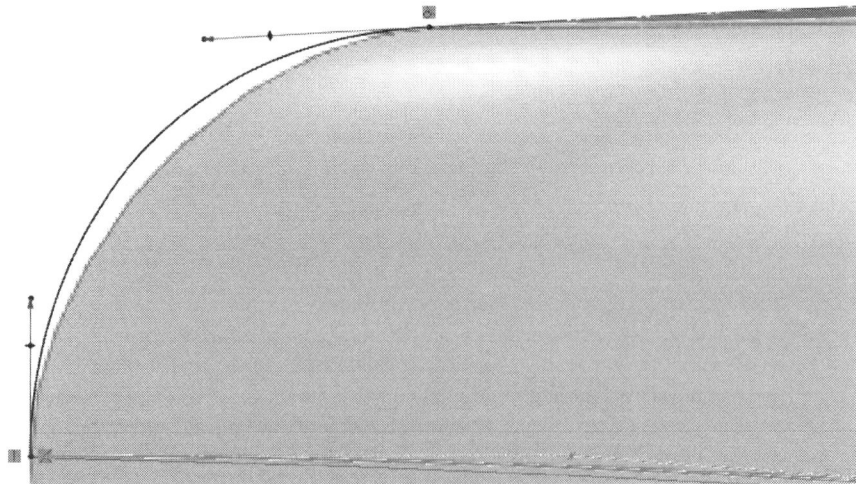

Figure 161 - Relations on Spline

You will notice that the Diamond Part in both Spline handles has changed in colour from Blue to Black showing that the Angular Direction of each of the Spline Handles is now Fully Defined.

However, the arrow portion of both Spline Handles still remains blue in colour which means the Tangent Magnitude of both Spline Handles is still Under Defined.

Drag the arrow head handle of the Top Spline Handle to control tangency weight and move the spline closer to the sketch picture we are tracing. Do the same with the Bottom Spline Handle until you get the spline as close to the sketch picture as possible.

DIMENSIONING TO SPLINE HANDLES

You can add dimensions to spline handles for Tangent Weighting and or Tangent Radial Direction.

Click Smart Dimension (Dimensions/Relations toolbar) or Tools > Dimensions > Smart.

Select the Top Handle on the arrow end, and click to place the dimension.

Set the dimension in the Modify box as shown in the following image to 10.75mm.

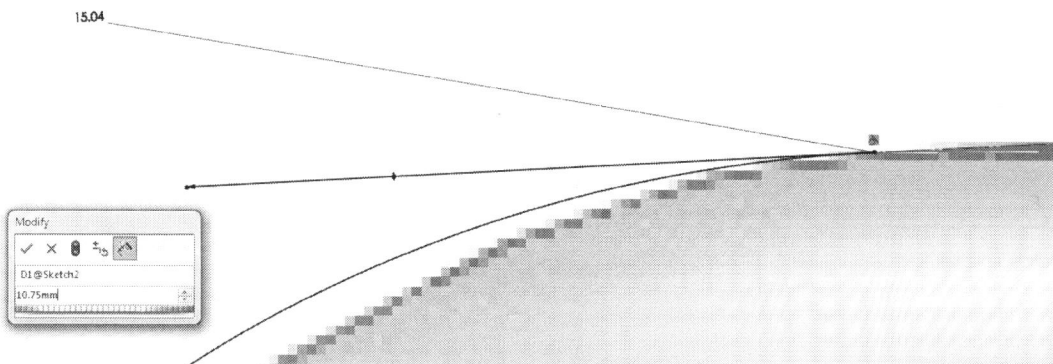

Figure 162 - Dimensioning to Spline Handles

Click Ok and your sketch should now appear as shown in the following image.

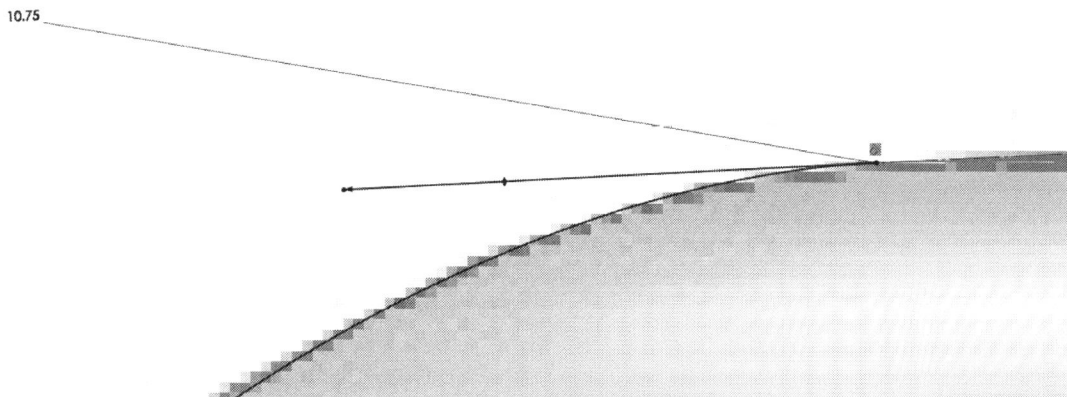

Figure 163 - Dimensioning to Spline Handles

93

Repeat the same process for the Bottom Spline Handle and add a dimension of 11.50mm. Your Sketch should now appear as shown in the following image.

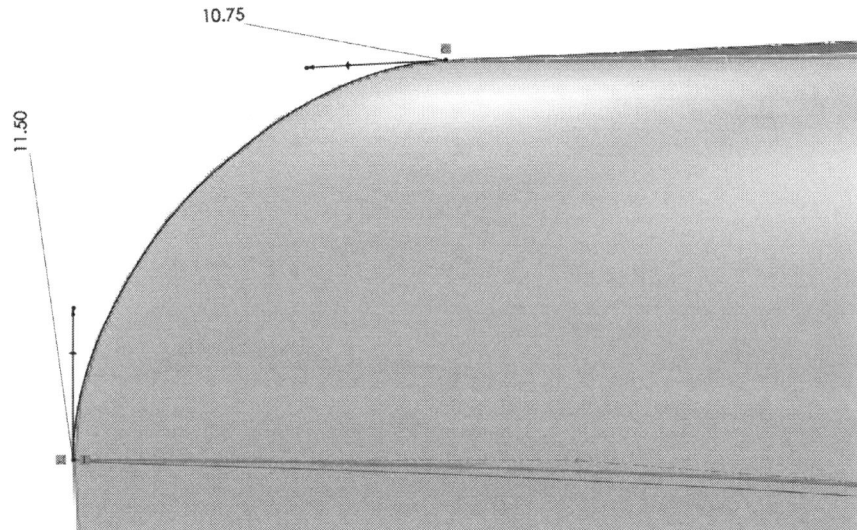

Figure 164 - Fully Defined Two Point Spline

You will notice that all parts of both spline handles are now black in colour showing that they are now Fully Defined.

Click Spline (Sketch toolbar) or Tools > Sketch Entities > Spline. The pointer changes to ⟲. Click to place the first point coincident with the end point of the curve under sketch AAA (Point 3). Drag out the first segment and click the next point coincident with the bottom end point of one of the Two Point Spline we created in sections above (Point 1). Hit the Escape Key on your keyboard to complete the spline as shown in the following image.

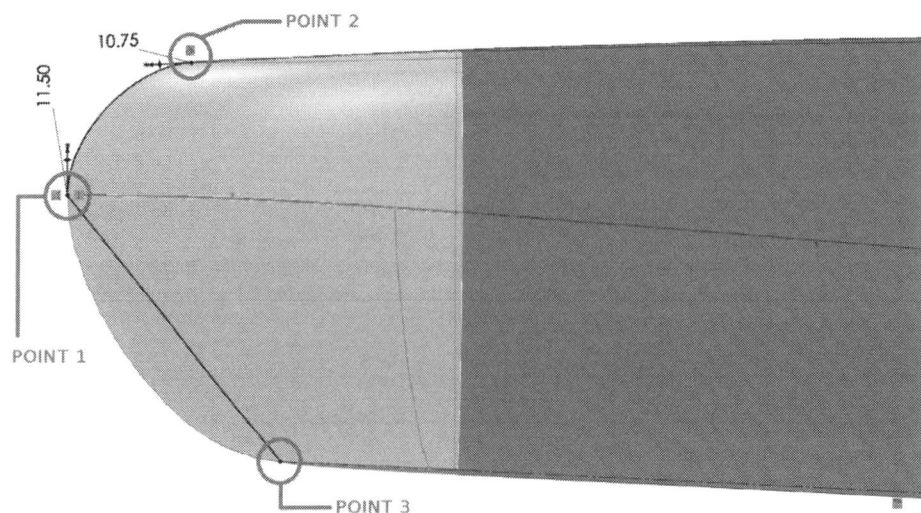

Figure 165 - Two Point Spline

Move Spline Handles on this second Two point Spline, add relations and dimension it as shown in the following image using the same procedure we used with the first spline.

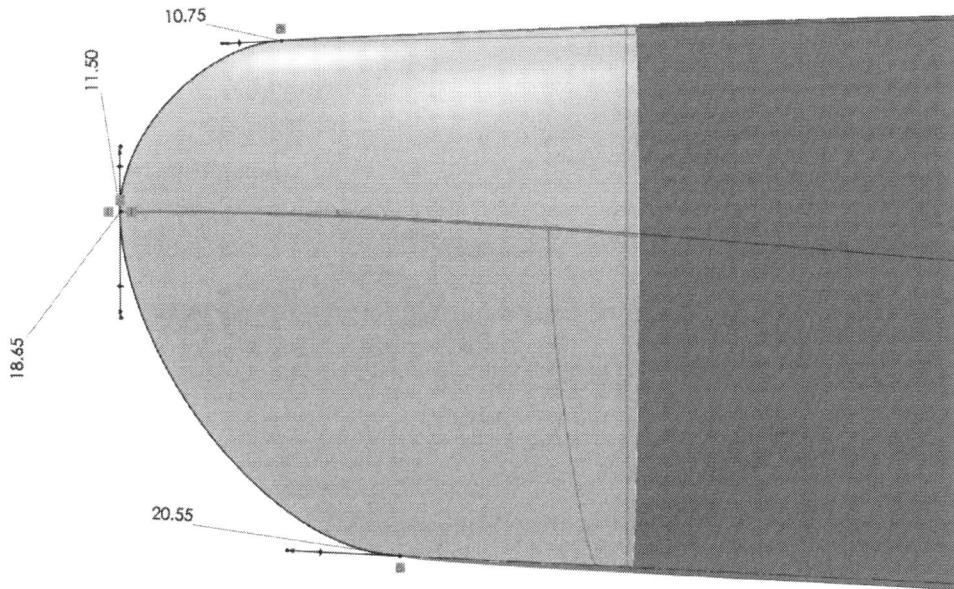

Figure 166 - Two Point Spline Adding Relations and Dimensioning

Exit the Sketch Mode. Save your part.

Your part should now look as shown in the following image with sketches AAA, BBB, CCC and DDD hidden.

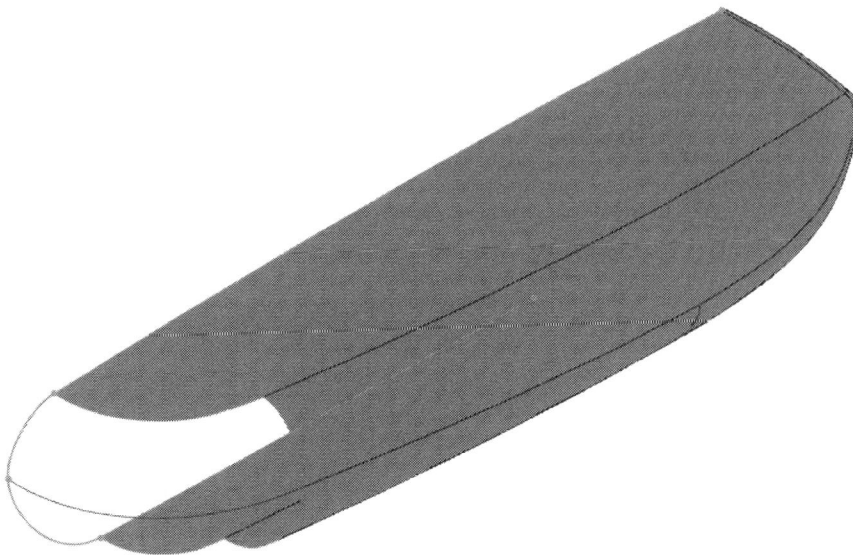

Figure 167 - Part current status

BOUNDARY SURFACE

The Boundary Surface feature creates a surface that spans two or more profiles in one or two directions - **Direction 1** and or **Direction 2**. Sketch curves, edges, faces, or other sketch entities can be used to control the shape of a boundary feature.

To create a Boundary Surface Click Boundary Surface (Surface toolbar) or Insert > Surface > Boundary Surface.

ACCESSSING THE SELECTION MANAGER

Right Click anywhere in the Graphics Area and choose SelectionManager from the popup menu as shown in the following image.

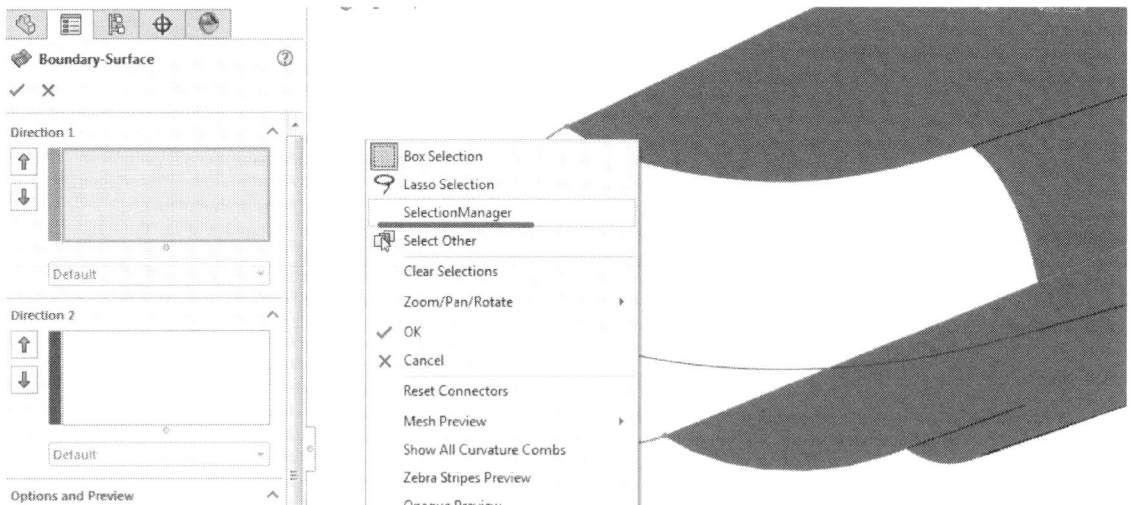

Figure 168 - Accessing the SelectionManager

The SelectionManager appears.

In the SelectionManager, click the pushpin so that the SelectionManager remains available. Choose the Select Group Option as shown in the following image.

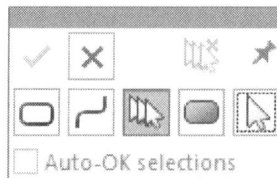

Figure 169 - SelectionManager - Select Group Option

In the PropertyManager under Direction 1, select a sketch entity as shown in the following image and Click Ok in the SelectionManager.

96

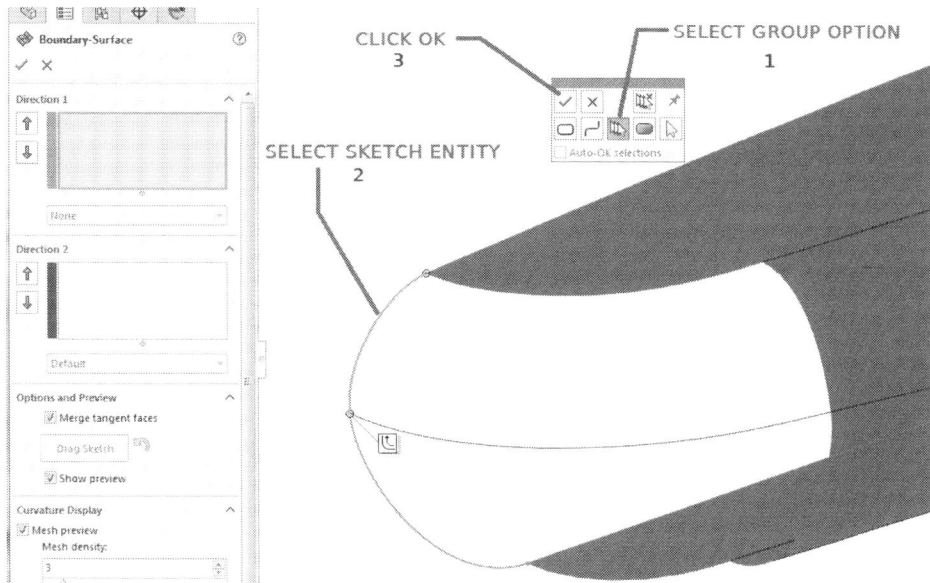

Figure 170 - Choosing a Direction 1 Profile

Open Group<n> appears under Direction 1 in the Boundary Surface PropertyManager. For the curve influence option select **Normal to Profile** which applies a tangency constraint normal to the sketch entity as shown in the following image.

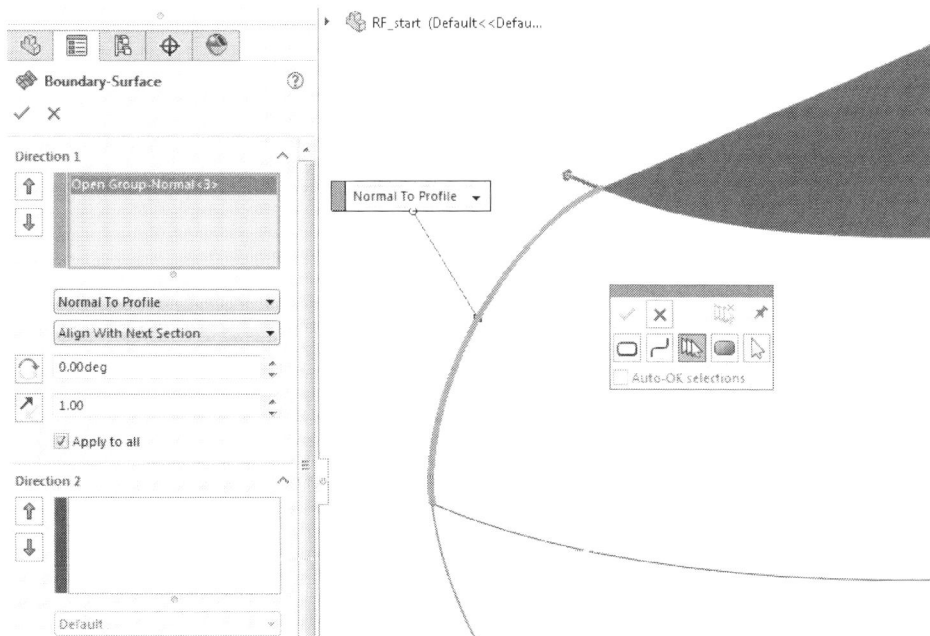

Figure 171 - Boundary Surface - Curve Influence

In the PropertyManager still under Direction 1, select an edge as shown in the following image and Click Ok in the SelectionManager. Open Group<n> appears under Direction 1 in the Boundary Surface PropertyManager. For the curve influence option select **Tangency To Face** which makes the adjacent face tangent at the selected edge as shown in the following image.

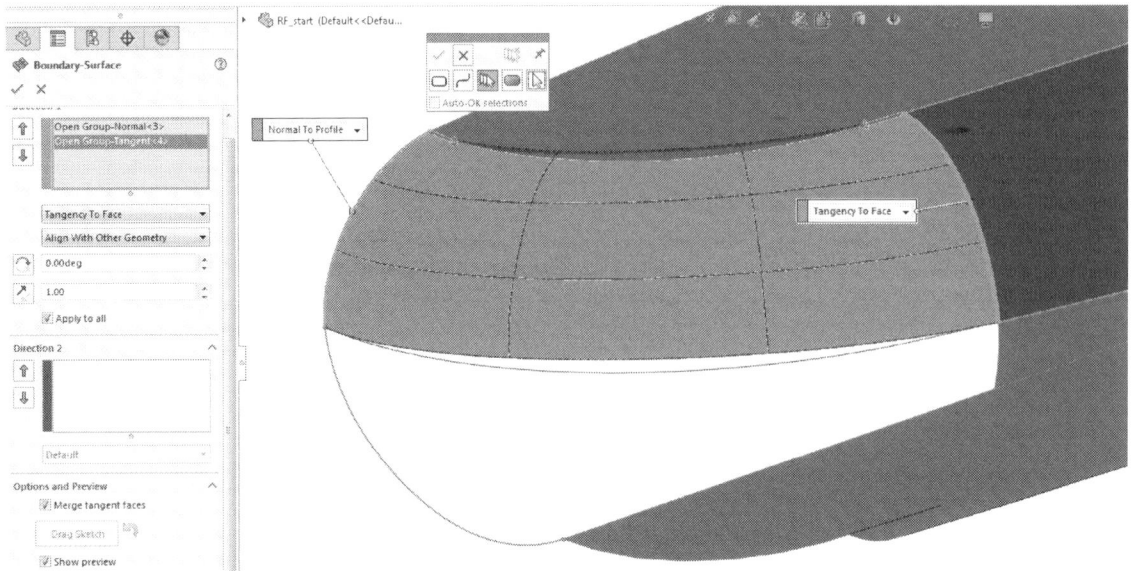

Figure 172 - Choosing a Direction 1 Profile

In the PropertyManager under Direction 2, select an edge as shown in the following image and Click Ok in the SelectionManager. Open Group<n> appears under Direction 2 in the Boundary Surface PropertyManager. For the curve influence option select **Tangency To Face** which makes the adjacent face tangent at the selected edge as shown in the following image.

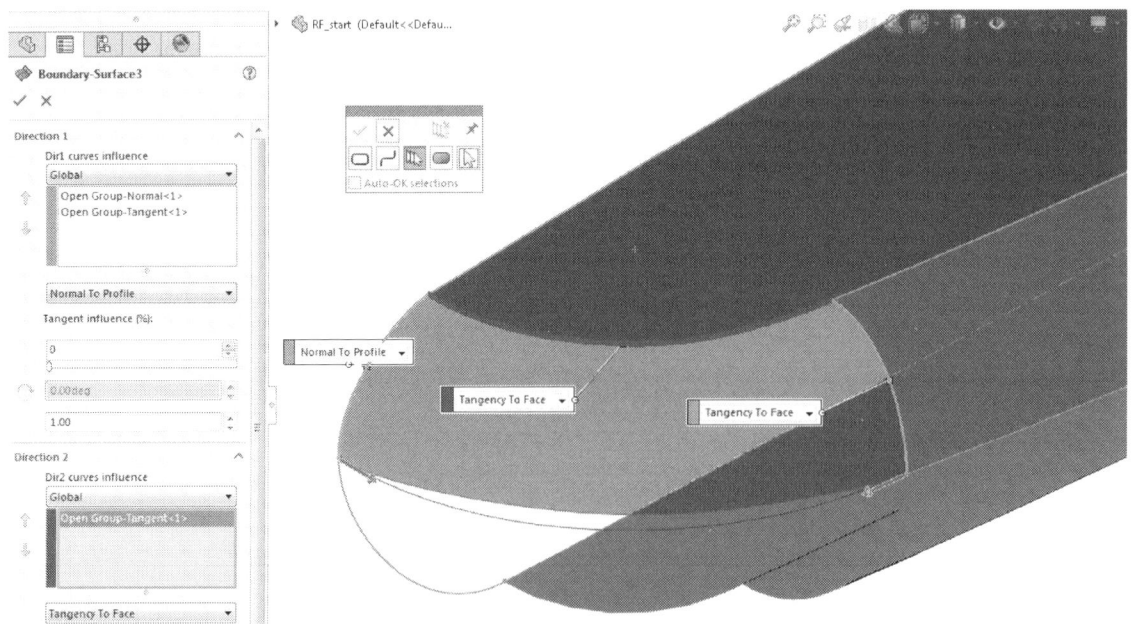

Figure 173 - Choosing a Direction 2 Profile and applying a Tangency To Face Curve Influence

In the PropertyManager still under Direction 2, select the Curve in the Flyout Feature Manager Design Tree as shown in the following image. Curve1 appears under Direction 2 in the Boundary

Surface PropertyManager. For the curve influence option select **None** as shown in the following image.

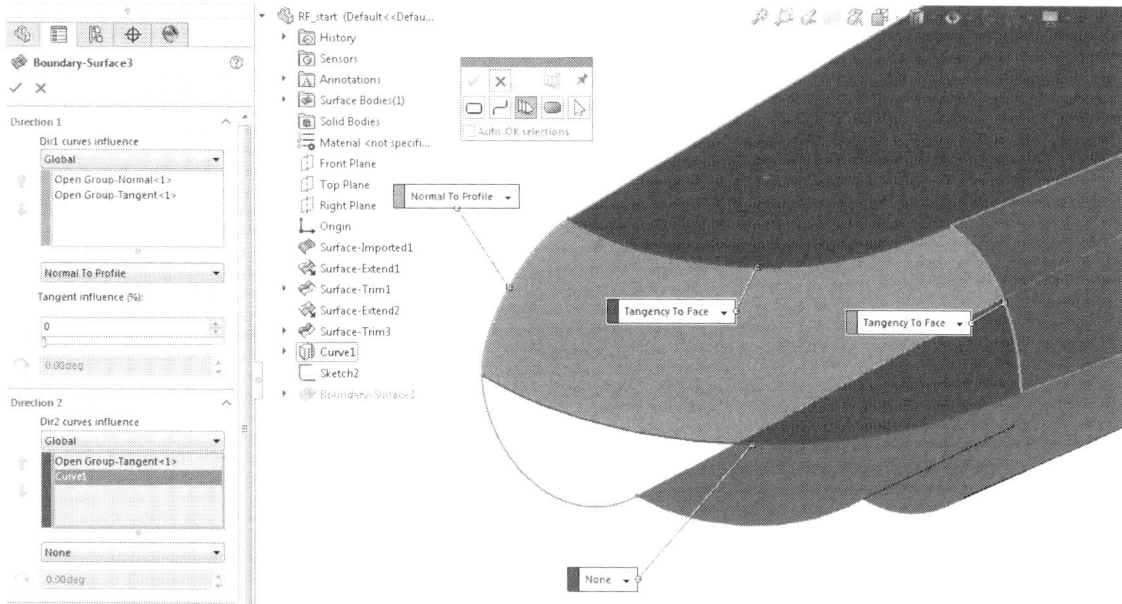

Figure 174 - Choosing a Curve as a Direction 2 Profile

TIP: The availability of curves influence options depends on the geometry of the curves you select for a direction.

TIP: The two directions - **Direction 1** and **Direction 2** are interchangeable and give the same results regardless of whether you select the first set or second set of curves as Direction 1 or Direction 2.

TIP: If you choose curves in two directions - **Direction 1** and **Direction 2** - all the curves in Direction 1 must intersect all the curves in Direction 2 to form a closed 2D or 3D boundary.

TIP: You don't have to have selections in both directions (**Direction 1** and **Direction 2)** to create a Boundary Surface - you can create a Boundary Surface with selections in one direction only - see an example in the following image. However, a minimum of two selections are required. You may open the part file named ONE DIRECTION BOUNDARY SURFACE.sldprt from the Chapter 4 Downloaded Folder. Notice the use of Derived Sketches in that part file. A **Derived Sketch** is basically a copy of another sketch that belongs to the same part or a copy of a sketch from another sketch in the same assembly. When you derive a sketch from an existing sketch, you are assured that the two sketches will retain the characteristics that they share in common. Changes that you make to the original sketch are reflected in the derived sketch. Instead of using a Derived Sketch, you may also create, save and use a linked **Sketch Block**

(select the <u>Link to File</u> checkbox when inserting the Sketch Block) which will be more versatile in terms of being able to be used in different parts and or assemblies.

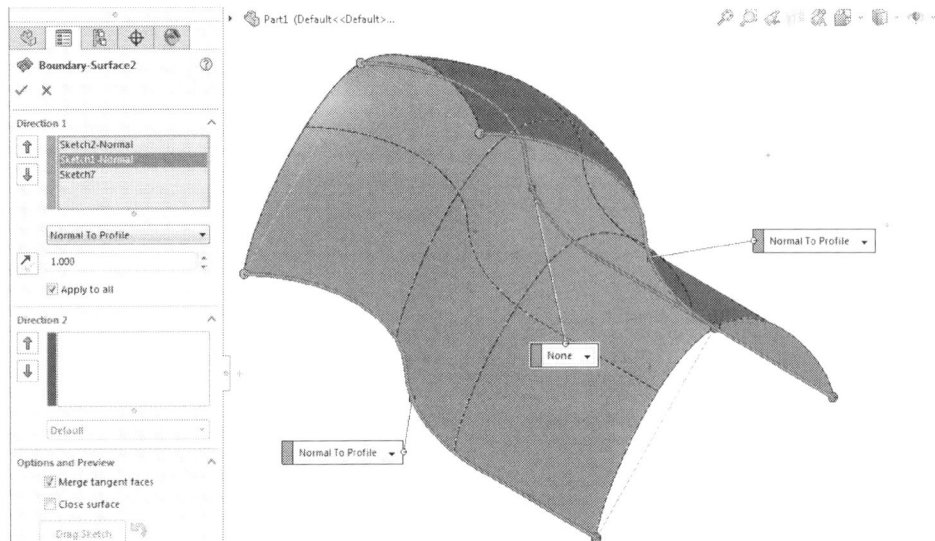

Figure 175 - Boundary Surface using Selections in one Direction Only

Save and close the example part - ONE DIRECTION BOUNDARY SURFACE.sldprt.

CURVATURE DISPLAY

In the Boundary Surface PropertyManager under Curvature Display, there is three options for Mesh Preview with the ability to control mesh density, Zebra Stripes *(see the following image)* and Curvature Combs.

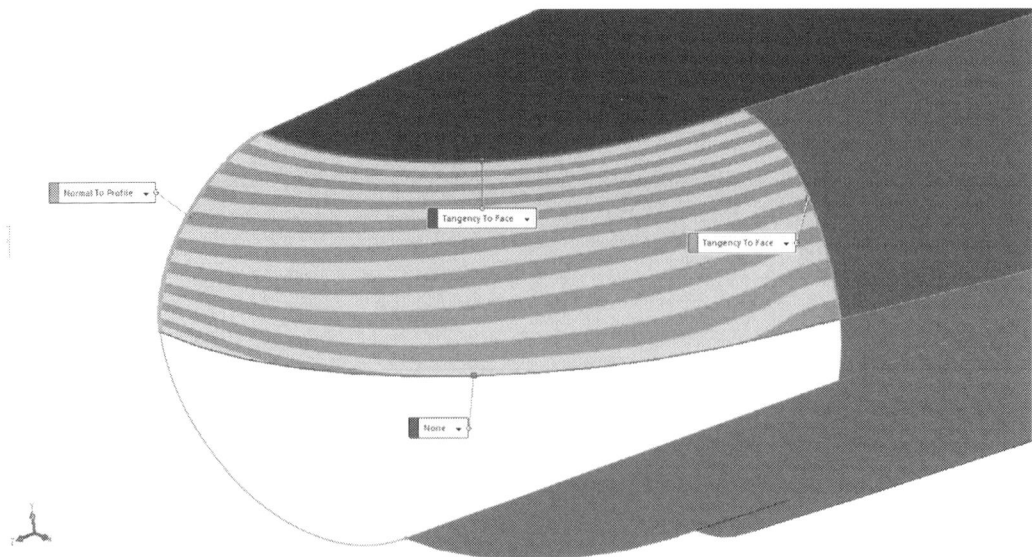

Figure 176 - Curvature Display - Zebra Stripes

The aforementioned Curvature Display Options are qualitative tools used to basically check the quality of the surface being created such as ensuring that there are no abrupt changes in curvature e.t.c.

TIP : Curvature is an inverse of radius i.e. C = 1/R where C: - Curvature and R: - Radius. Thus the bigger the radius the smaller the curvature and the smaller the radius the bigger the curvature.

Click Ok and your part should now look as shown in the following image.

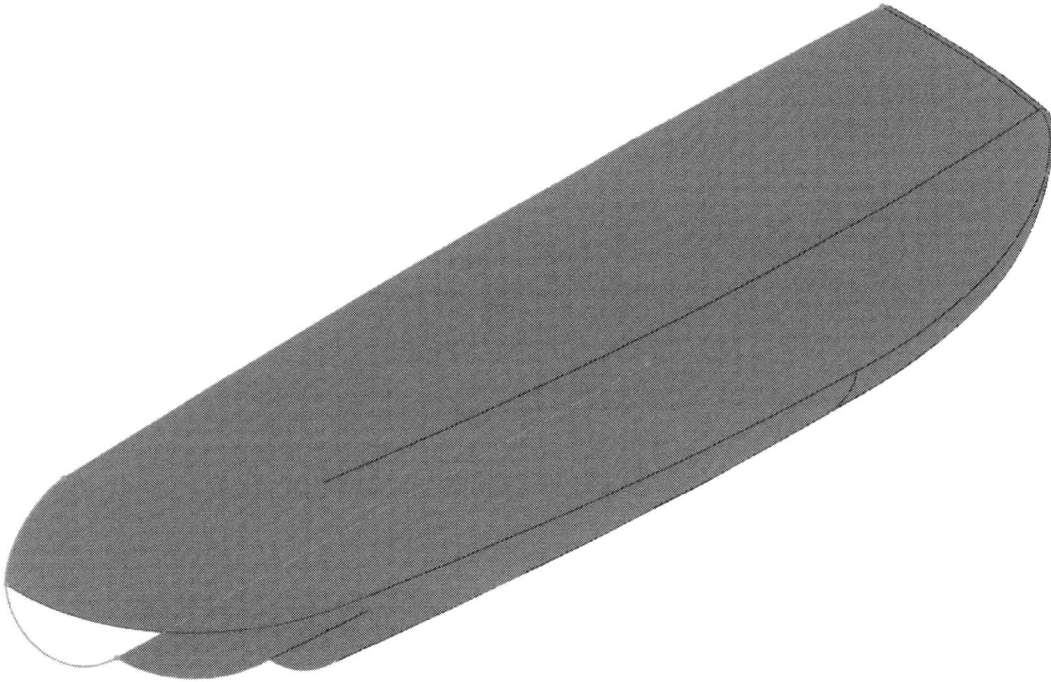

Figure 177 - Part current status

USING THE KNIT SURFACE TOOL

The Knit Surface Tool is used to combine two or more surfaces into one.

Click Knit Surface on the Surfaces toolbar, or click Insert > Surface > Knit.

In the PropertyManager, under Selections:

Select surfaces for Surfaces and Faces to Knit in the graphics area or in the Fly Out Feature Manager Design Tree as shown in the following image.

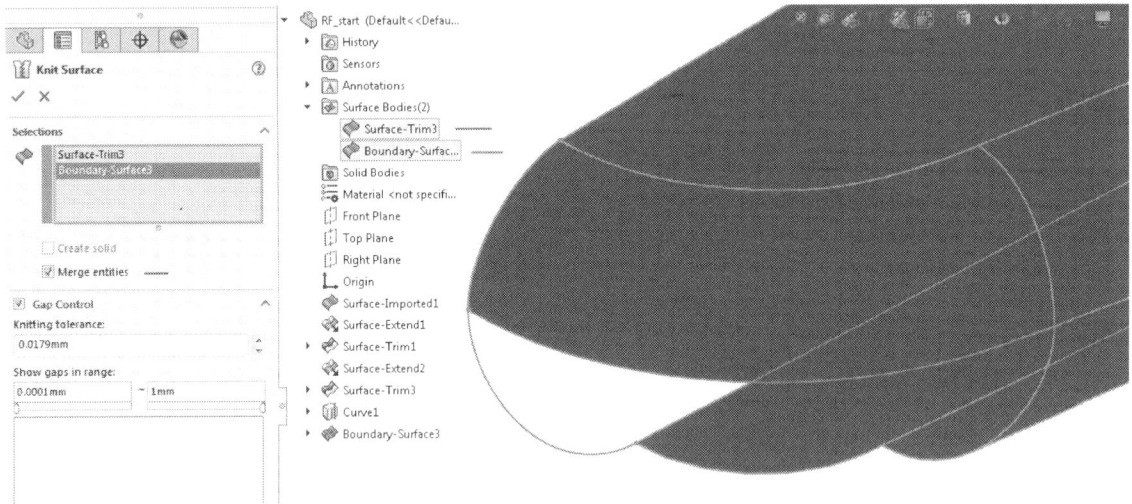

Figure 178 - Knit Surface

Select the Merge Entities Checkbox.

Click OK.

The two surface bodies are now combined into one surface body named Surface-Knit<n> - the change is also reflected in the Surface Bodies Folder in the Feature Manager Design Tree as shown in the following image from two surface bodies to one surface body.

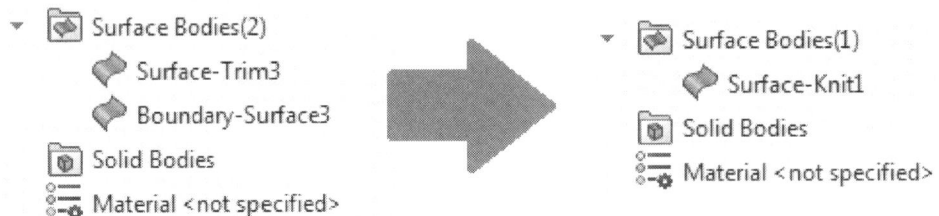

Figure 179 - Knit Surface change in the Surface Bodies Folder in the Feature Manager Design Tree

If the Surface Bodies Folder is not visible in your Feature Manager Design Tree, Click Tools > Options > System Options > FeatureManager and under Hide/show tree items select Show from the Surface Bodies Drop Down Box.

ZEBRA STRIPES

Zebra Stripes are used to visually determine what type of boundary condition exists between surfaces. They are one of the subjective Model Quality Evaluation Tools available in Solidworks that offer visual cues about the quality of a given geometry.

There are three **Boundary Conditions** zebra stripes can help you identify namely:

Contact (C0 Continuity) - the stripes do not match at the boundary.

Figure 180 - Zebra Stripes C0 Continuity

Tangent (C1 Continuity) - the stripes match but there is an abrupt change in direction or a sharp corner.

Figure 181 - Zebra Stripes C1 Continuity

Curvature Continuous (C2 Continuity) - the stripes continue smoothly across the boundary.

Figure 182 - Zebra Stripes C2 Continuity

VIEWING A PART WITH ZEBRA STRIPES

Click Zebra Stripes (View toolbar) or click View > Display > Zebra Stripes. In the PropertyManager, set options, then click OK. Your part should appear as shown in the following image.

Figure 183 - Viewing a part with Zebra Stripes

To turn off zebra stripes, click Zebra Stripes (View toolbar) or click View > Display > Zebra Stripes again to clear the zebra stripes.

VIEWING SELECTED FACES WITH ZEBRA STRIPES

You can improve the accuracy of your display by selecting only those faces that you want displayed with zebra stripes. To view a face or a combination of faces with zebra stripes, right-click the faces in the graphics area and select Zebra Stripes as shown in the following image - hold the Ctrl Button on your keyboard, select the two faces in the graphics area, release the Ctrl button and right click anywhere in the graphics area then select zebra stripes.

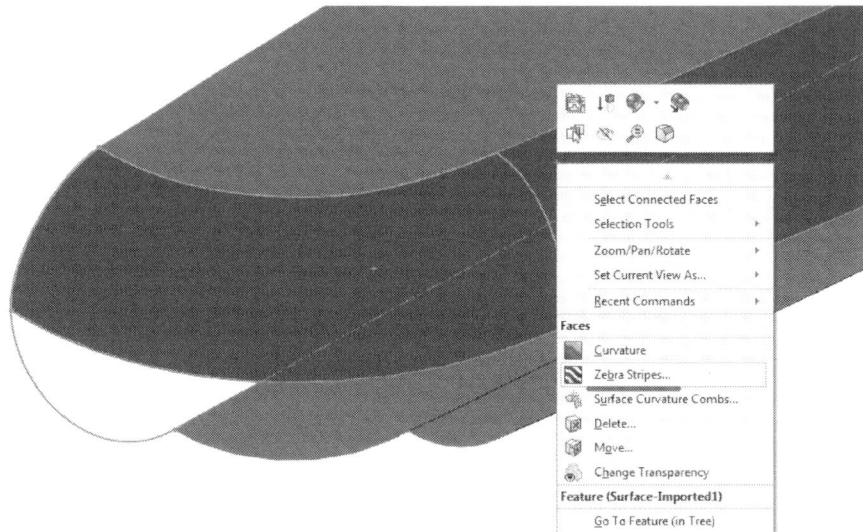

Figure 184 - Viewing selected faces with Zebra Stripes

Your part should now appear as shown in the following image.

Figure 185 - Viewing selected faces with Zebra Stripes

104

Click Zebra Stripes (View toolbar) or click View > Display > Zebra Stripes again to clear or turn off the zebra stripes.

Save your part.

CREATING THE SECOND BOUNDARY SURFACE

Click Boundary Surface (Surface toolbar) or Insert > Surface > Boundary Surface.

Right Click anywhere in the Graphics Area and choose SelectionManager from the popup menu as shown in the following image.

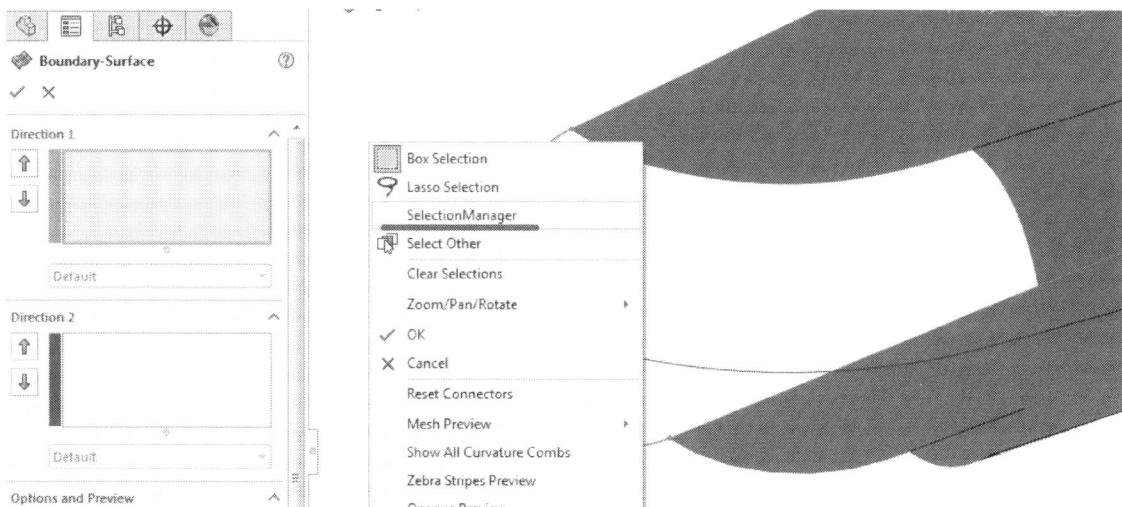

Figure 186 - Accessing the SelectionManager

The SelectionManager appears.

In the SelectionManager, click the pushpin so that the SelectionManager remains available. Choose the Select Group Option as shown in the following image.

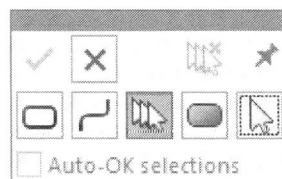

Figure 187 - SelectionManager - Select Group Option

In the PropertyManager under Direction 1, select a sketch entity as shown in the following image and Click Ok in the SelectionManager.

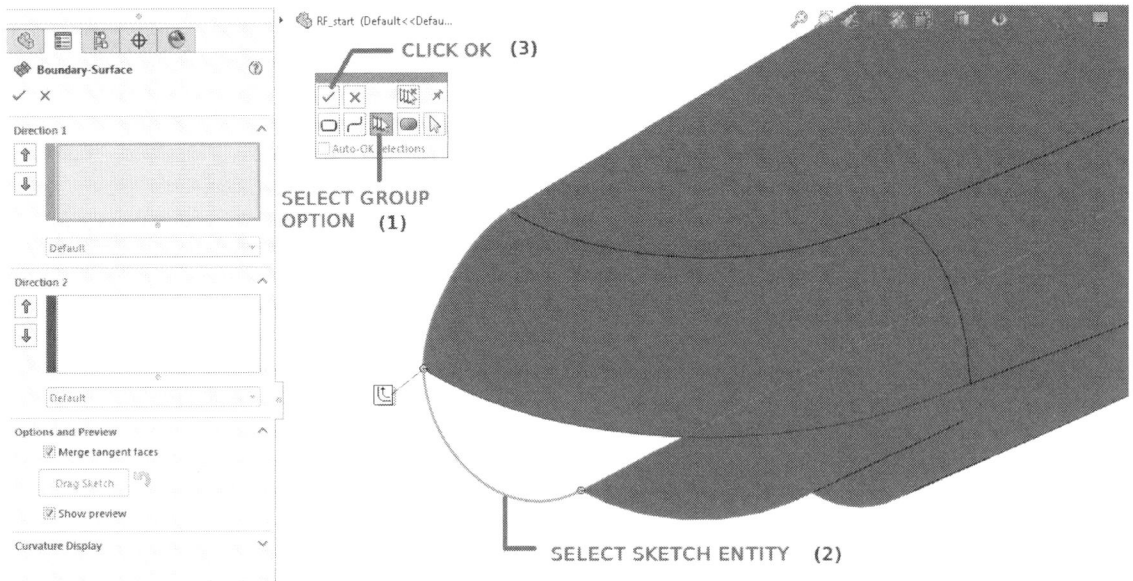

Figure 188 - Choosing a Direction 1 Profile

Open Group<n> appears under Direction 1 in the Boundary Surface PropertyManager. For the curve influence option select **Normal to Profile** which applies a tangency constraint normal to the sketch entity as shown in the following image.

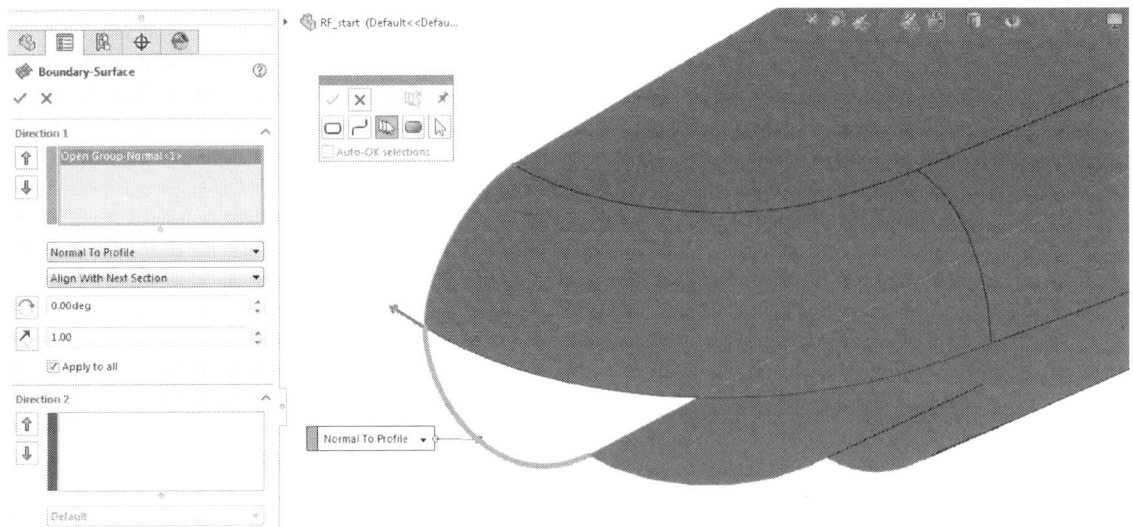

Figure 189 - Boundary Surface - Curve Influence

In the PropertyManager still under Direction 1, select an edge as shown in the following image and Click Ok in the SelectionManager. Open Group<n> appears under Direction 1 in the Boundary Surface PropertyManager. For the curve influence option select **Tangency To Face** which makes the adjacent face tangent at the selected edge as shown in the following image.

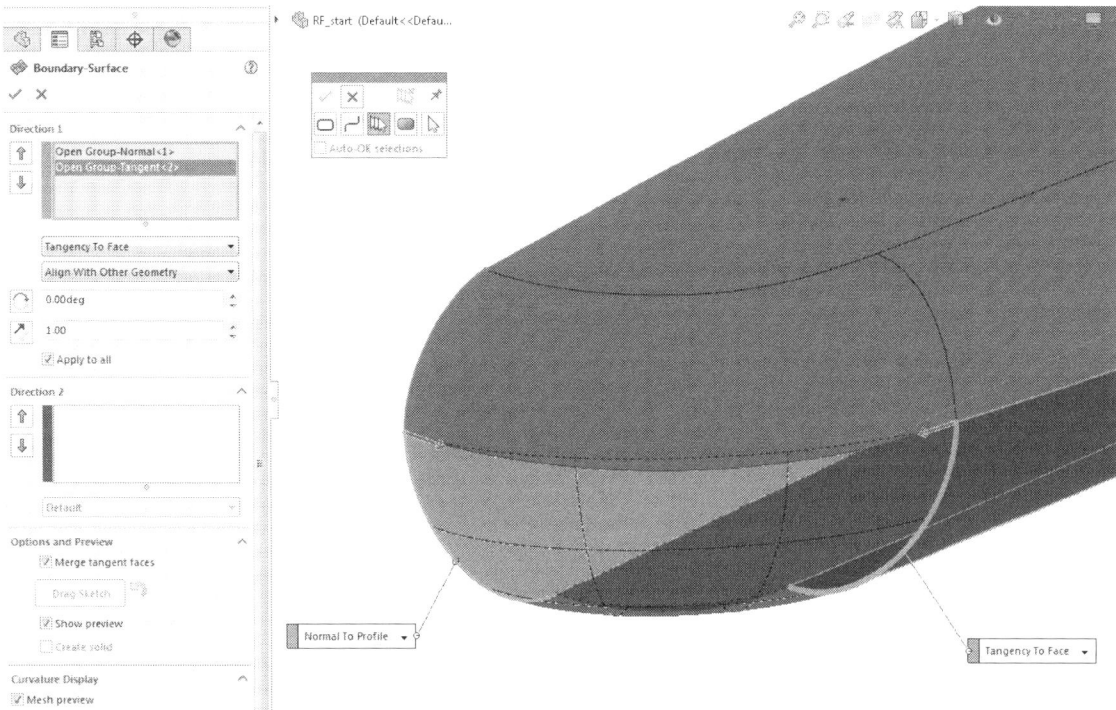

Figure 190 - Choosing a Direction 1 Profile

In the PropertyManager under Direction 2, select an edge as shown in the following image and Click Ok in the SelectionManager. Open Group<n> appears under Direction 2 in the Boundary Surface PropertyManager. For the curve influence option select **Tangency To Face** which makes the adjacent face tangent at the selected edge as shown in the following image.

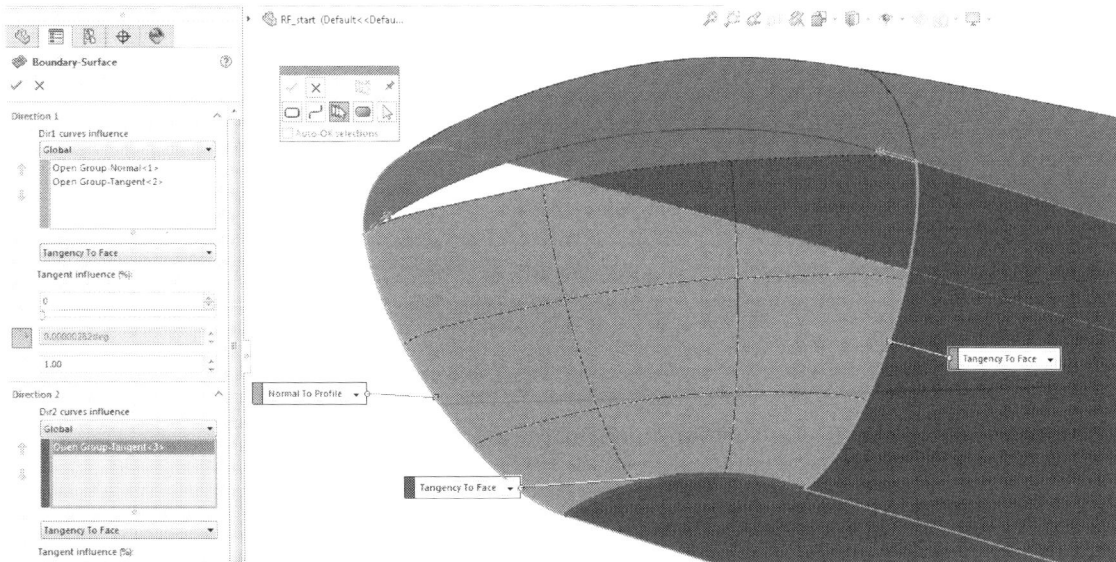

Figure 191 - Choosing a Direction 2 Profile and applying a Tangency To Face Curve Influence

107

TIP : Make sure you select the whole edge since it is not continuous but broken into two entities. When you select the first segment of the edge the icon shown in the following image appears.

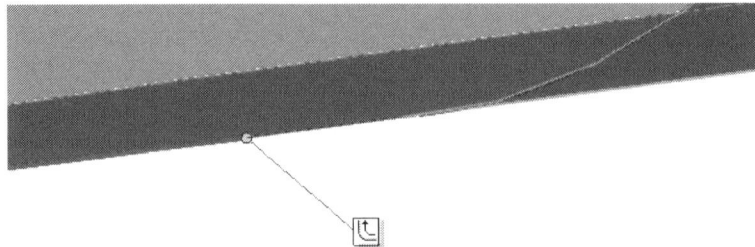

TIP : Click inside the small box to extend the selection to the second segment of the as shown in the following image then Click OK on the Selection Manager.

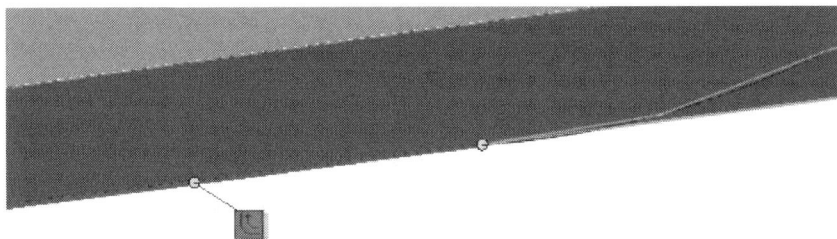

In the PropertyManager still under Direction 2, select the Curve in the Flyout Feature Manager Design Tree as shown in the following image. Curve1 appears under Direction 2 in the Boundary Surface PropertyManager. For the curve influence option select **None** as shown in the following image.

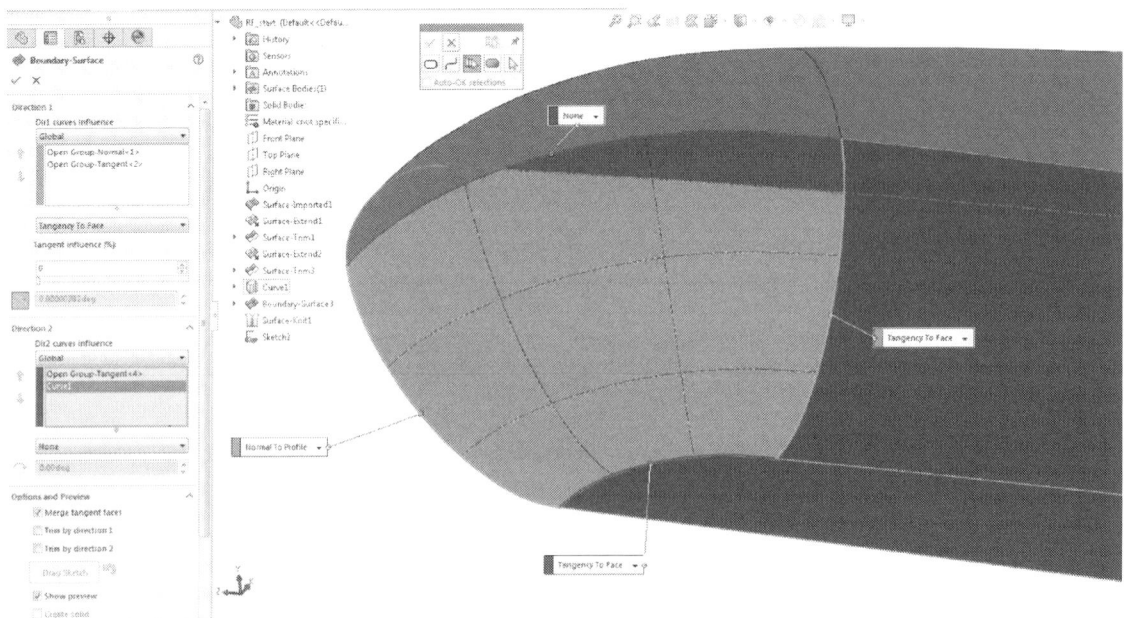

Figure 192 - Choosing a Curve as a Direction 2 Profile

108

Click OK on the Boundary Surface Property Manager to finish and close the Boundary Surface command.

Your part should now look as shown in the following image.

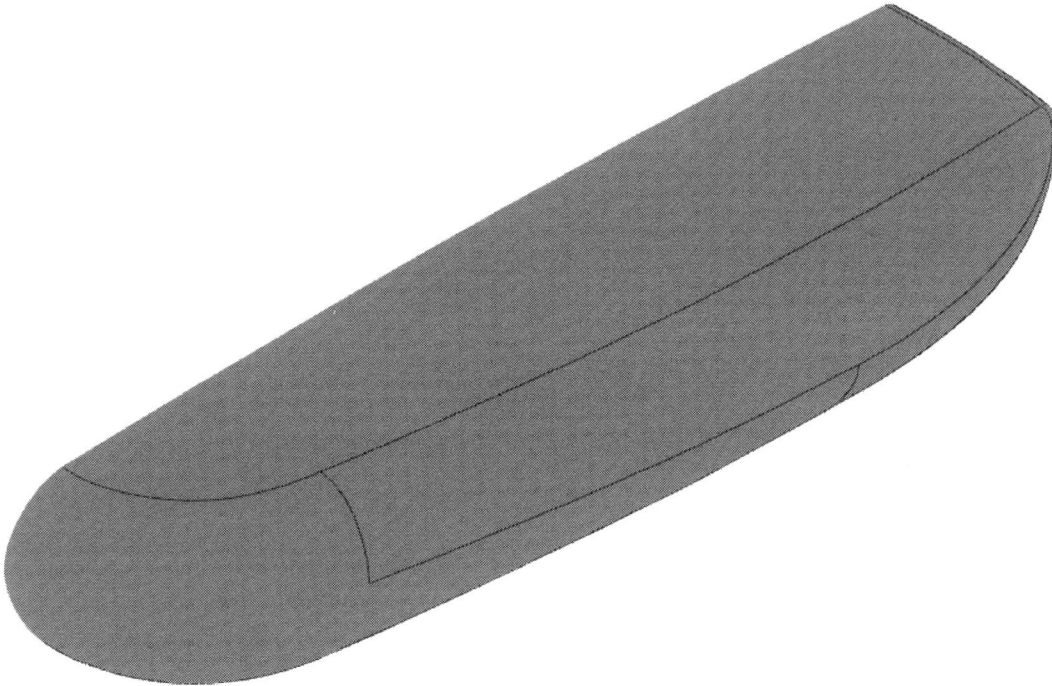

Figure 193 - Part Current Status

USING THE KNIT SURFACE TOOL

Click Knit Surface on the Surfaces toolbar, or click Insert > Surface > Knit.

In the PropertyManager, under Selections:

Select surfaces for Surfaces and Faces to Knit in the graphics area or in the Fly Out Feature Manager Design Tree as shown in the following image.

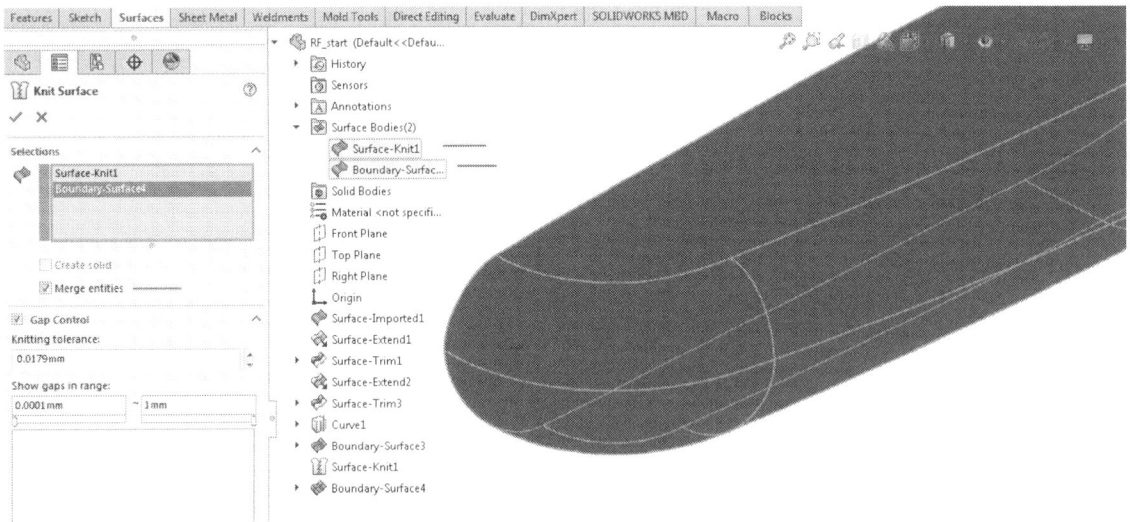

Figure 194 - Knit Surface

Select the Merge Entities Checkbox.

Click OK.

Your part should now look as shown in the following image.

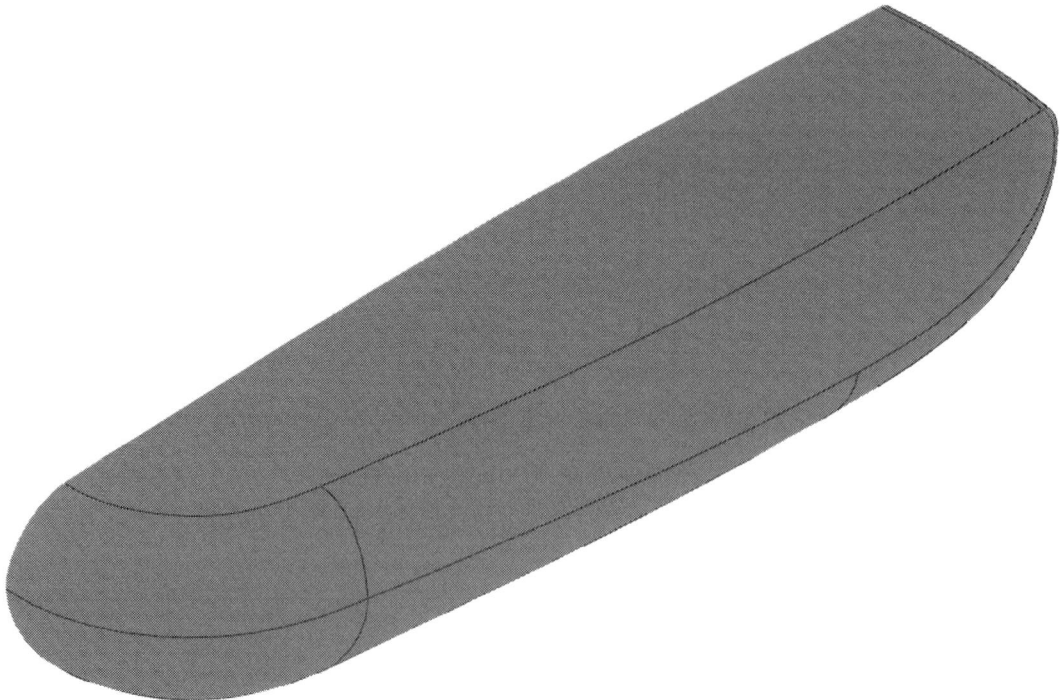

Figure 195 - Part Current Status

Your part should now look as shown in the following image.

Use the Zebra Stripes to Evaluate your model and identify different boundary conditions on different edges. Rotate the model while the Zebra Stripes are applied. Turn Off the Zebra Stripes once satisfied.

CURVATURE DISPLAY

Click Curvature (Evaluate Toolbar), or click View > Display > Curvature.

The curvature of the model is displayed in color. When you point to a model surface, a spline, or a curve, the curvature value and the radius of curvature are displayed beside the pointer. Curvature Display is used to visualize if the curvature of a model changes abruptly or if there are surprise areas of small curvature in the middle of a face.

Your part should look as shown in the following image - **NB:** You may have to observe the colours in the graphics area on your computer if you are using the black and white printed version of this book.

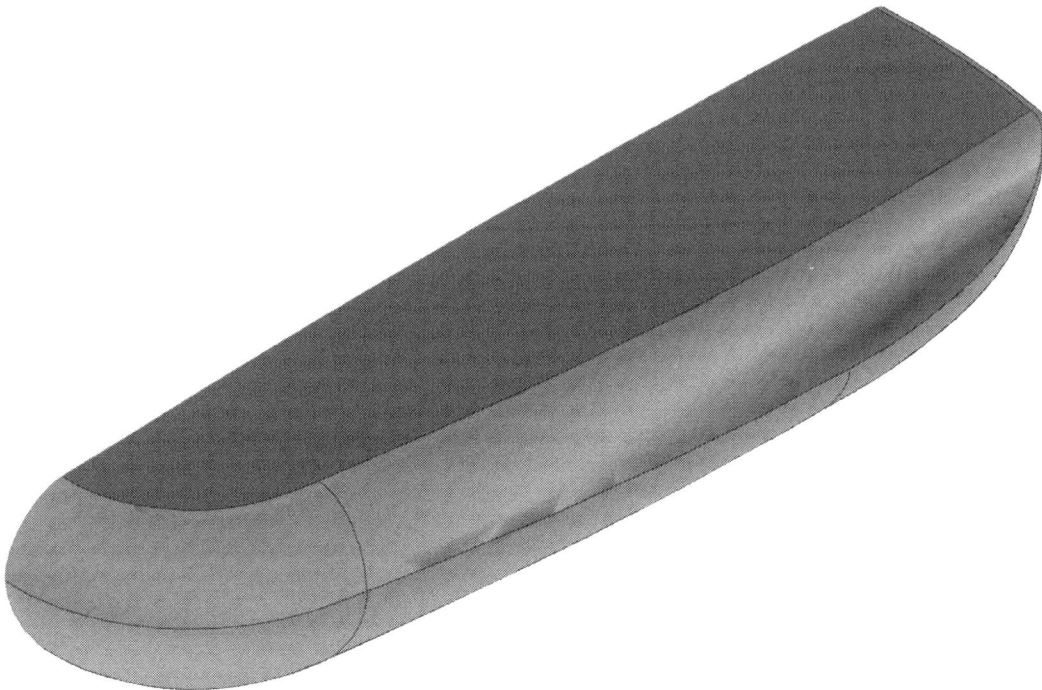

Figure 196 - Curvature Display

To remove the color, clear View > Display > Curvature.

Save your part.

DEVIATION ANALYSIS

The Deviation Analysis Tool is one of the objective Model Quality Evaluation Tools available in Solidworks that offer quantitative results in terms of the angle between faces. It actually measures how far (in degrees) two surfaces are from being tangent to one another along the length of an edge shared by adjacent faces.

To apply a Deviation Analysis:

Click Deviation Analysis on the Tools toolbar, or Tools > Evaluate > Deviation Analysis.

The Deviation Analysis PropertyManager appears.

Under Analysis Parameters, select the edges in the graphics area as shown in the following image.

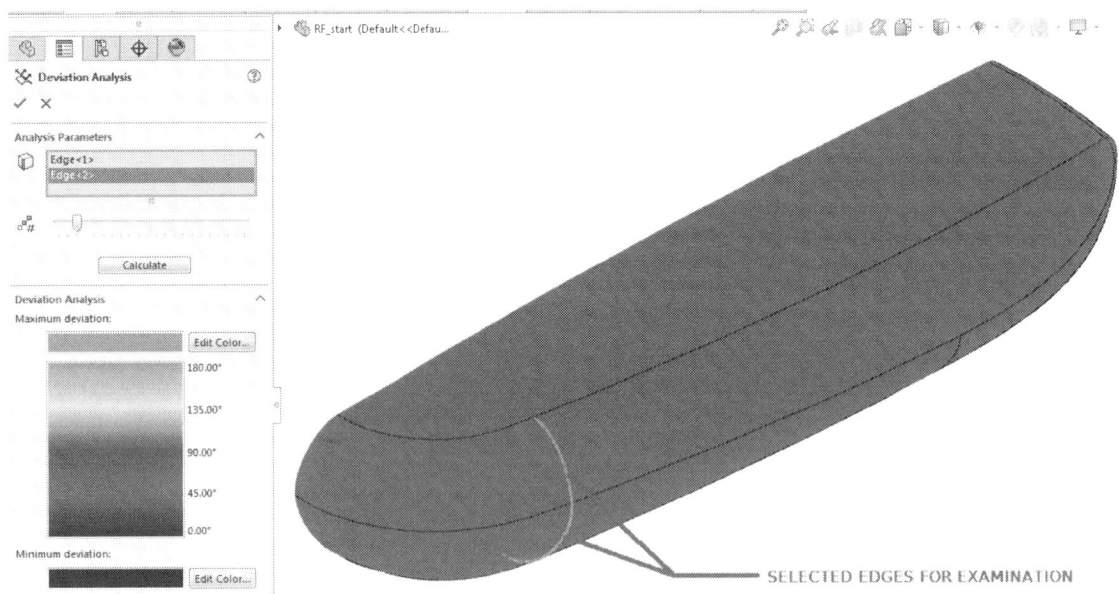

Figure 197 - Deviation Analysis

Use the slider to select the Number of Sample Points to include in the analysis.

The number of points is based on the size of the window client area. If you select more than one edge, the sample points are distributed across the selected edges, proportional to the edge length.

Click Calculate.

Your part should now appear as shown in the following image.

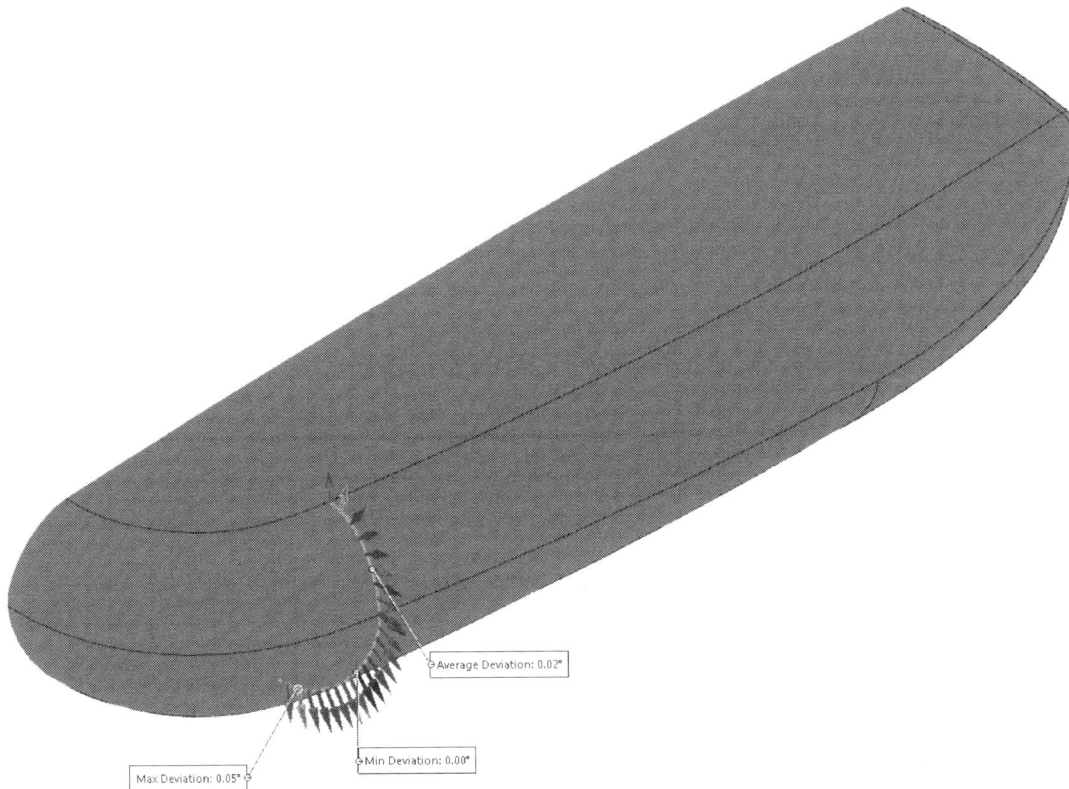

Figure 198 - Deviation Analysis

The maximum deviation is indicated as 0.05 Degrees which you may or may not consider tangent enough depending on your application, design intent or design requirements. What this also shows us is that though we applied a tangent relation when we created these two boundary faces, the result is almost always never 100 percent tangent.

Click OK and Save your part.

In summary, we have looked at three ways to evaluate model shape and continuity, which are Zebra Stripes, Curvature Display and Deviation Analysis. Though these tools do not fix any errors for you, and don't even automatically point to problem areas, they do help you visualize certain properties of the geometry which may be difficult if not impossible to identify otherwise which I believe is an indispensable tool even in the CSWPA-SU exam. Besides, as an Engineer or CAD Designer it's always critical to check your work before handing it over for manufacturing or CNC Programming for CNC Machining e.t.c. The last thing you want is downstream operations struggling or complaining about having to fix your work to make it suitable for further processing hence the old adage - measure twice cut once always applies in CAD Work so it's important to know the various tools in your arsenal to do just that in any given scenario. And also to make it a habit to always double check your work which is also critical in the exam.

You are required to measure and provide the total surface area of the recreated surfaces in square millimeters.

Click Measure (Tools toolbar) or Tools > Evaluate > Measure.

Select the two recreated surfaces (1 and 2) as shown in the following image.

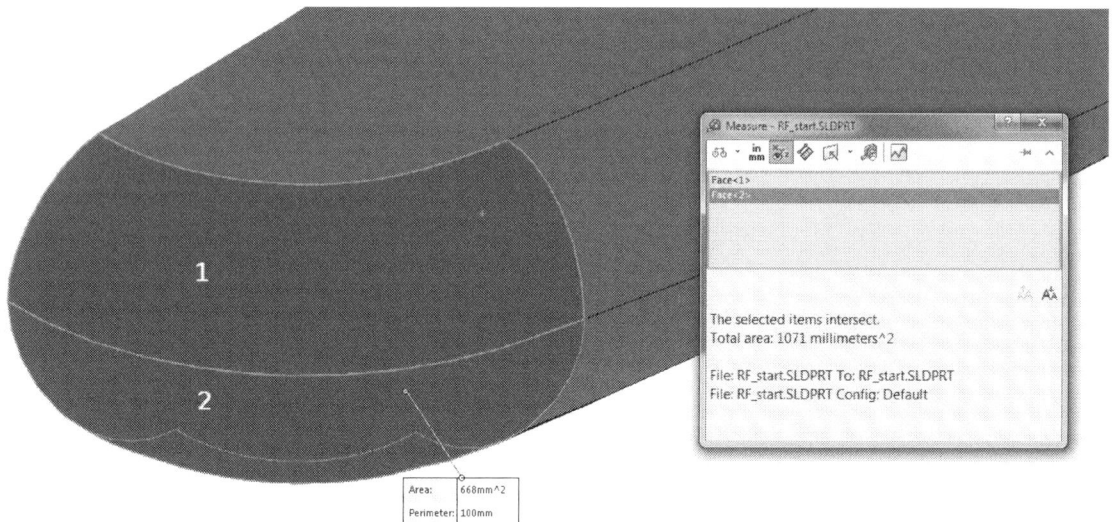

Figure 199 - Surface Area of Recreated Surfaces

The answer is 1071 square millimeters. **NB:** Your answer may be slightly different due to the nature of Surface Modeling but should be within +/- 5mm^2 from <u>my personal experience</u>. You may research acceptable tolerances online or by asking your VAR.

CHAPTER 5

This Chapter is a continuation from Chapter Four.

You are required to:

- Mirror existing faces / surfaces about the YZ plane.

- Knit the faces / surfaces together to create a solid.

- Measure the volume of the final solid.

You are then required to measure the volume of the final solid in cubic millimeters and your answer should be to two decimal places. No material is assigned.

MIRRORING BODIES IN A PART

Click Mirror (Features toolbar) or Insert > Pattern/Mirror > Mirror. In the Mirror Property Manager, under Mirror Face/Plane - Select the YZ Plane or Right Plane. Under Bodies to Mirror - Select the Surface Body in the Graphics Area or in the Surface Bodies folder on the Flyout Feature Manager Design Tree. Under Options check the Knit Surfaces Checkbox as shown in the following image.

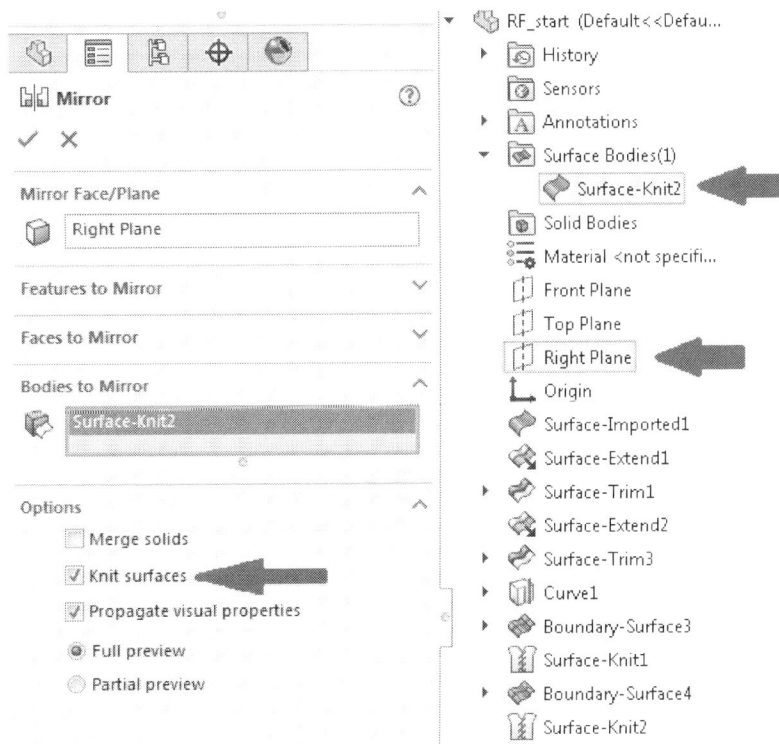

Figure 200 - Mirror Property Manager - Mirroring Bodies in a Part

The preview in the Graphics Area should appear as shown in the following image if you have the Full Preview Radio Button selected under Options in the Mirror Property Manager.

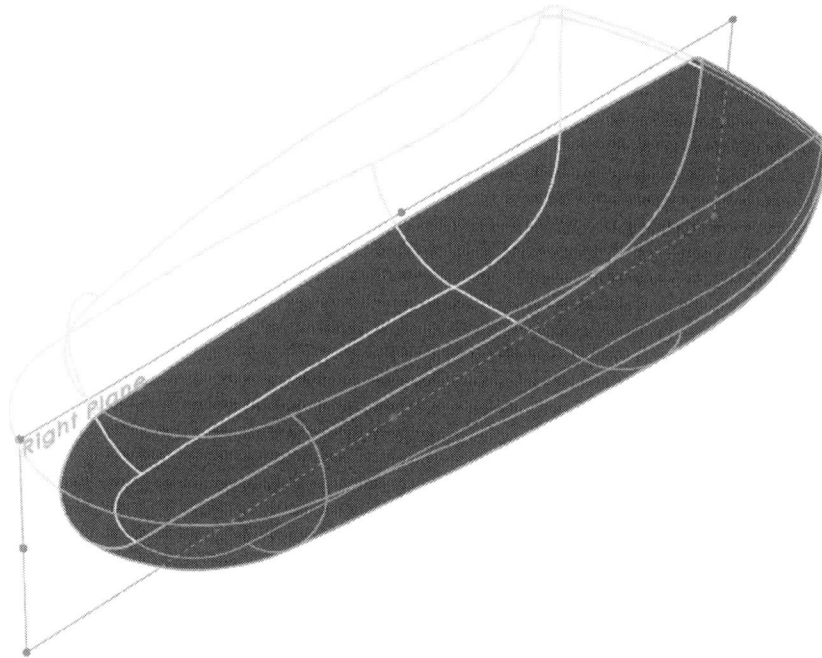

Figure 201 - Mirror Body Preview

In the PropertyManager, click ✓ (OK). Your Part should now appear as shown in the following image.

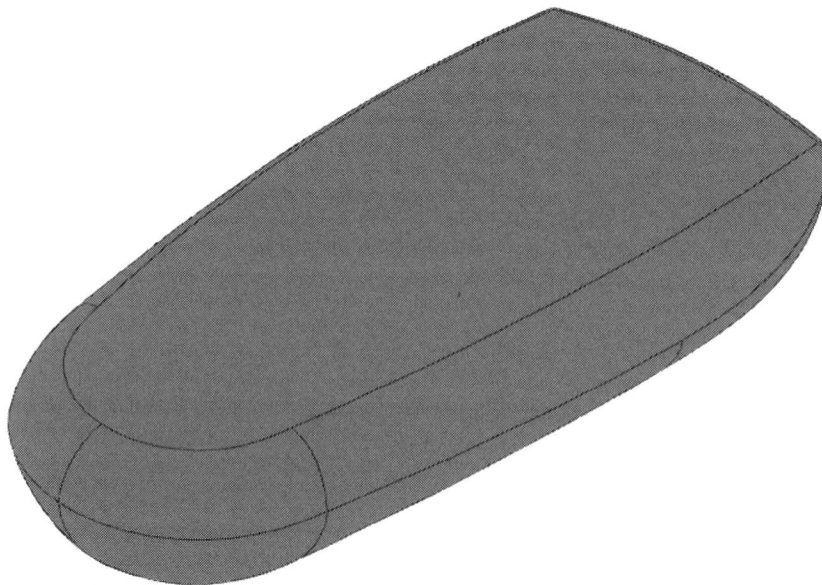

Figure 202 - Part Current Status

Though the part now has surfaces forming a fully enclosed volume, it still doesn't have the properties of a solid such as mass or volume - thus it is still a Surface Body and not a Solid Body. You will notice that even the Mass Properties command button is grayed out under the Evaluate Tab or if you Click Tools > Evaluate > Mass Properties. You will also notice that in the Feature Manager Design Tree that we have one Surface Body in the Surface Bodies Folder and nothing in the Solid Bodies Folder.

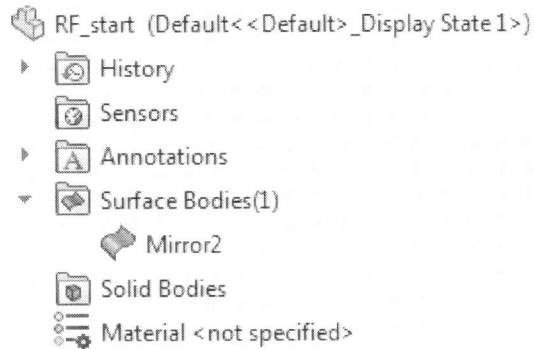

Figure 203 - Surface Bodies Folder

You may also take cross sections to see that the body is hollow as shown in the following image where I have taken a Front Plane or XY Plane Cross Section.

Figure 204 - Front Plane Cross Sectional View - Surface Body

117

USING THE THICKEN FEATURE

The Thicken feature is used to:

1. Add thickness to an open surface by offsetting the faces and connecting their edges.

2. To solidify an enclosed volume made of surfaces.

Click Thicken on the Surfaces Toolbar, or click Insert > Boss/Base > Thicken.

In the Thicken Feature Property Manager select the surface body in the Surface Bodies Folder in the Flyout Feature Manager Design Tree as shown in the following image. To create a solid, click the Create solid from enclosed volume Checkbox. This option is available only if you created a volume that is fully enclosed by surfaces like we what we have done.

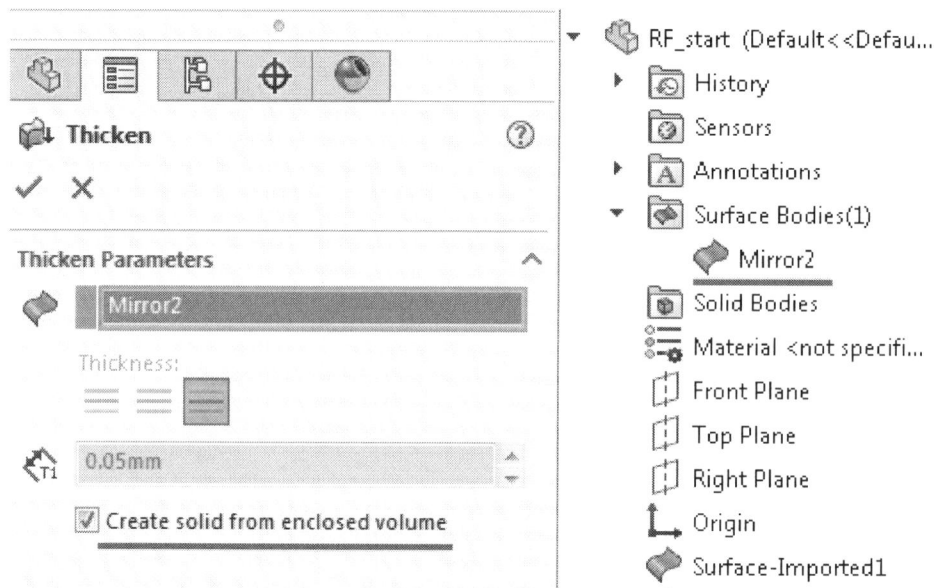

Figure 205 - Using the Thicken Feature to create a Solid from a volume that is fully enclosed by surfaces

Click OK.

Save your part.

Your part should now look as shown in the following image.

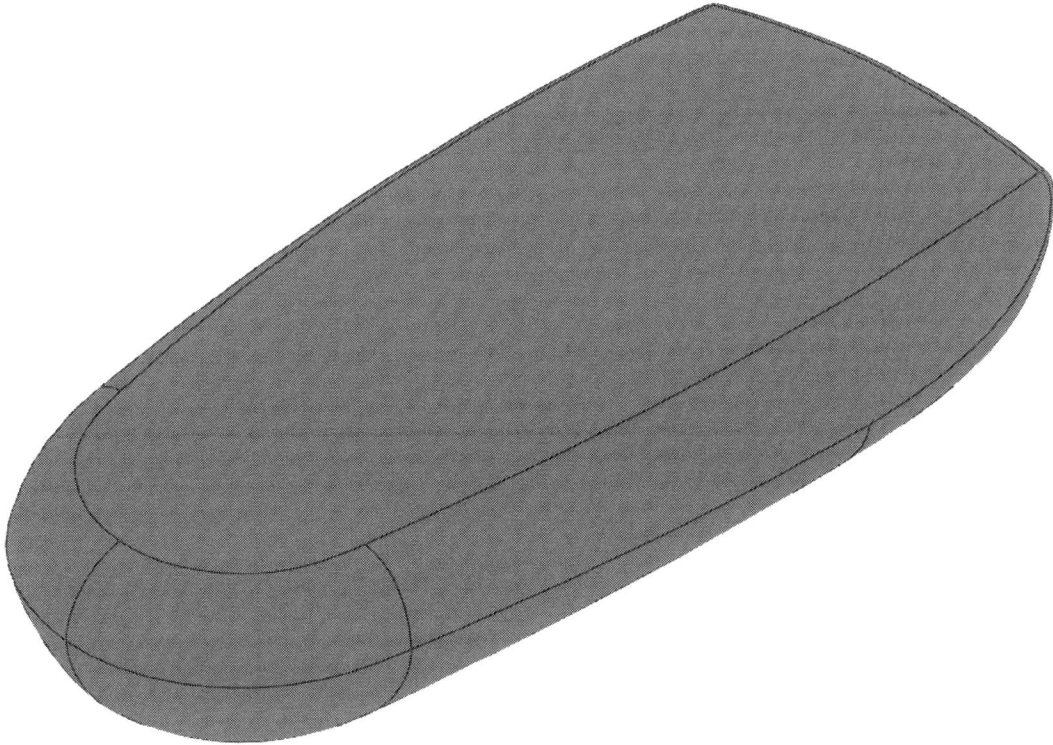

Figure 206 - Part Current Status

You will notice that in the Feature Manager Design Tree that we now have one Solid Body in the Solid Bodies Folder and nothing in the Surface Bodies Folder - the change is depicted in the following image.

Figure 207 - Feature Manager Design Tree Surface and Solid Body Folders

Let's take a across section to see that the body is not hollow anymore as shown in the following image where I have taken a Front Plane or XY Plane Cross Section.

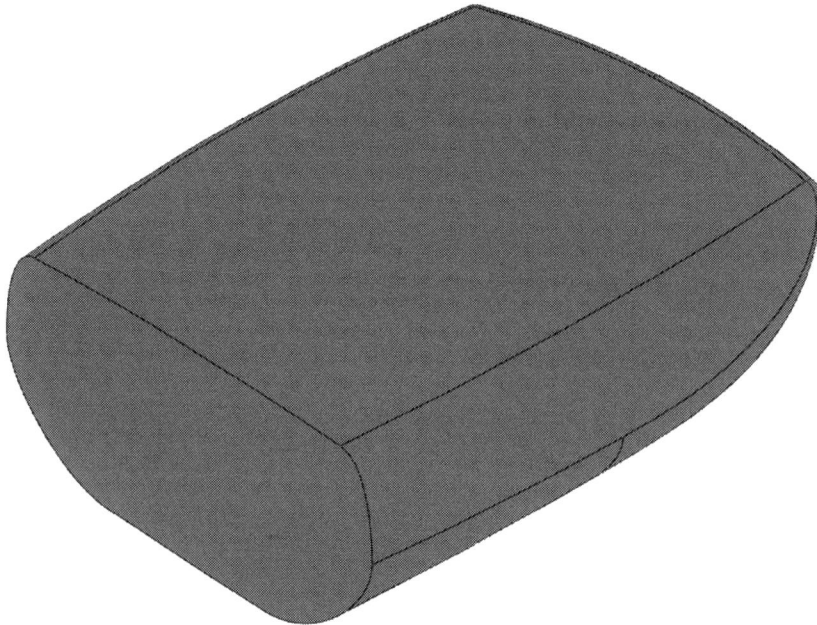

Figure 208 - Front Plane Cross Sectional View - Solid Body

CHAPTER 5 ANSWER

MEASURING VOLUME

Click Mass Properties (Tools toolbar) or Tools > Evaluate > Mass Properties.

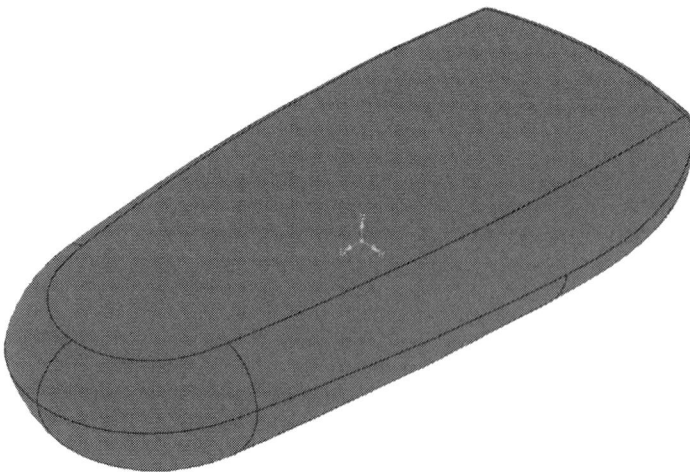

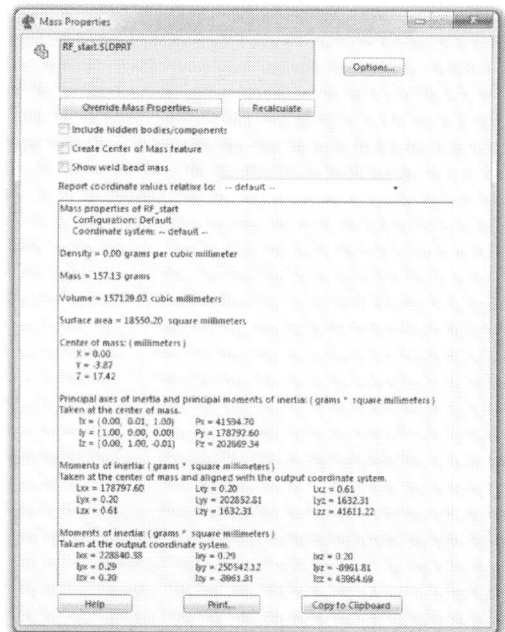

Figure 209 - Mass Properties

The Mass Properties Menu appears showing that the volume of the solid body is 157129.03 cubic millimeters as shown in the following image. Again your answer may be slightly different but as long as the <u>Mass</u> is within +/- 0.1 grams *(with no material applied)* it should be alright - **NB:** Again, this is just from my personal experience.

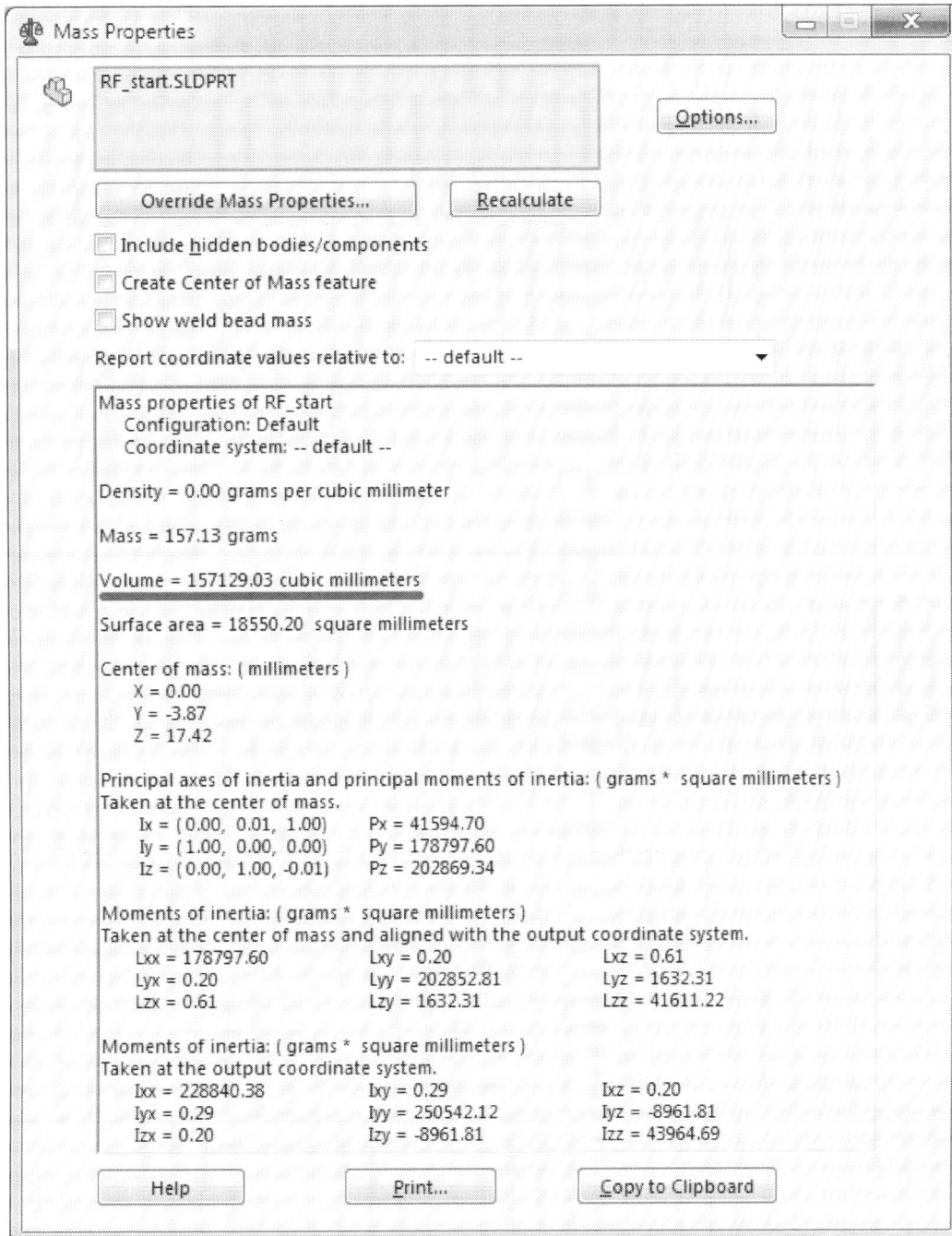

```
Mass Properties                                                    ⬜ ◻ ✕

⚖  RF_start.SLDPRT

                                                              Options...

        Override Mass Properties...        Recalculate

     ☐ Include hidden bodies/components

     ☐ Create Center of Mass feature

     ☐ Show weld bead mass

     Report coordinate values relative to:  -- default --          ▼

     Mass properties of RF_start
         Configuration: Default
         Coordinate system: -- default --

     Density = 0.00 grams per cubic millimeter

     Mass = 157.13 grams

     Volume = 157129.03 cubic millimeters

     Surface area = 18550.20  square millimeters

     Center of mass: ( millimeters )
         X = 0.00
         Y = -3.87
         Z = 17.42

     Principal axes of inertia and principal moments of inertia: ( grams * square millimeters )
     Taken at the center of mass.
         Ix = ( 0.00,  0.01,  1.00)     Px = 41594.70
         Iy = ( 1.00,  0.00,  0.00)     Py = 178797.60
         Iz = ( 0.00,  1.00, -0.01)     Pz = 202869.34

     Moments of inertia: ( grams * square millimeters )
     Taken at the center of mass and aligned with the output coordinate system.
         Lxx = 178797.60        Lxy = 0.20           Lxz = 0.61
         Lyx = 0.20             Lyy = 202852.81      Lyz = 1632.31
         Lzx = 0.61             Lzy = 1632.31        Lzz = 41611.22

     Moments of inertia: ( grams * square millimeters )
     Taken at the output coordinate system.
         Ixx = 228840.38        Ixy = 0.29           Ixz = 0.20
         Iyx = 0.29             Iyy = 250542.12      Iyz = -8961.81
         Izx = 0.20             Izy = -8961.81        Izz = 43964.69

        Help              Print...           Copy to Clipboard
```

Figure 210 - Mass Properties - Volume

Close the Mass Properties Dialog Box. Save your part. Close the part.

MIRRORING BODIES IN A PART

Instead of selecting the knit surfaces option when we mirrored the surface body in sections above, we could have just mirrored the surface body without selecting the Knit Surfaces option as shown in the following image - open the part in the Chapter 5 Downloaded Folder RF_start CH05 EXAMPLE.SLDPRT.

Click Mirror (Features toolbar) or Insert > Pattern/Mirror > Mirror. In the Mirror Property Manager, under Mirror Face/Plane - Select the YZ Plane or Right Plane. Under Bodies to Mirror - Select the Surface Body in the Graphics Area or in the Surface Bodies folder on the Flyout Feature Manager Design Tree. Under Options leave the Knit Surfaces Checkbox unchecked as shown in the following image.

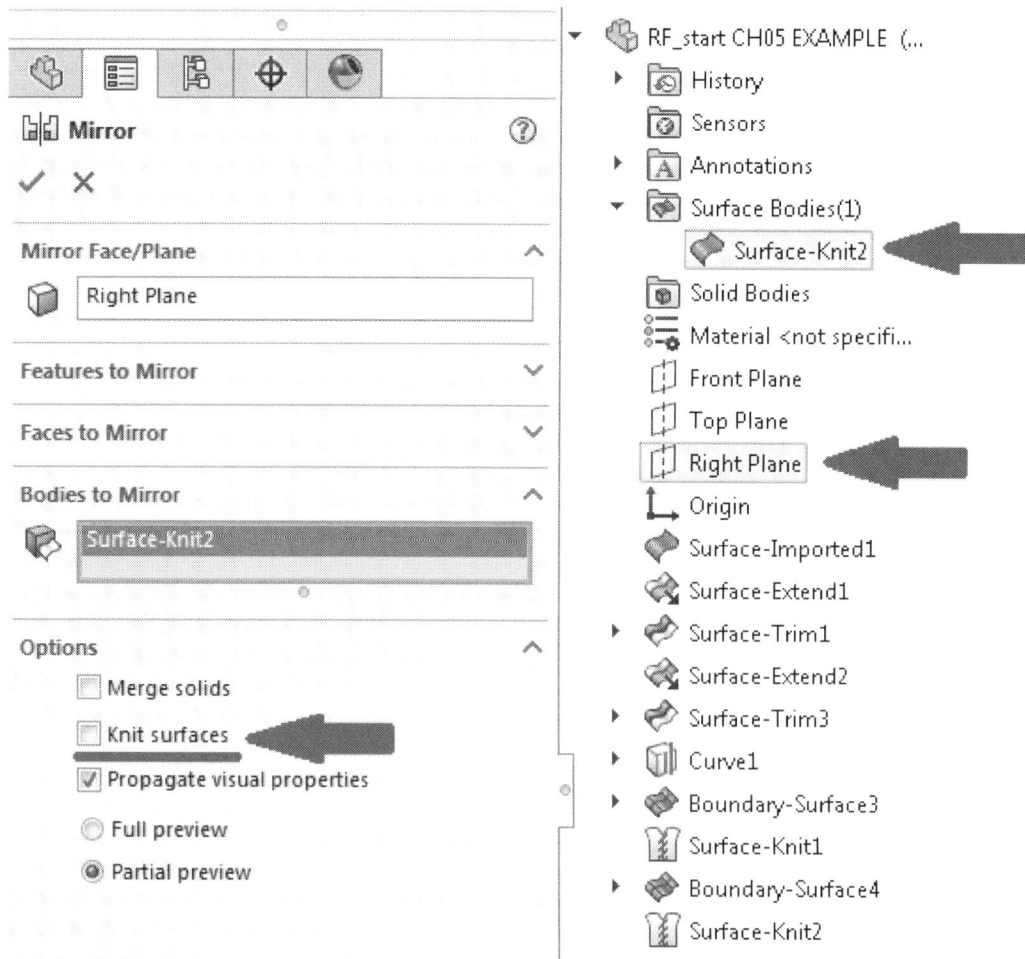

Figure 211 - Mirroring a surface body in a part without Knitting the Mirrored Surfaces

Click OK and your part should now look as shown in the following image.

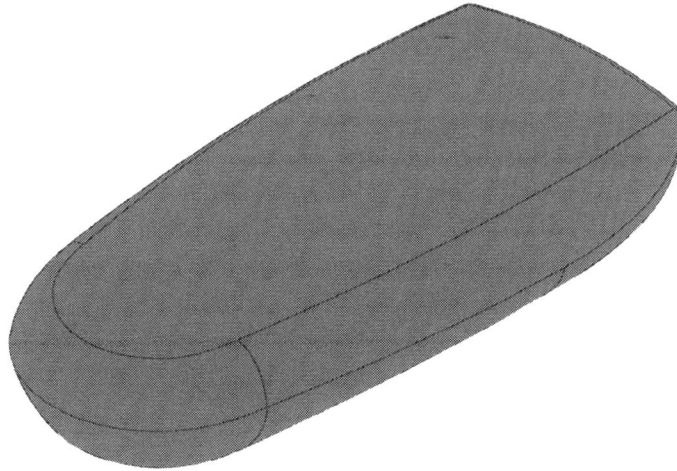

Figure 212 - Mirrored surface body in a part

If you check in the Feature Manager Design Tree you will notice that we now have two surface bodies in the Surface Bodies Folder as shown in the following image.

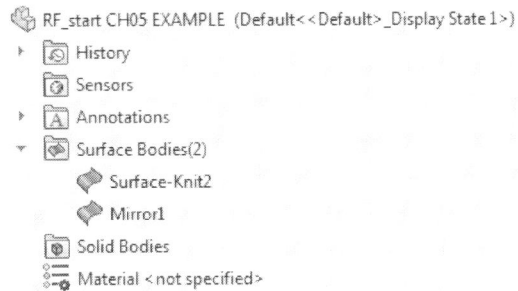

Figure 213 - Surface Bodies Folder in the Feature Manager Design Tree

KNITTING SURFACES

The Knit Surface tool is used to combine two or more faces and or surfaces into one. One thing to note about the Knit Surface tool is that it may be used to create a solid body when the knit surfaces form a closed volume as is the case in our example here.

To knit the two surface bodies in our part:

Click Knit Surface on the Surfaces toolbar, or click Insert > Surface > Knit.

In the PropertyManager, under Selections:

Select faces and surfaces for Surfaces and Faces to Knit - you may select the two surface bodies in our part directly from the Graphics Area or you may select them from the Surface Bodies Folder in the Flyout Feature Manager Design Tree.

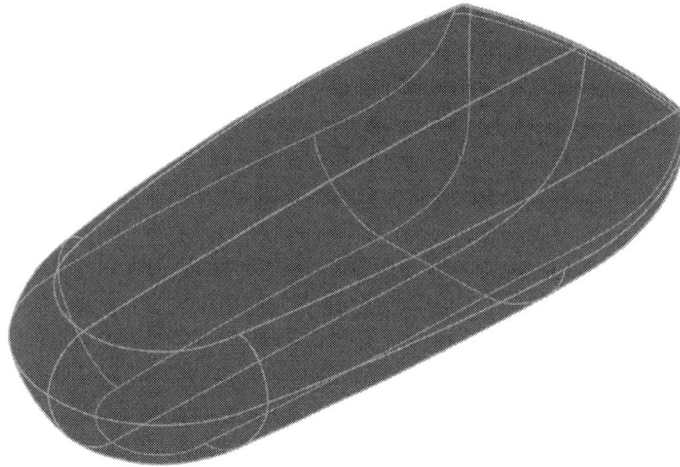

Figure 214 - Surface to Knit

Select Create solid to create a solid model from enclosed surfaces. Select Merge entities to merge faces with the same underlying geometry. Select Gap Control to view edge pairs that might introduce gap problems, and to view or edit the knitting tolerance or gap range. View the Knitting tolerance. Modify it if required. The gap range depends on the knitting tolerance. Only gaps within the selected gap range are listed. You can modify the gap range if required. Your Knit Surface Property Manager should have selections and options check as shown in the following image.

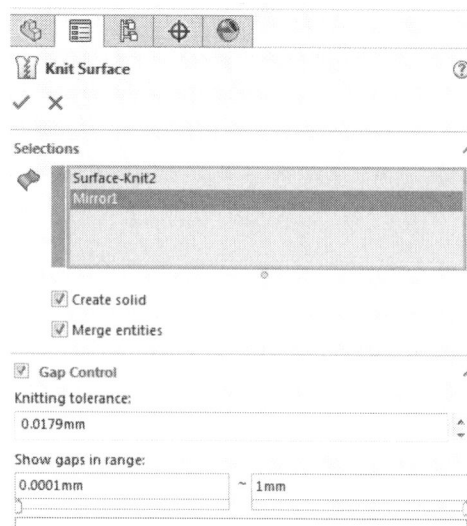

Figure 215 - Knit Surface Property Manager

Click OK. You will notice that in the Feature Manager Design Tree Folders we now have no Surface Bodies and one Solid Body in the Solid Bodies Folder. Taking a cross-section across any of the planes also shows that the part is now really a solid. Your part should now look as shown below - thus we have achieved the same result using a slightly different route to the one described in the previous section before the answer in this Chapter.

In the following image is a side by side comparison of the Feature Manager Design Trees of these two parts which are basically the same except for the last two features.

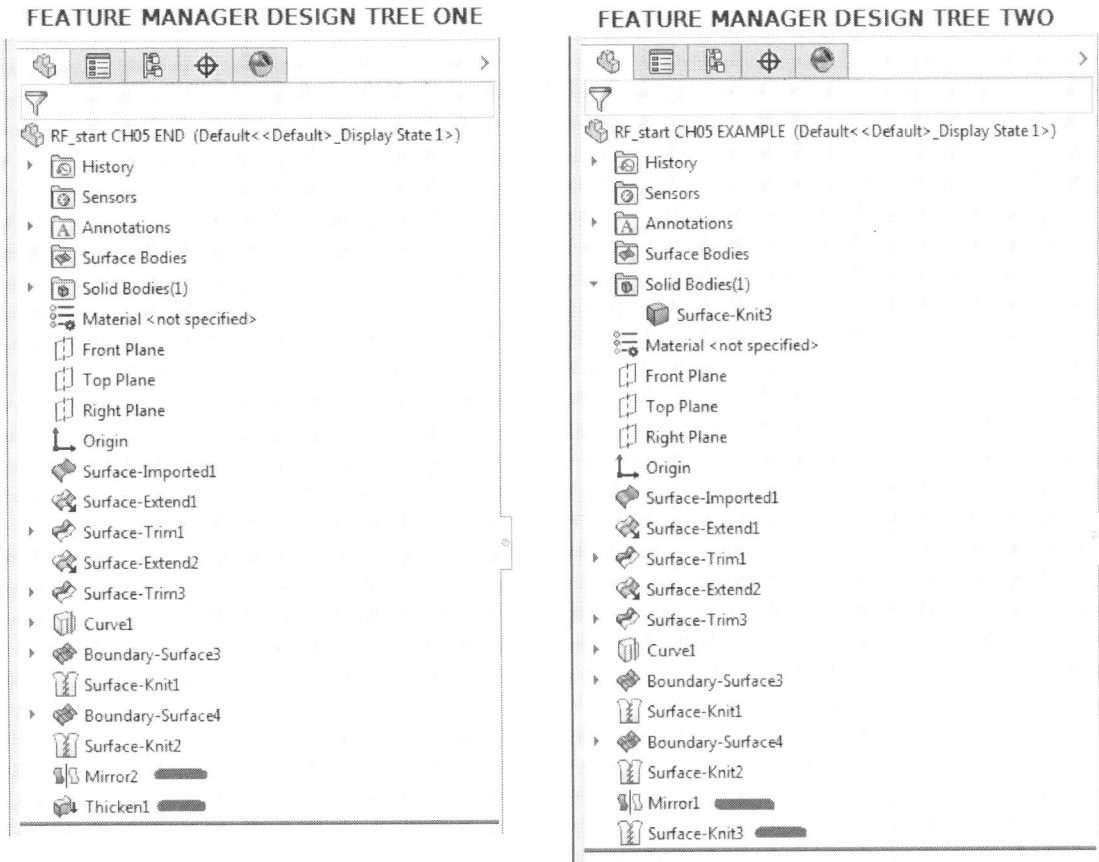

Figure 216 - Feature Manager Design Trees of two parts

Which method is better? It's difficult to say, so sometimes it's just a matter of personal preference. But what we can do is compare the rebuild time of these two parts to see which method is better especially if say you are building a very complex part with hundreds of surfaces and faces.

TILE VERTICALLY

Tile Horizontally or Tile Vertically arranges open Solidworks document windows so they are all visible.

With the two parts open - that is RF_start CH05 END.SLDPRT and RF_start CH05 EXAMPLE END.SLD PRT click Tile Vertically (Standard toolbar) or Window > Tile Vertically to display the windows vertically as shown in the following image.

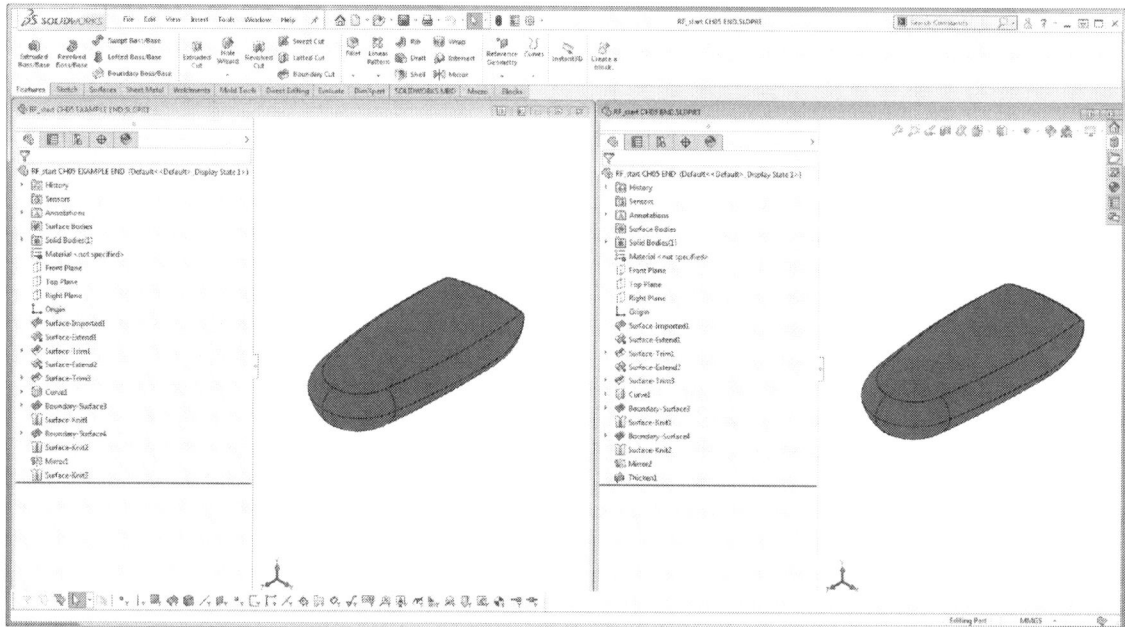

Figure 217 - Tile Vertically

PERFORMANCE EVALUATION

Performance Evaluation is a tool that displays the amount of time it takes to rebuild each feature in a part.

Click anywhere in the Graphics Area of the LHS Window or RF_start CH05 EXAMPLE END.SLD PRT part document.

Click Performance Evaluation (Evaluation Toolbar) or Tools > Evaluate > Performance Evaluation. The Performance Evaluation dialog box appears with a list of all features and their rebuild times in descending order.

- Feature Order. Lists each item in the FeatureManager Design Tree: features, sketches, and derived planes. Use the shortcut menu to edit feature definition, suppress features, and other actions.
- Time %. Displays the percentage of the total part rebuild time to regenerate each item.

126

- Time(s). Displays the amount of time in seconds that each item takes to rebuild.

Click Feature Order. This sorts the features to match the FeatureManager Design Tree. Click Time(s) to sort by each feature's rebuild time again.

Click one of the following:

- Print. Prints the Feature Statistics to an external printer or a pdf printer.
- Copy. Copies the Feature Statistics so you can paste them into another file such as a word document or excel spreadsheet.

Refresh.

The total rebuild time of RF_start CH05 EXAMPLE END.SLD PRT is 3.90 seconds as shown in the following image under the Performance Evaluation dialog box.

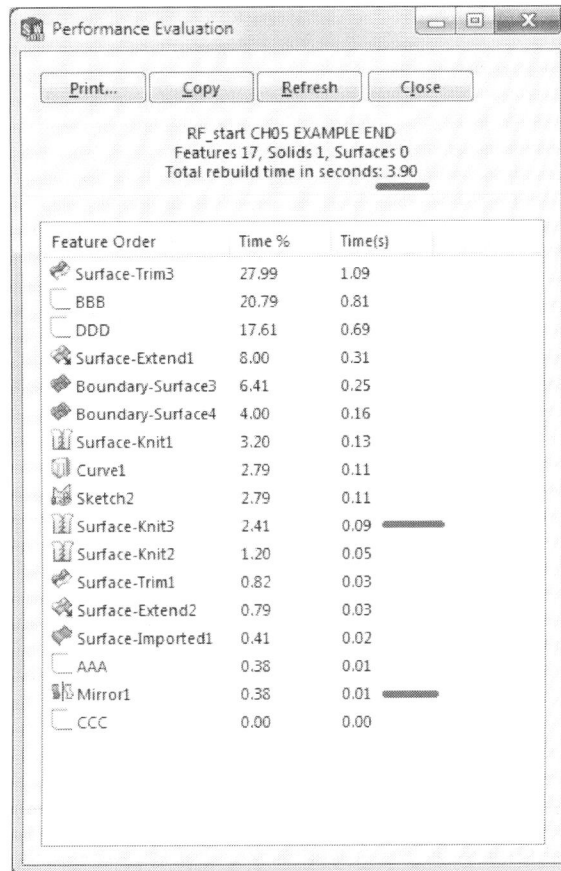

Figure 218 - Performance Evaluation Dialog Box

Click Close.

Click anywhere in the Graphics Area of the RHS Window or RF_start CH05 END.SLD PRT part document.

Click Performance Evaluation (Evaluation Toolbar) or Tools > Evaluate > Performance Evaluation. The Performance Evaluation dialog box appears with a list of all features and their rebuild times in descending order as shown in the following image - the total rebuild time of RF_start CH05 END.SLD PRT is thus 4.01 seconds.

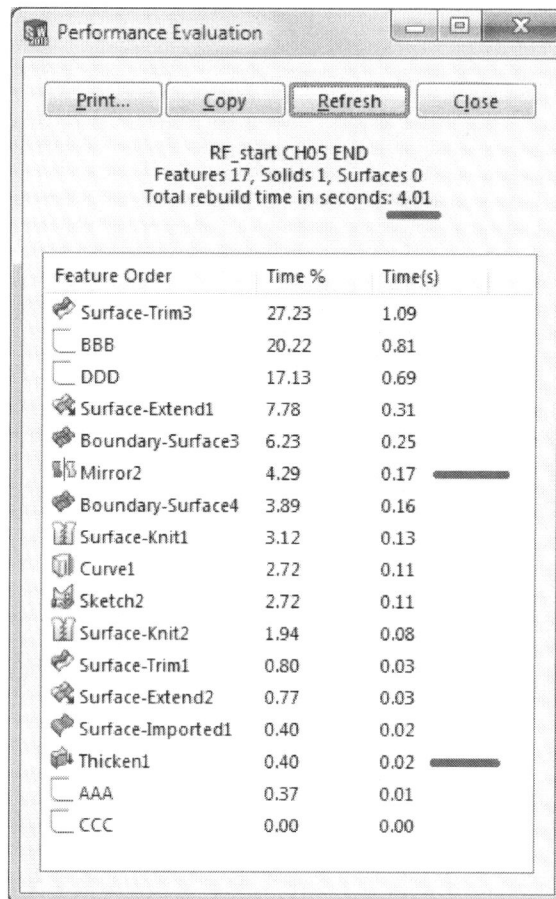

Feature Order	Time %	Time(s)
Surface-Trim3	27.23	1.09
BBB	20.22	0.81
DDD	17.13	0.69
Surface-Extend1	7.78	0.31
Boundary-Surface3	6.23	0.25
Mirror2	4.29	0.17
Boundary-Surface4	3.89	0.16
Surface-Knit1	3.12	0.13
Curve1	2.72	0.11
Sketch2	2.72	0.11
Surface-Knit2	1.94	0.08
Surface-Trim1	0.80	0.03
Surface-Extend2	0.77	0.03
Surface-Imported1	0.40	0.02
Thicken1	0.40	0.02
AAA	0.37	0.01
CCC	0.00	0.00

Figure 218 - Performance Evaluation Dialog Box

Click Close.

If we put the two Performance Evaluation Dialog Boxes side by side as shown in the following image you will notice that the Mirror feature with the Knit Surfaces Option selected (Mirror 2 under Method One) seems to be one of the main culprits contributing to a longer rebuild time of 0.17 seconds compared to a Mirror with the Knit Surfaces Option not selected (Mirror 1 under Method Two) with a rebuild time of 0.01 seconds. However, the Thicken feature to convert a surface with an enclosed volume into a solid seems to take less rebuild time of 0.01

seconds compared to the alternative method of Knit Surfaces with the Create solid option selected which takes a rebuild time of 0.13 seconds.

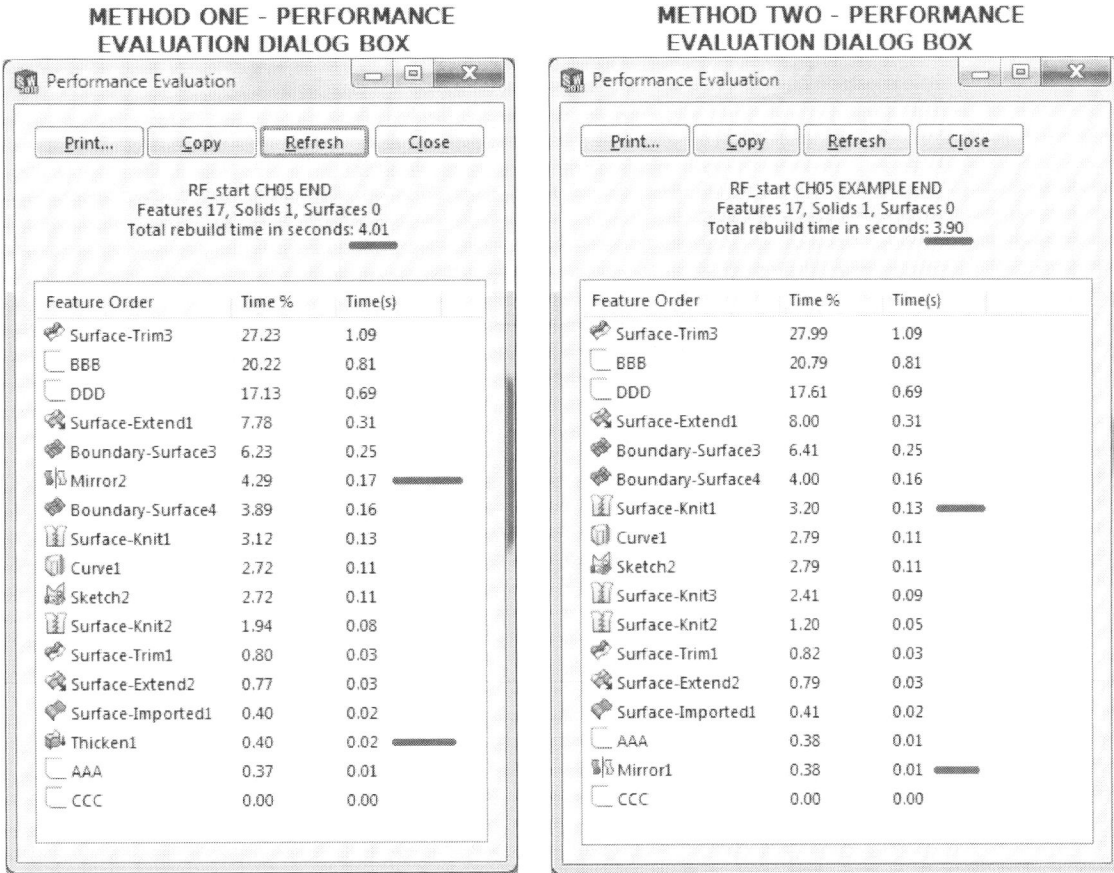

METHOD ONE - PERFORMANCE
EVALUATION DIALOG BOX

Performance Evaluation		
Print... Copy Refresh Close		
RF_start CH05 END		
Features 17, Solids 1, Surfaces 0		
Total rebuild time in seconds: 4.01		

Feature Order	Time %	Time(s)
Surface-Trim3	27.23	1.09
BBB	20.22	0.81
DDD	17.13	0.69
Surface-Extend1	7.78	0.31
Boundary-Surface3	6.23	0.25
Mirror2	4.29	0.17
Boundary-Surface4	3.89	0.16
Surface-Knit1	3.12	0.13
Curve1	2.72	0.11
Sketch2	2.72	0.11
Surface-Knit2	1.94	0.08
Surface-Trim1	0.80	0.03
Surface-Extend2	0.77	0.03
Surface-Imported1	0.40	0.02
Thicken1	0.40	0.02
AAA	0.37	0.01
CCC	0.00	0.00

METHOD TWO - PERFORMANCE
EVALUATION DIALOG BOX

Performance Evaluation		
Print... Copy Refresh Close		
RF_start CH05 EXAMPLE END		
Features 17, Solids 1, Surfaces 0		
Total rebuild time in seconds: 3.90		

Feature Order	Time %	Time(s)
Surface-Trim3	27.99	1.09
BBB	20.79	0.81
DDD	17.61	0.69
Surface-Extend1	8.00	0.31
Boundary-Surface3	6.41	0.25
Boundary-Surface4	4.00	0.16
Surface-Knit1	3.20	0.13
Curve1	2.79	0.11
Sketch2	2.79	0.11
Surface-Knit3	2.41	0.09
Surface-Knit2	1.20	0.05
Surface-Trim1	0.82	0.03
Surface-Extend2	0.79	0.03
Surface-Imported1	0.41	0.02
AAA	0.38	0.01
Mirror1	0.38	0.01
CCC	0.00	0.00

Figure 220 - Performance Evaluation Dialog Boxes - Comparison

The second method (METHOD TWO) under the example part therefore wins this battle since it has a rebuild time of 3.9 seconds compared to the first method which has a rebuild time of 4.01 seconds. However, like I mentioned before this is mainly critical when building complex parts with multiple faces and is definitely not something to worry much about in the exam when dealing with such simple parts.

Save and close the two parts and move on to Chapter Six.

CHAPTER 6

In this Chapter we will start working with a new imported part. The part is basically an incomplete Bicycle Frame with imported tube geometry surfaces. You are required to create smooth blended surfaces to form the transition between three tubes that make up the front part of this bicycle frame - namely the Top Tube, Head Tube and the Down Tube. The final result should be as shown in the following three images.

Figure 221 - Bicycle Frame Expected Result Image One

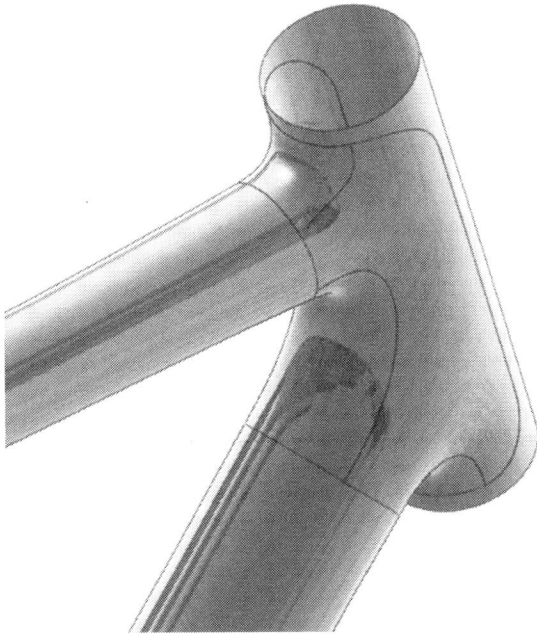

Figure 222 - Bicycle Frame Expected Result Image Two

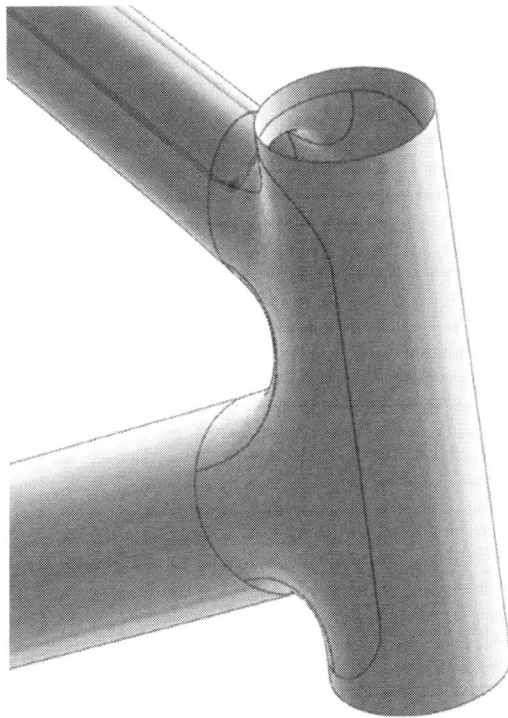

Figure 223 - Bicycle Frame Expected Result Image Three

Download the part named BICYCLE FRAME.SLDPRT from this Google drive location - *http://bit.ly/CSWPA-SU* or Scan the QR Code shown in the following image:

If you experience any problems with downloading any files you may send an email to *cswpasmebook@gmail.com* with the title of the book indicated in your email subject.

Save the part to your computer. When you open the downloaded part in Solidworks it should look as shown in the following image.

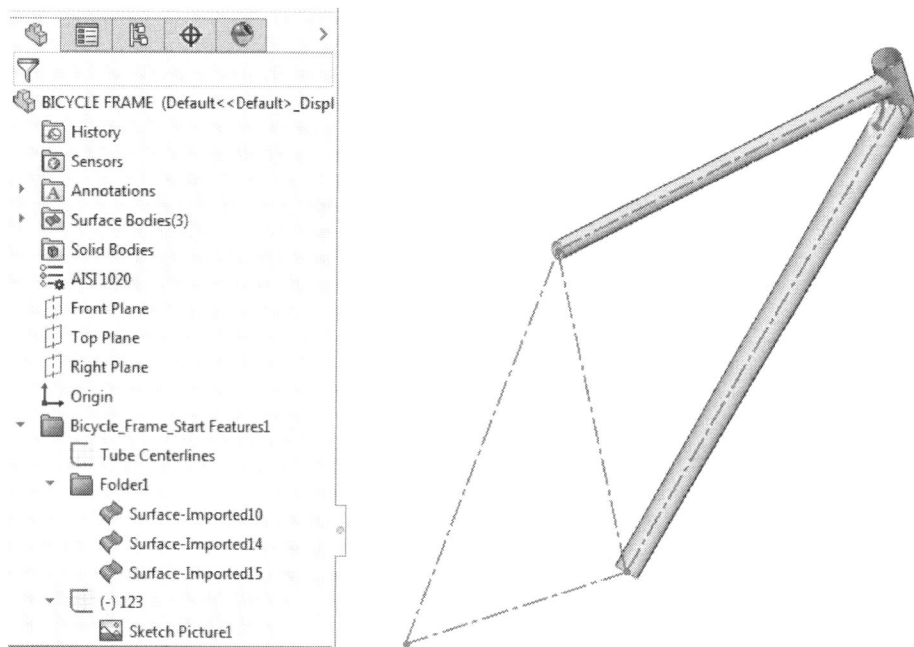

Figure 224 - Bicycle Frame Downloaded Part

In this first Chapter towards creating the surface blends shown in sections above, you are required to:

- Trim the frame surfaces to the dimensions shown in the following image. **NB:** There are two different fillet sizes in the trimmed image below - thus 20mm and 15mm.

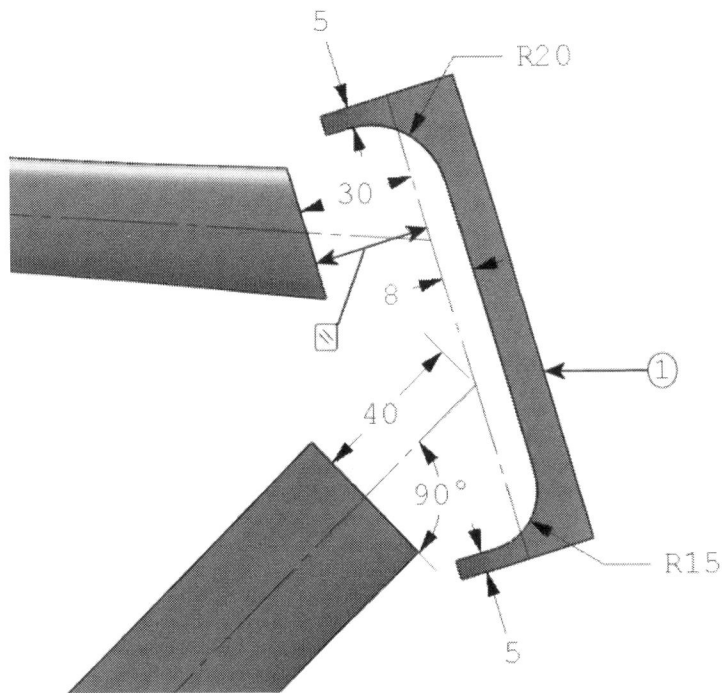

Figure 225 - Bicycle Frame Trim Dimensions

- Use the dimensions of the surfaces in the following image to plan correctly the trimming of the frame surfaces.

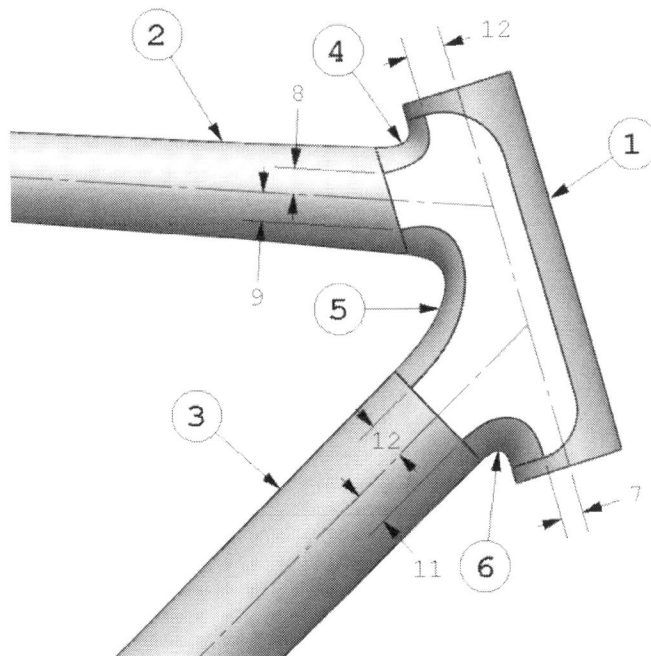

Figure 226 - Bicycle Frame Trim Dimensions

133

Figure 227 - Bicycle Frame Trim Dimensions

- Measure the area of the front surface (Head Tube) after trimming
- Select the range that contains the measured surface area of Surface1 in square millimeters from the options given below:
  - a. 5490 - 5590
  - b. 5850 - 5950
  - c. 6065 - 6165
  - d. 5740 - 5840

CREATING A 2D SKETCH TO USE AS A TRIM TOOL

Hide all sketches *(if the sketches are not already hidden)* by Right clicking on each sketch in the Feature Manager Design Tree and selecting Hide.

Start a sketch on the Right Plane or YZ Plane. Show the Tube Centerlines sketch only.

Click Convert Entities (Sketch toolbar) or Tools > Sketch Tools > Convert Entities. Select the sketch entities or lines from the Tube Centerlines Sketch in the Graphics Area as shown in the following image then Click OK.

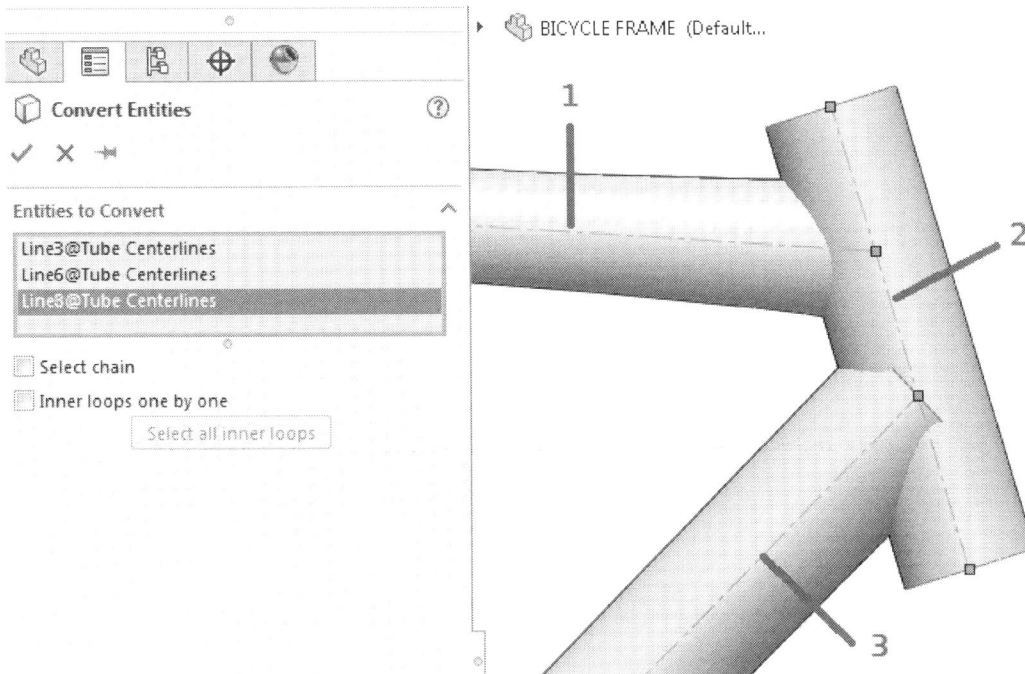

Figure 228 - Using Convert Entities

Hide the Tube Centerlines Sketch *(Right Click on the Sketch in the Feature Manager Design Tree and select Hide)*. Click anywhere in the Graphics Area and then press Ctrl + A on your keyboard to select all sketch entities in the sketch. From the Properties Manager, select the For Construction checkbox then Click OK.

Your part should now look as shown in the following image with all inactive sketches hidden.

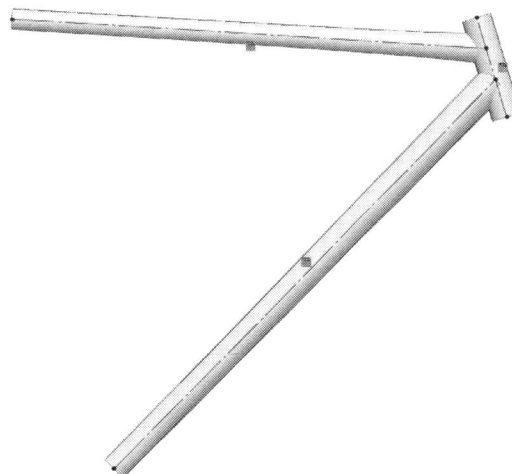

Figure 229 - Converting sketch entities to Construction Lines - For Construction

Offset the centerline on the centre of Head Tube 8mm to the right and the 30mm to the left as shown in the following image.

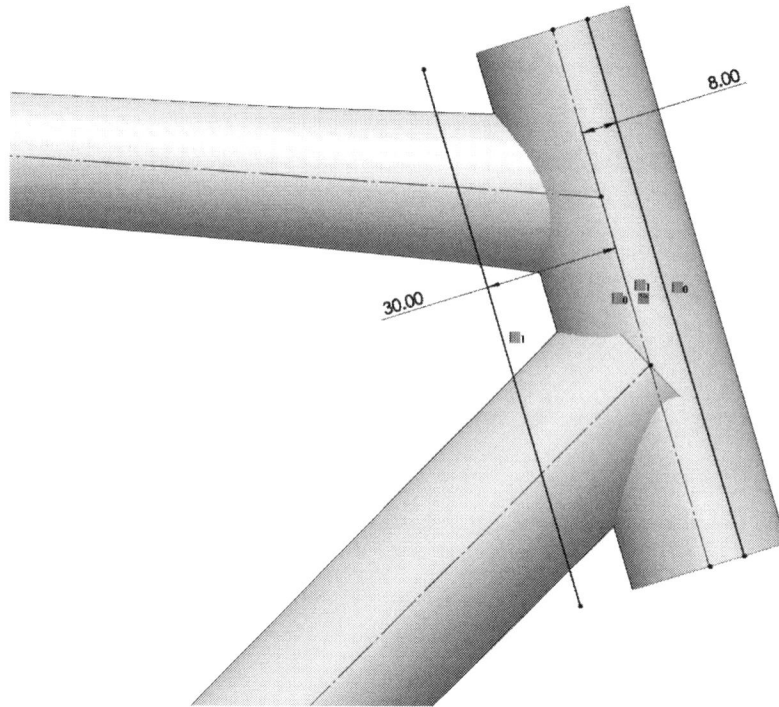

Figure 230 - Using the Offset Entities - Offsetting Sketch Entities

Offset the bottom and top edges of the Head Tube 5mm down and 5mm up as shown in the following image.

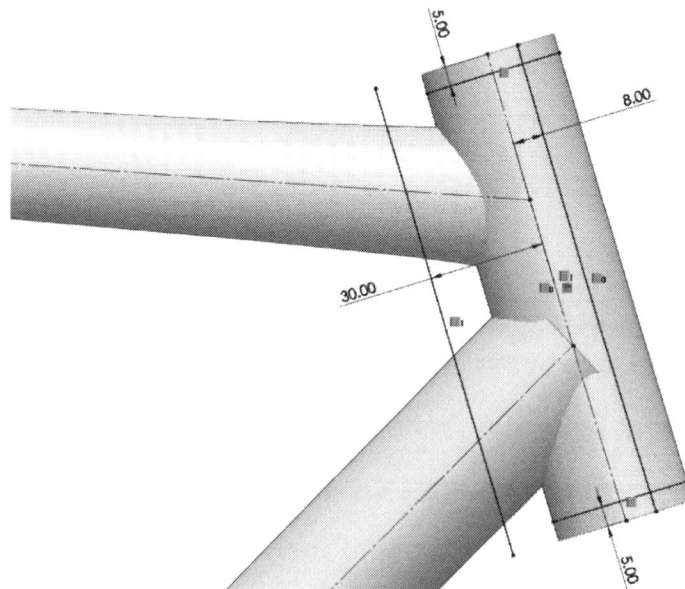

Figure 231 - Using the Offset Entities - Offsetting Selected Model edges

Offset the centerline on the centre of Down Tube 25mm up as shown in the following image.

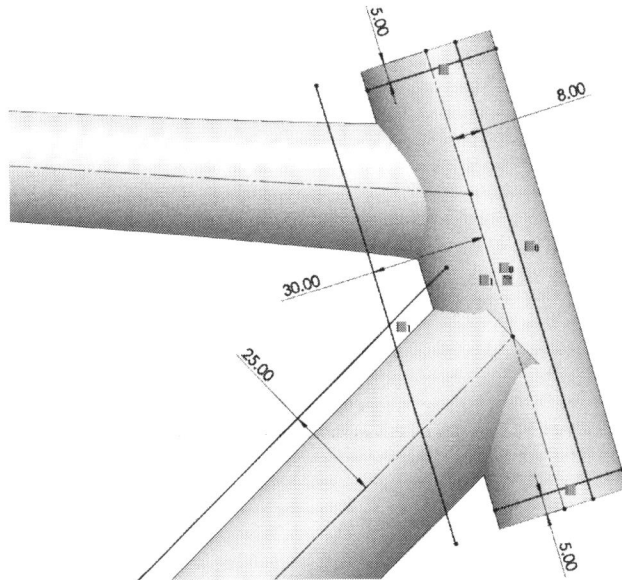

Figure 232 - Using the Offset Entities - Offsetting Sketch Entities

Draw a line as shown in the following image - take note of the Coincident and Perpendicular Sketch Relations *(added to the line manually or automatically)* and the 40mm Dimension.

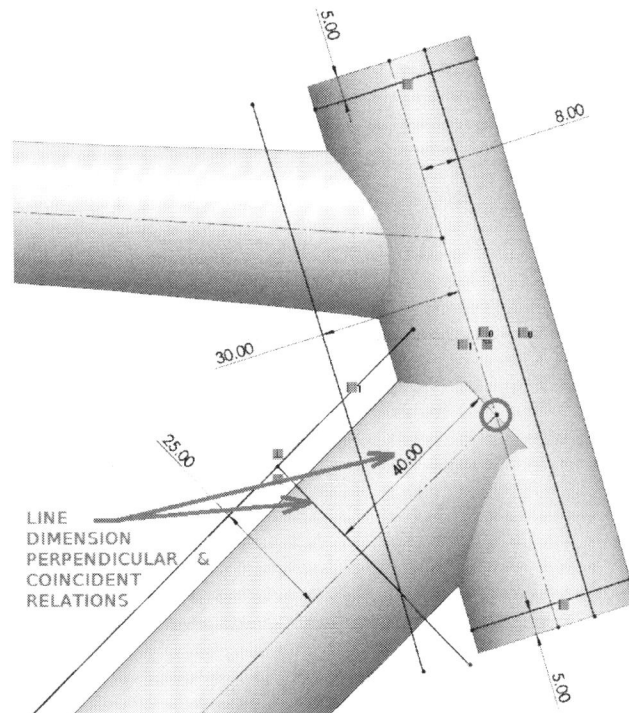

LINE
DIMENSION
PERPENDICULAR &
COINCIDENT
RELATIONS

Figure 233 - Line with Coincident and Perpendicular Sketch Relations

Using the <u>Trim with Corner Tool</u> - click Trim Entities (Sketch toolbar) or Tools > Sketch Tools > Trim - In the PropertyManager, under Options, select Corner - trim the sketch entities to achieve the result shown in the following image.

137

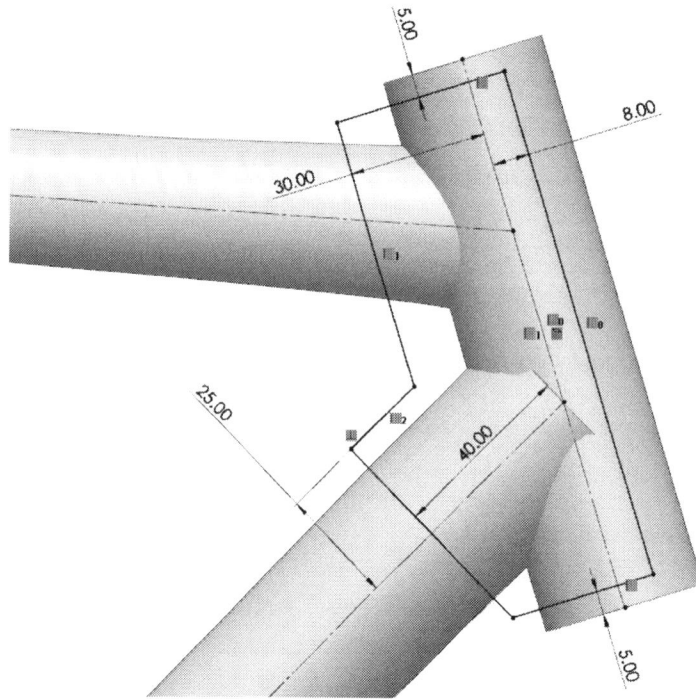

Figure 234 - Line with Coincident and Perpendicular Sketch Relations

Create a 15mm Fillet at the Bottom Right Corner as shown in the following image - click Sketch Fillet on the Sketch toolbar, or Tools > Sketch Tools > Fillet.

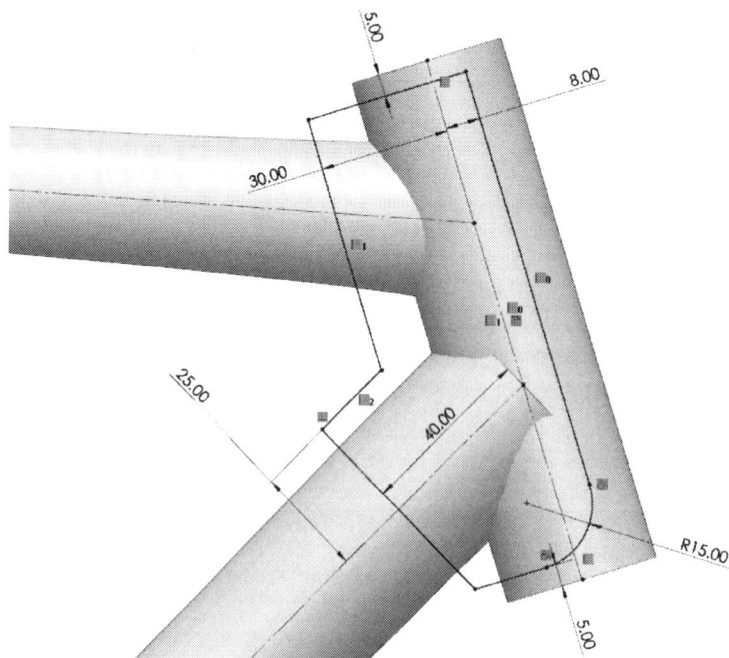

Figure 235 - Sketching Sketch Fillets

Create a 20mm Fillet at to the Top Right Corner as shown in the following image - click Sketch Fillet on the Sketch toolbar, or Tools > Sketch Tools > Fillet.

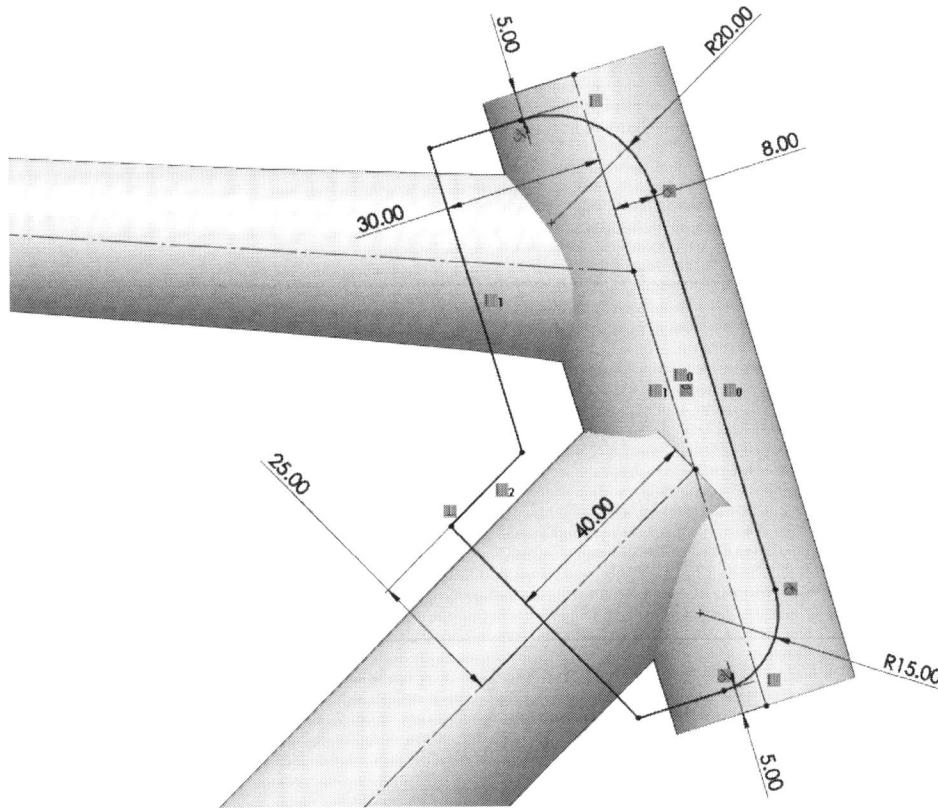

Figure 236 - Sketching Sketch Fillets

Our sketch is done, we can now use it as a trim tool.

TRIMMING MULTIPLE SURFACES USING A SKETCH AS A TRIM TOOL

Click Trim Surface on the Surfaces toolbar, or click Insert > Surface > Trim.

- In the PropertyManager, under Trim Type, select a Standard.
- Under Selections, Trim Tool - choose the sketch we just created from the Graphics Window or from the Flyout Feature Manager Design Tree.
- Select a Trim Action - Keep Selections Radio Button
- Under Surface Split Options, Select the Natural Radio Button. make sure the Split All checkbox is not selected.
- Select surfaces in Pieces to Keep as shown in the following image

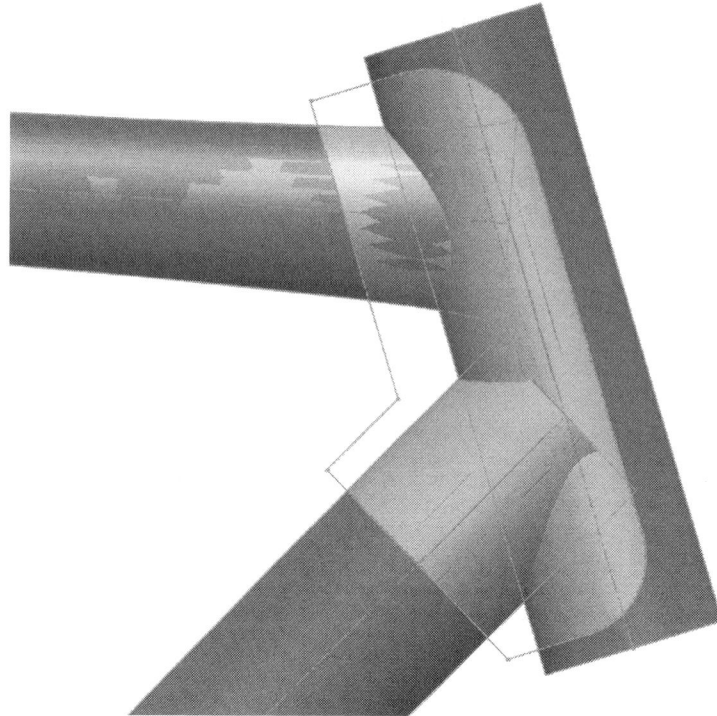

Figure 237 - Trim Surface - Surfaces in Pieces to Keep

Your Trim Surface Property Manager should look as shown in the following image.

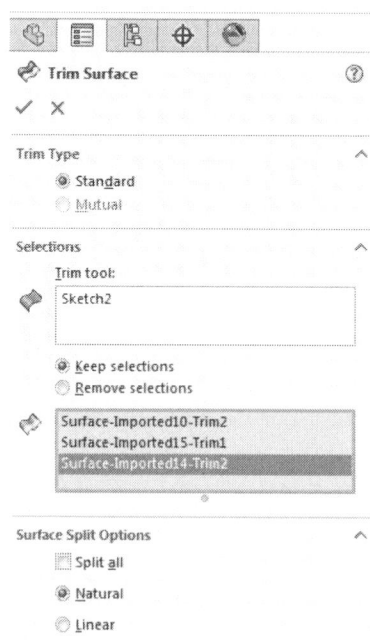

Figure 238 - Trim Surface Property Manager

Click OK. Your part should now look as shown in the following image.

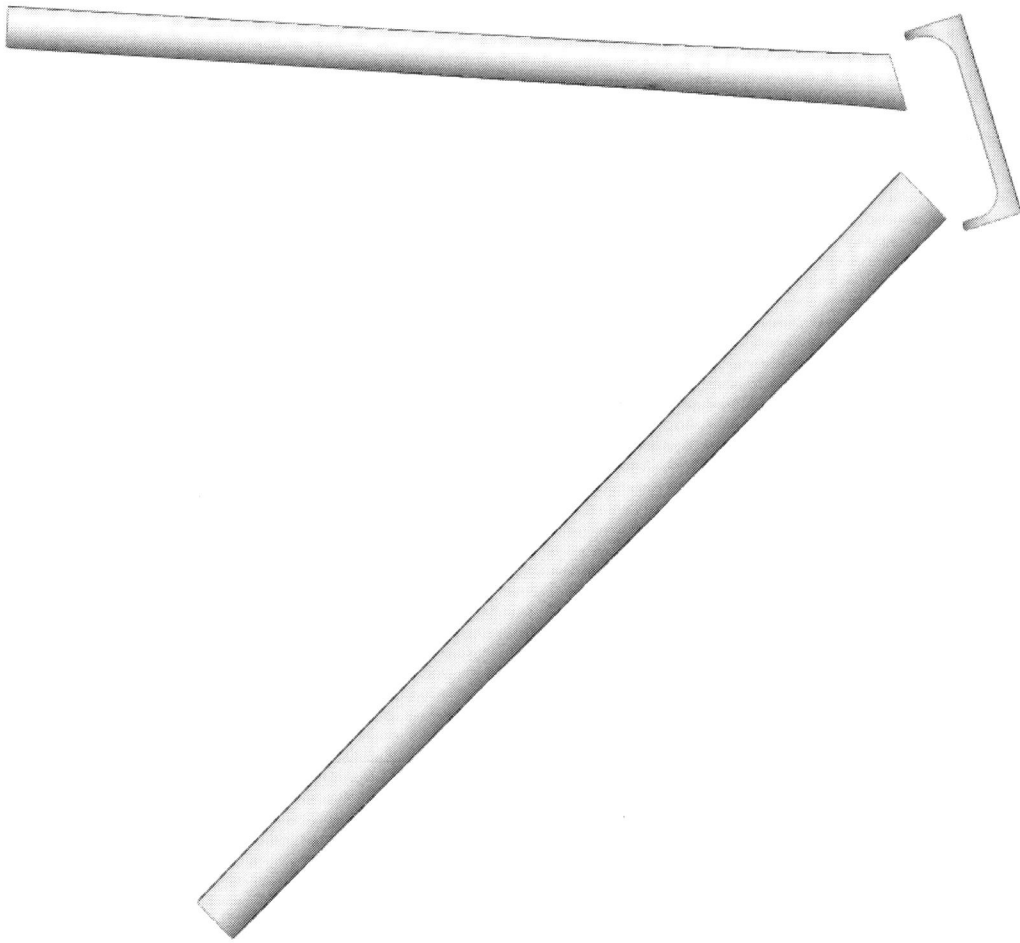

Figure 239 - Part Current Status

In this Chapter you are required to:

- Measure the area of the front surface (Head Tube) after trimming
- Select the range that contains the measured surface area of Surface1 in square millimeters from the options given below:
  a. 5490 - 5590

  b. 5850 - 5950

  c. 6065 - 6165

  d. 5740 - 5840

To measure the area of the front surface (Head Tube):

- Click Measure (Evaluate Toolbar) or Tools > Evaluate > Measure.
- Click on the Trimmed Head Tube Surface in the Graphics Area

The answer is thus 5793.83 square millimeters as shown in the following image.

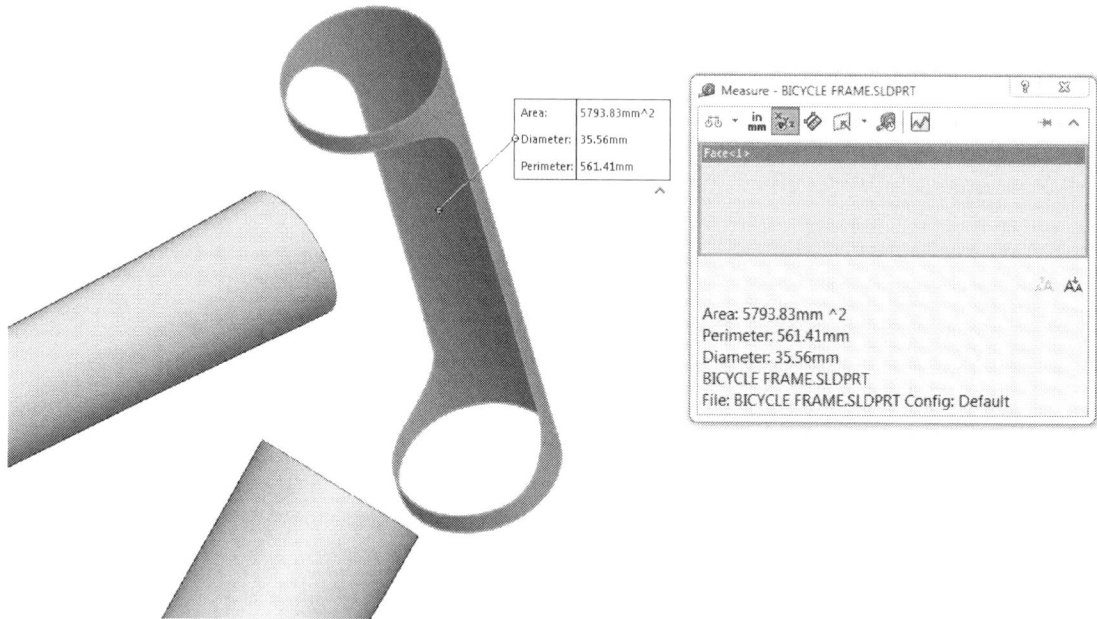

Figure 240 - Measuring Surface Area using the Measure Tool

However, you are required to select the range that contains the measured surface area of the trimmed Head Tube in square millimeters from a set of options given in sections above.
So the correct is range is on (d): **5740 - 5840**

since 5740 < **5793.83** < 5840.

Save your part and move on to Chapter Seven.

CHAPTER 7

In this Chapter we will continue working with the same part from Chapter Six.

You are required to:

- Show the sketch labeled "123" which has an outline and a Sketch Picture of the blend surface #5 to be created.
- Rotate the viewport to the default Right View.
- Create a blend Surface that as closely as possible matches the shape of the outline and sketch picture using the Boundary Surface Feature. The Boundary Surface created should use the Tangency to Face Option on both boundaries selected.
- Measure the area of the surface created and provide the answer in square millimeters.

See the following images that forms part of the problem set described above to be dealt with in this Chapter.

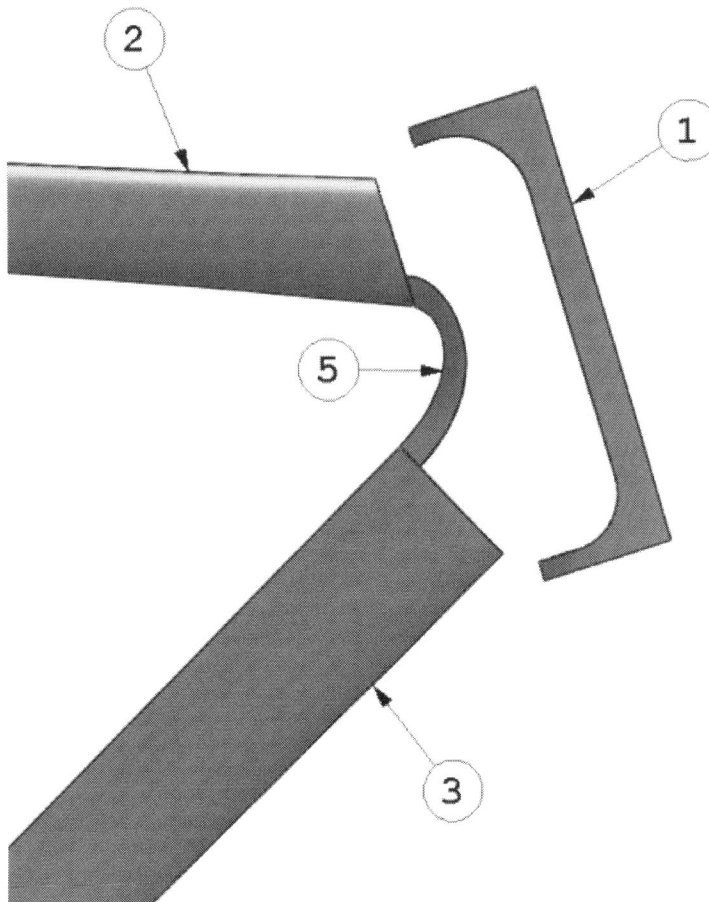

Figure 241 - Surface #5 to be created in this Chapter

143

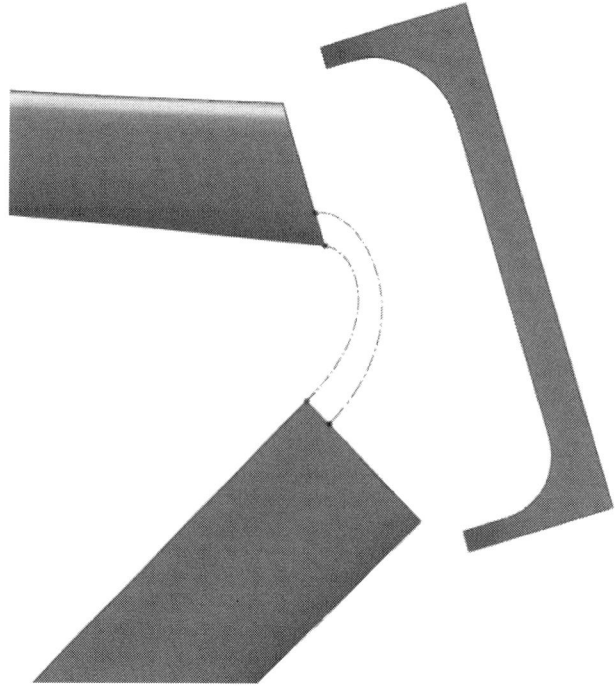

Figure 242 - Sketch "123" Outline

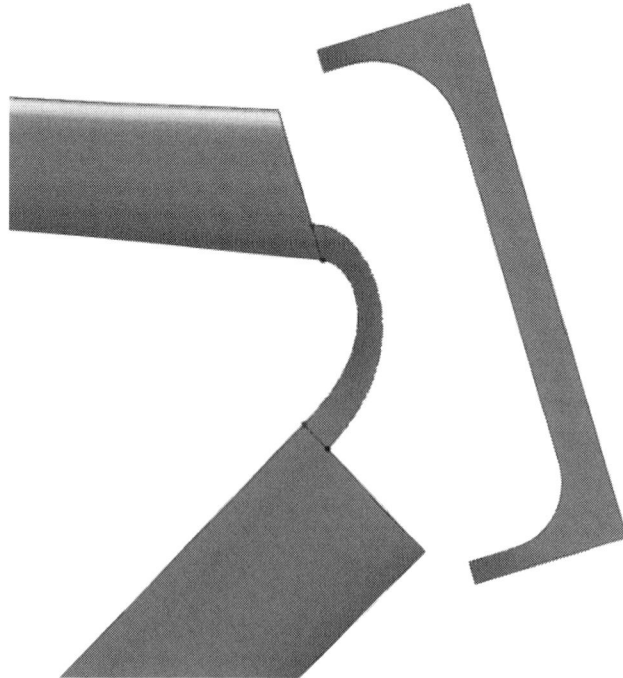

Figure 243 - Blend Surface #5 - Right View

144

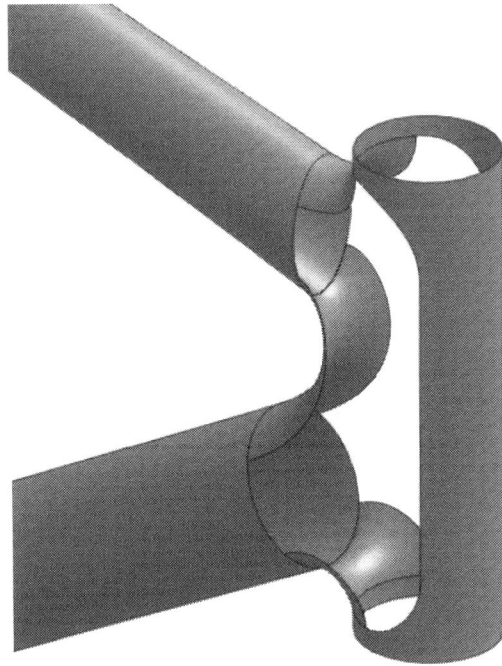

Figure 244 - Blend Surfaces - Isometric View

CREATING BOUNDARIES

You are required to create a boundary surface and as discussed in previous chapters, to create a boundary surface we need to have some boundaries. Looking at the surface we need to create we will definitely need at least four boundaries in the directions shown in the following image with the Head Tube surface hidden.

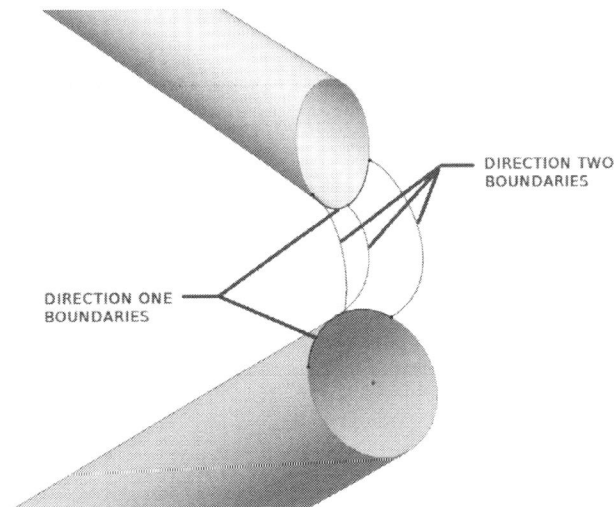

DIRECTION TWO
BOUNDARIES

DIRECTION ONE
BOUNDARIES

Figure 245 - Boundaries required for Surface #5 Boundary Surface

There are 3 ways to create the Direction One Boundaries as listed and illustrated below from Method One to Method Three in the following section.

CREATING DIRECTION ONE BOUNDARIES - METHOD ONE

The first method, Method One involves creating a 3D Sketch.

Click 3D Sketch (Sketch toolbar) or Insert > 3D Sketch.

Click Convert Entities (Sketch toolbar) or Tools > Sketch Tools > Convert Entities.

Select the two edges shown in image below.

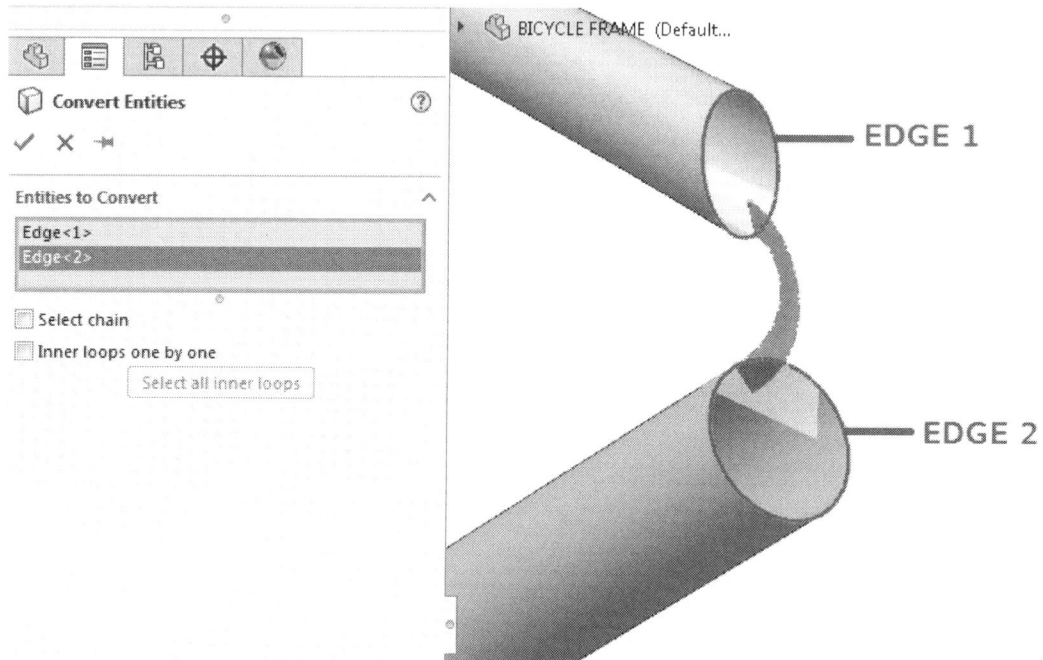

Figure 246 - Convert Entities

Click OK.

Draw two horizontal line as shown below then apply an Along X axis relation to each one of these two lines if the relation was not automatically applied - each of the lines should have relations as shown in the following image.

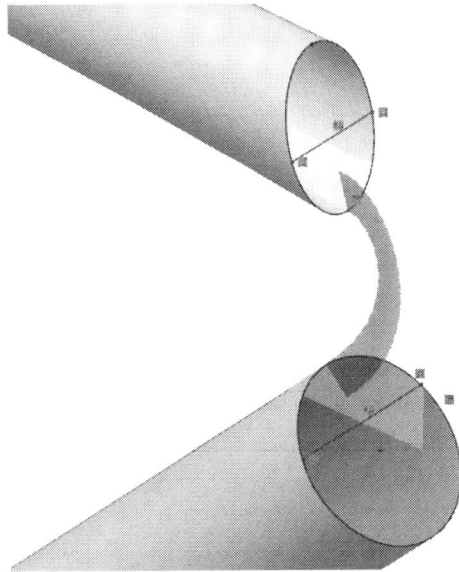

Figure 247 - Creating Direction One Boundaries

Add a coincident relation between the line on the Top Tube edge and the end point on sketch 123 as shown in the following image. Do the same for the line on the edge of the Down Tube as shown in the following image.

Figure 248 - Direction One Boundaries Creation

Click Trim Entities (Sketch toolbar) or Tools > Sketch Tools > Trim.

The pointer changes to ✂. Select each sketch entity you want trimmed to the closest intersection as shown in the following image.

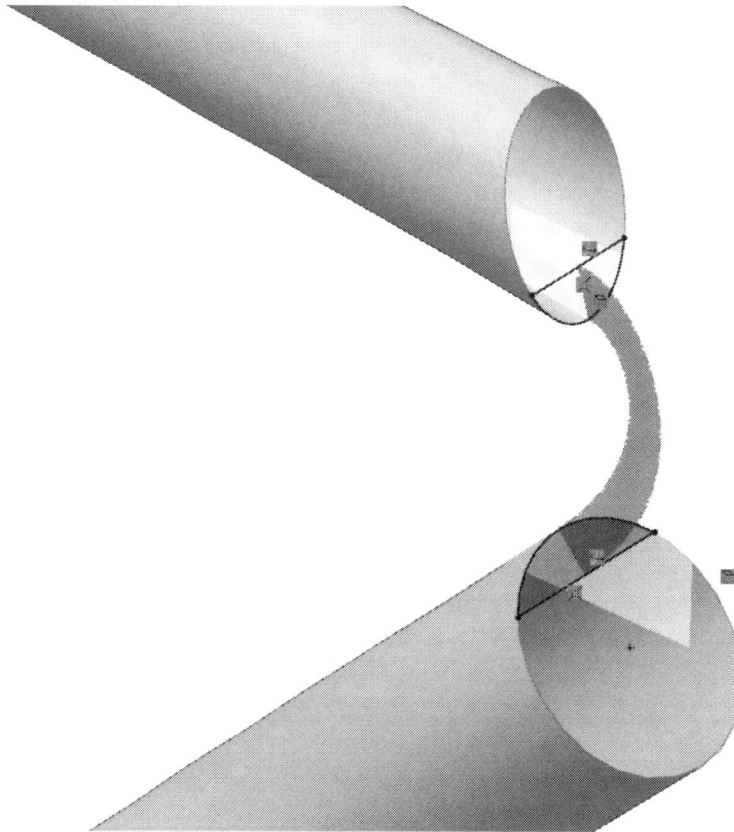

Figure 249 - Direction One Boundaries Creation

Click Trim Entities (Sketch toolbar) or Tools > Sketch Tools > Trim to deactivate the Trim Command.

Select the two straight lines by holding the Ctrl Key on your Keyboard then clicking on one of the lines followed by the other one. Select For Construction under Options in the Property Manager .

Options
☑ For construction

Click OK.

Your Parts should now look as shown in the following image.

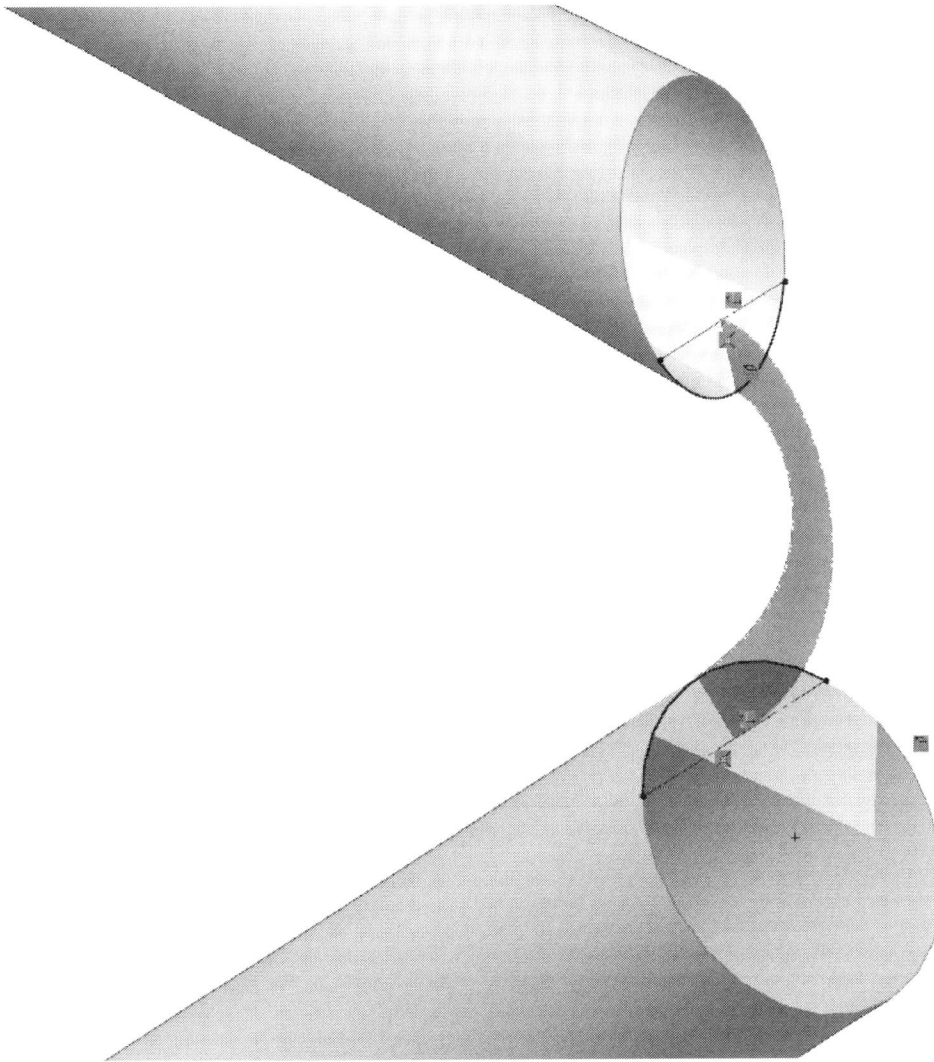

Figure 250 - Boundaries Creation

Exit the 3D Sketch.

Method One is done.

Save your part.

Suppress the 3D Sketch you just created and move on to Method Two.

CREATING DIRECTION ONE BOUNDARIES - METHOD TWO

The second method, Method Two involves using Split Lines. Select the Right Plane or YZ Plane in the FeatureManager Design Tree and click the sketch entity tool or click Sketch on the Sketch Toolbar.

Unhide the Tube Centerlines Sketch.

Draw two lines as shown in the following image with Coincident Relations to the end points of Sketch 123 and parallel relations to the Top Tube centerline and Down Tube centerline respectively as shown in the following image.

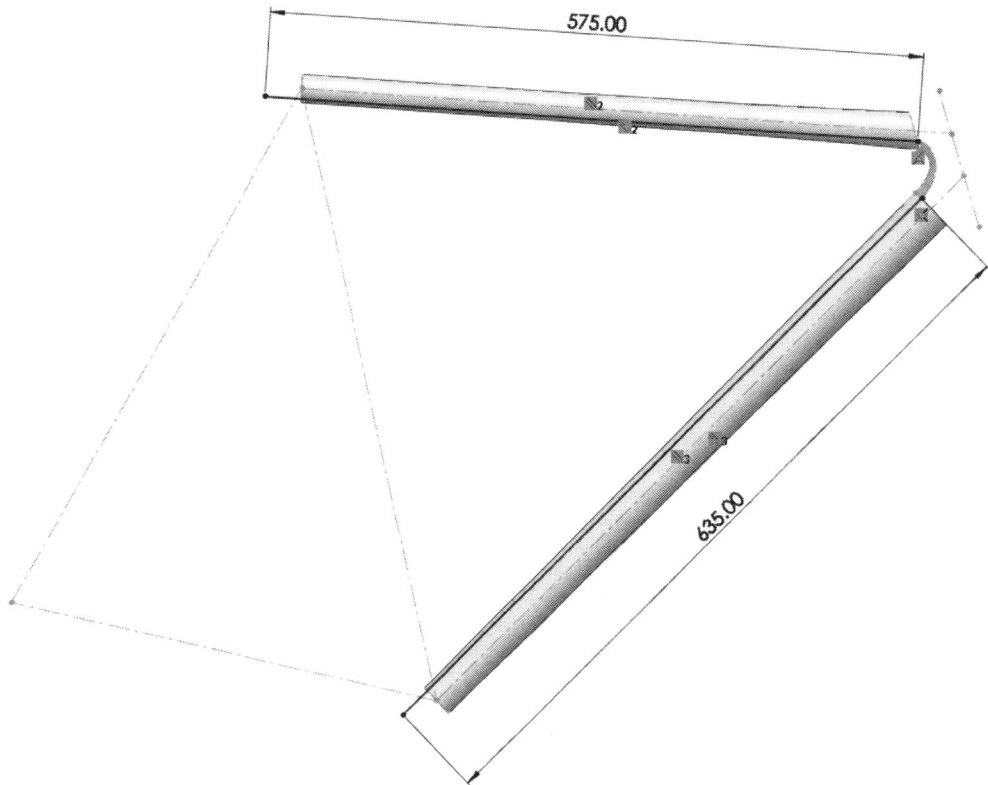

Figure 250 - Creating Direction One Boundaries - Method Two

Click Split Line (Curves toolbar) or Insert > Curve > Split Line. In the PropertyManager, under Type of Split, select Projection.

Under Selections, Current Sketch automatically appears since we haven't exited the sketch. Under Faces to Split select the two surface bodies of the Top Tube and the Down Tube.

Make sure Single direction is not checked since it will project the split line in one direction only.

Click OK.

Your part should now appear as shown in the following image.

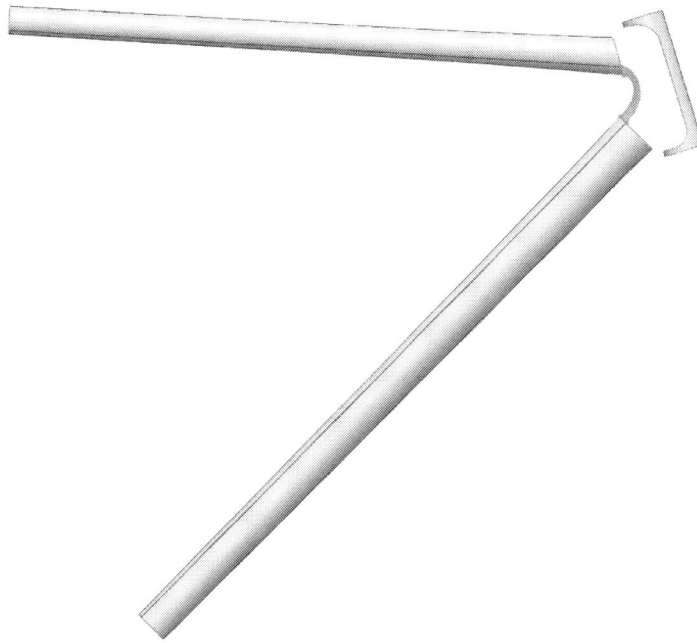

Figure 252 - Creating Direction One Boundaries - Method Two

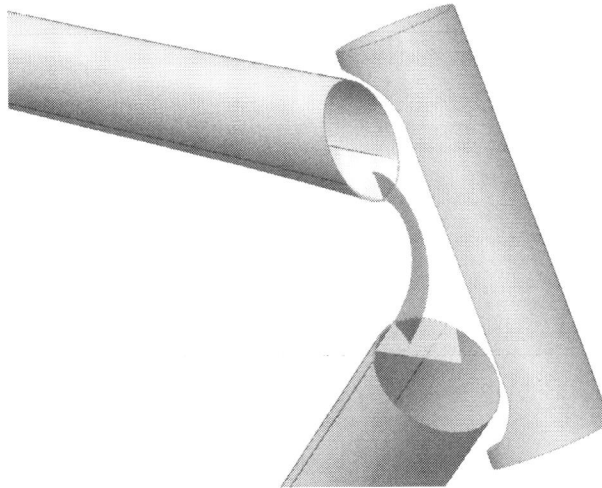

Figure 253 - Creating Direction One Boundaries - Method Two

Method Two is done. Save your part.

Suppress the Split Line Feature you just created and move on to Method Three.

CREATING DIRECTION ONE BOUNDARIES - METHOD THREE

The third method involves editing the sketch used as the Trim Tool in Chapter Six. The reason being that the sketch entities used were continuous in the trim tool sketch were continuous. So we can split the sketch entities at the points we want our boundaries to be as described in the following section.

Right Click the sketch used as the Trim Tool that has been absorbed by the Surface Trim Feature in the FeatureManager Design Tree and select Edit Sketch as shown in the following image.

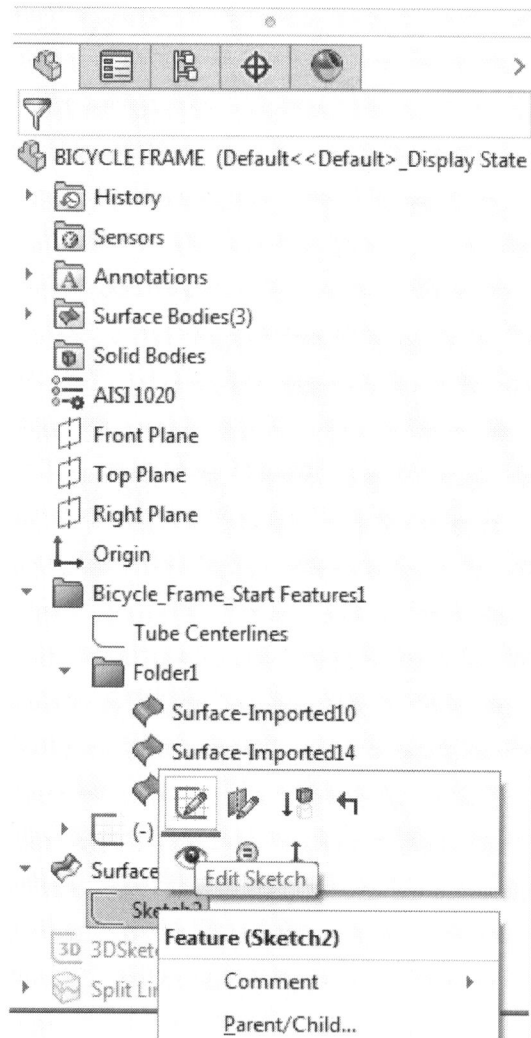

Figure 254 - Editing a Sketch

USING SPLIT ENTITIES

Split Entities is used to split a sketch entity to create two sketch entities. Conversely, you can delete a split point to combine two sketch entities into a single sketch entity.

Click Split Entities 〳 (Sketch toolbar) or Tools > Sketch Tools > Split Entities.

The pointer changes to .

Click the sketch entities (straight lines) in the Graphics Area at the locations *(Point A and Point B)* as shown in the following image where we want the splits to occur. Each of the sketch entities splits into two entities, and a split point is added between the two sketch entities.

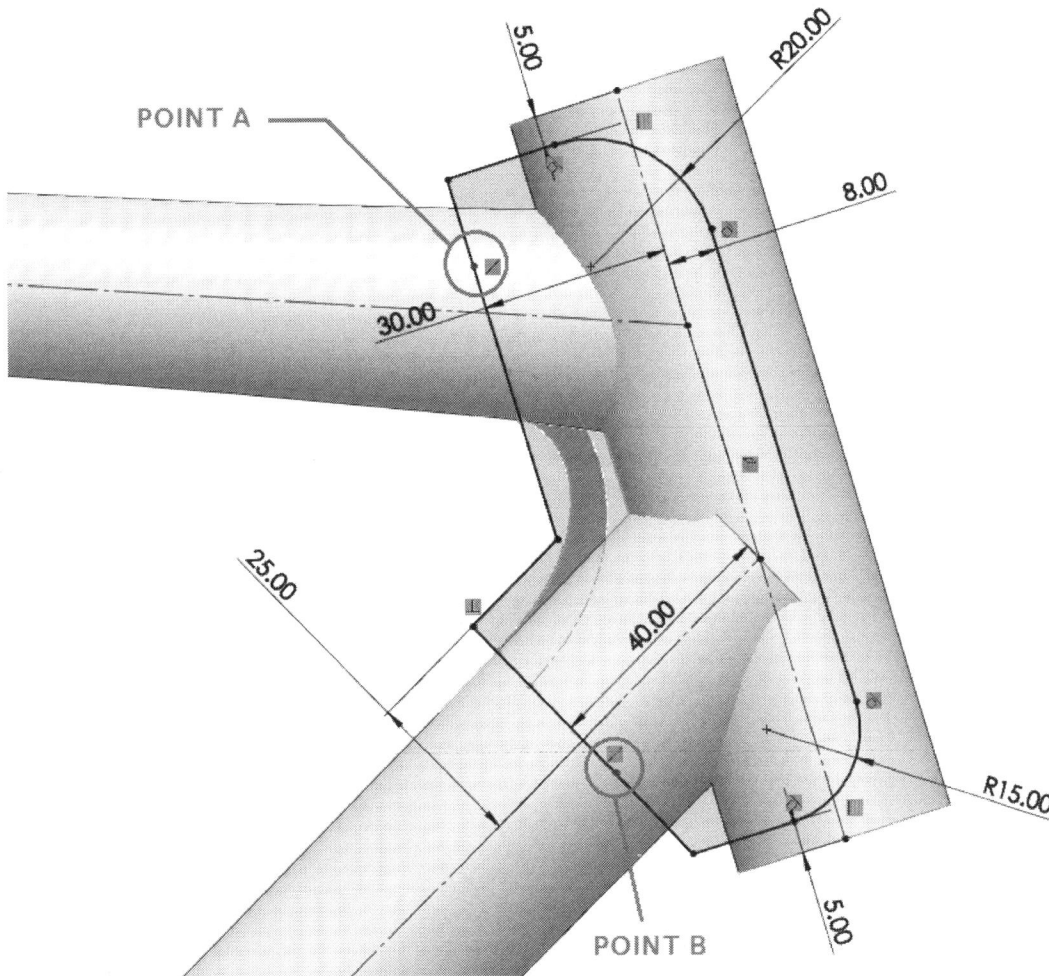

Figure 255 - Editing a Sketch - Using Split Entities

Press the Escape Key on your Keyboard or Click Split Entities (Sketch toolbar) or Tools > Sketch Tools > Split Entities to deactivate the Split Entities Command.

153

CHANGING DIMENSIONS TO DRIVEN DIMENSIONS AND ADDING SKETCH RELATIONS

We now want to add coincident sketch relations between the newly created split points and end points on Sketch "123" but if we try to do so our sketch becomes over defined. Hence, we need to change the 30mm and 40mm dimensions from Driving Dimensions to Driven Dimensions - thus the two dimensions will then be driven by the new coincident relations.

Right click on the 30mm Dimension and select Driven then Click Close Dialog or OK in the Dimension Property Manager. You will notice that the dimension changes in colour from Black *(Driving)* to Grey *(Driven)* and the line segments also change from Black *(Fully Defined)* to Blue *(Under Defined)*.

Right click on the 40mm Dimension and select Driven then Click Close Dialog or OK in the Dimension Property Manager. You will notice that the dimensions changes in colour from Black *(Driving)* to Grey *(Driven)* and the line segments also change from Black *(Fully Defined)* to Blue *(Under Defined)*.

Your sketch should now look as shown in the following image after changing the 30mm and 40mm dimensions from Driving Dimension to Driven Dimensions.

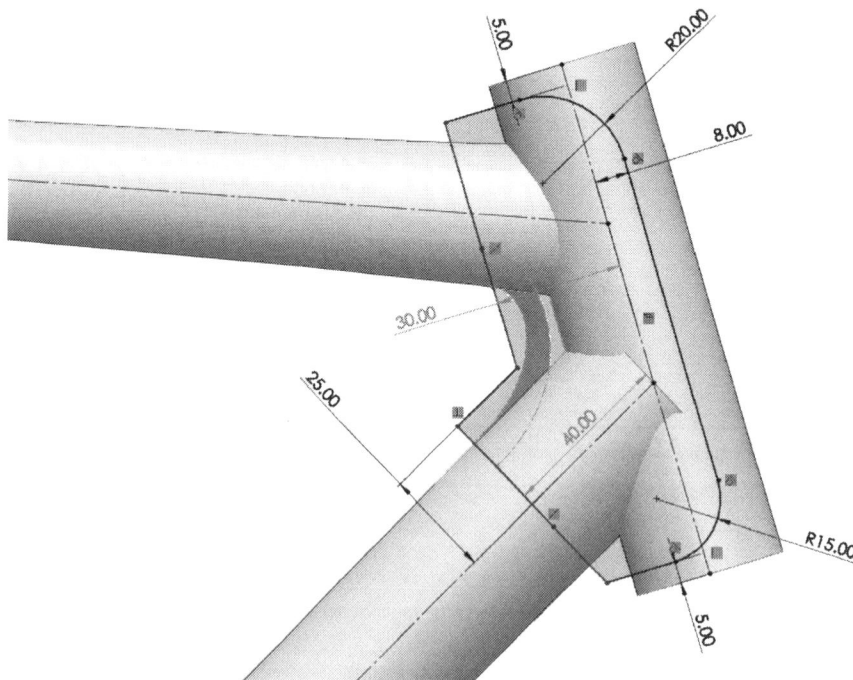

Figure 256 - Changing Dimensions from Driving Dimensions to Driven Dimensions

Now let's add coincident sketch relations between the split points and end points on sketch "123". Click Add Relations (Dimensions/Relations toolbar) or Click Tools > Relations > Add.

154

Under Selected Entities, select the top left point on sketch "123" and split point on the 30mm dimension line segments as shown in the following image.

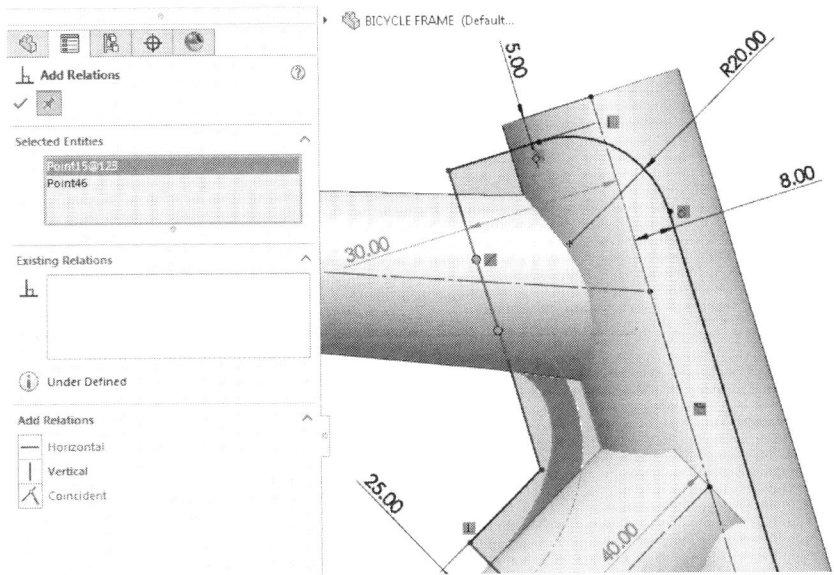

Figure 257 - Adding Sketch Relations

Click Coincident under Add Relations in the Add Relations Property Manager. Click OK. Your part should now look as shown in the following image - **NB:** The line segments become fully define again.

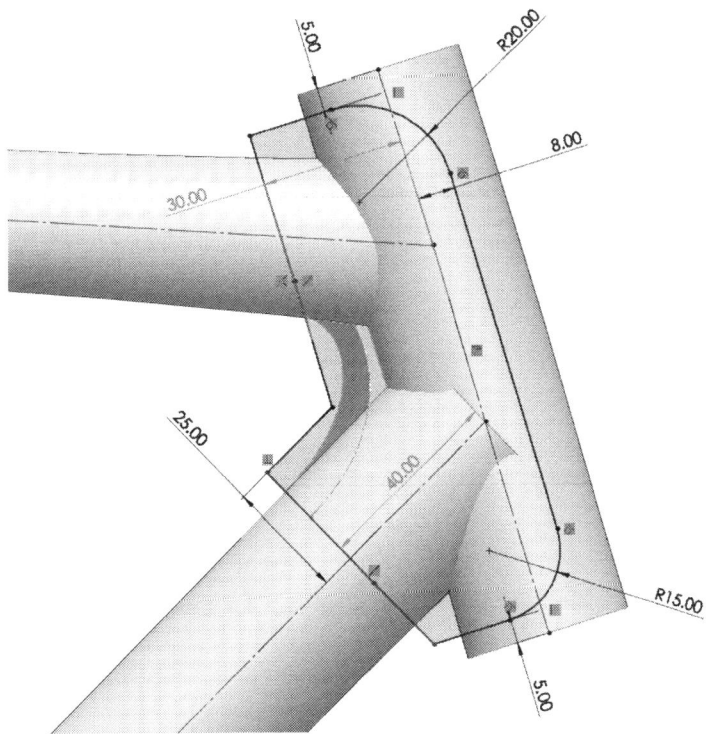

Figure 258 - Adding Sketch Relations

Repeat the same process and add a relation between the split point on the 40mm Dimension line segments and the bottom right point on sketch "123" as shown in the following image.

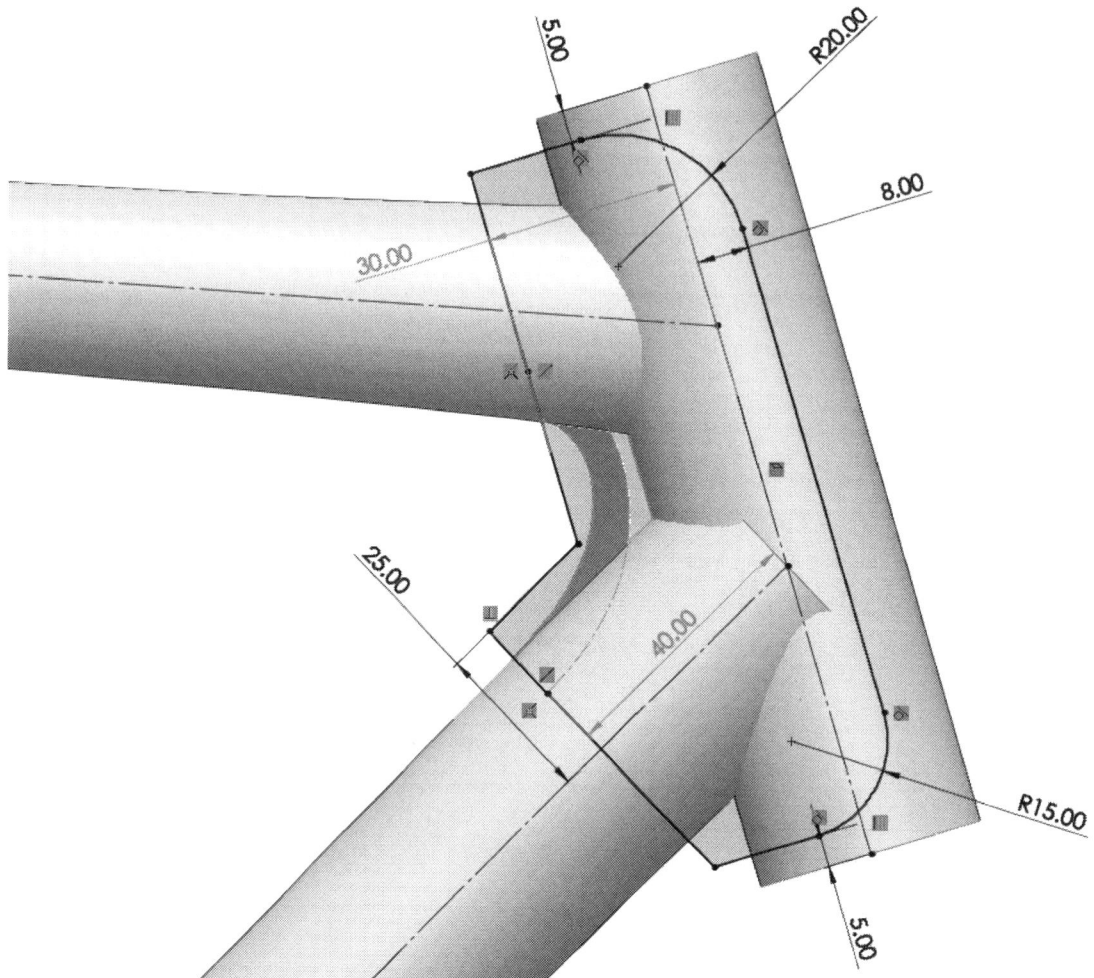

Figure 258 - Adding Sketch Relations

Click Coincident under Add Relations in the Add Relations Property Manager. Click OK. Your part should now look as shown in the following image.

Right-click and select Exit Sketch ⤶ or press the letter "D" on your keyboard and select Exit Sketch or OK to accept and finish the edit sketch command.

Save your part.

You will notice that no change is visually noticeable but when you click or select any of the trimmed edges you will notice that they no longer form a continuous loop - the change is depicted in the following image with the Head Tube Hidden.

156

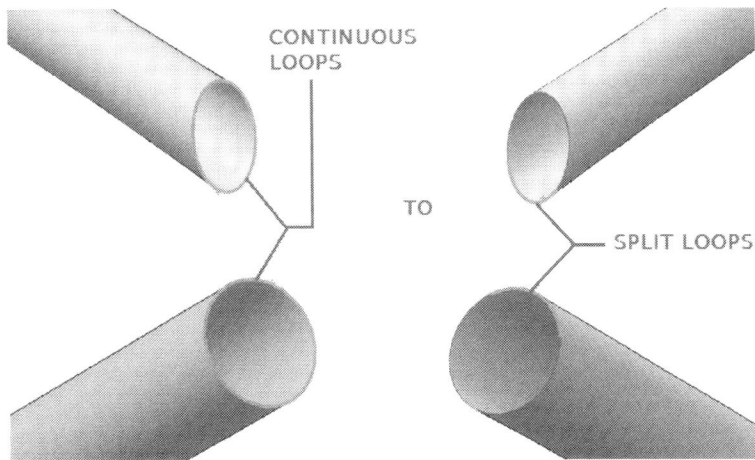

Figure 260 - Creating Direction One Boundaries - Method Three

Method Three is done.

Save your part.

Now, let's move on to create Direction Two Boundaries.

CREATING DIRECTION TWO BOUNDARIES

Now, let's move on to create Direction Two Boundaries. Click 3D Sketch (Sketch toolbar) or Insert > 3D Sketch. Click Spline (Sketch toolbar) or Tools > Sketch Entities > Spline. The pointer changes to 🖉. Click to place the first point on one of the split points on the trimmed edge of the Top Tube as shown in the following image.

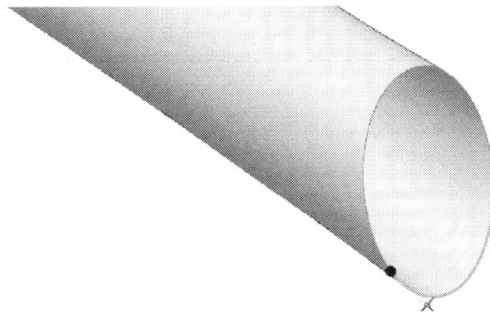

Figure 261 - Spline first point

Drag out the first segment then click the second point as shown in the following image.

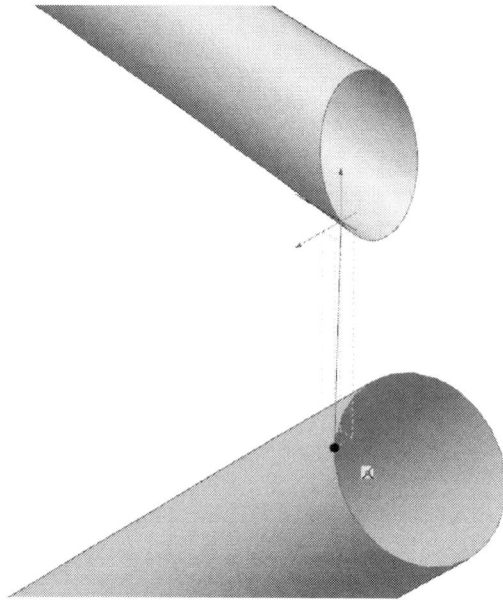

Figure 262 - Spline second point

Press the escape key on your keyboard to end the Spline Command. Your Part should now look as shown in the following image with all other sketches hidden.

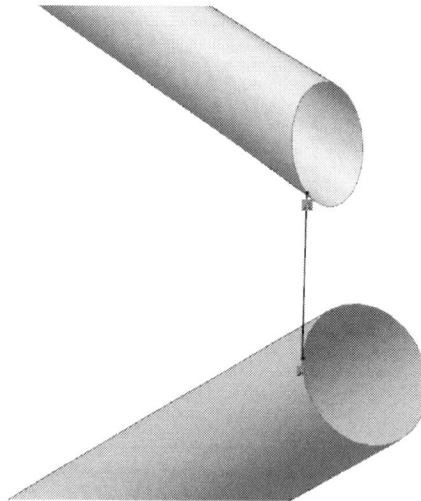

Figure 263 - Two Point Spline

Click on the Right Plane in the Feature Manager Design Tree and select Normal To as shown in the following image to align the model normal to the Right Plane or YZ Plane.

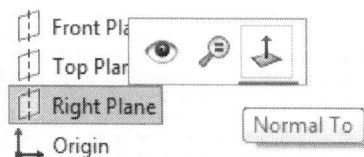

Figure 264 - Changing View Orientation using Normal To

TIP : You can also use the View Selector to see and select model views in context. Press Ctrl + Spacebar or click View Selector in the Orientation dialog box and select the Right Plane - a preview is shown in the top right corner of the Graphics area when you move your mouse pointer around different faces of the View Selector.

Your model should now look as shown in the following image.

Figure 265 - Model Current Status

Show sketch "123" by right clicking on it in the Feature Manager Design Tree and selecting Show. Your part should now look as shown in the following image - zoomed in to the area of interest.

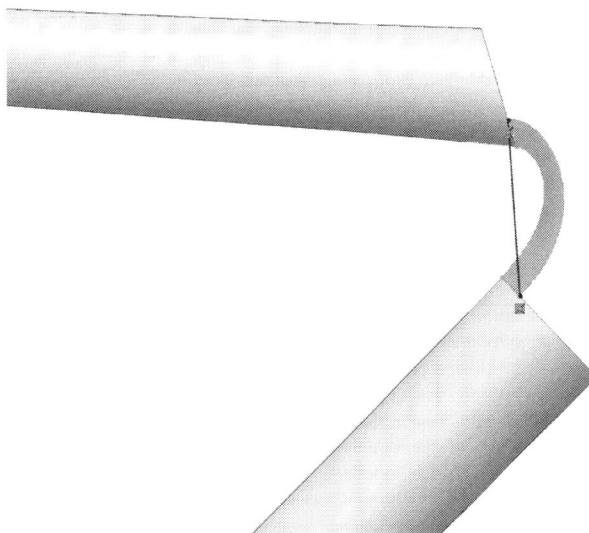

Figure 266 - Model Current Status

Now, strictly do not change the view orientation until you are done with the process of manipulating the spline's Spline Handles i.e. until the spline is as close as possible to the top spline of sketch "123".

Click on the spline to reveal its Spline handles which will appear in grey as shown in the following image.

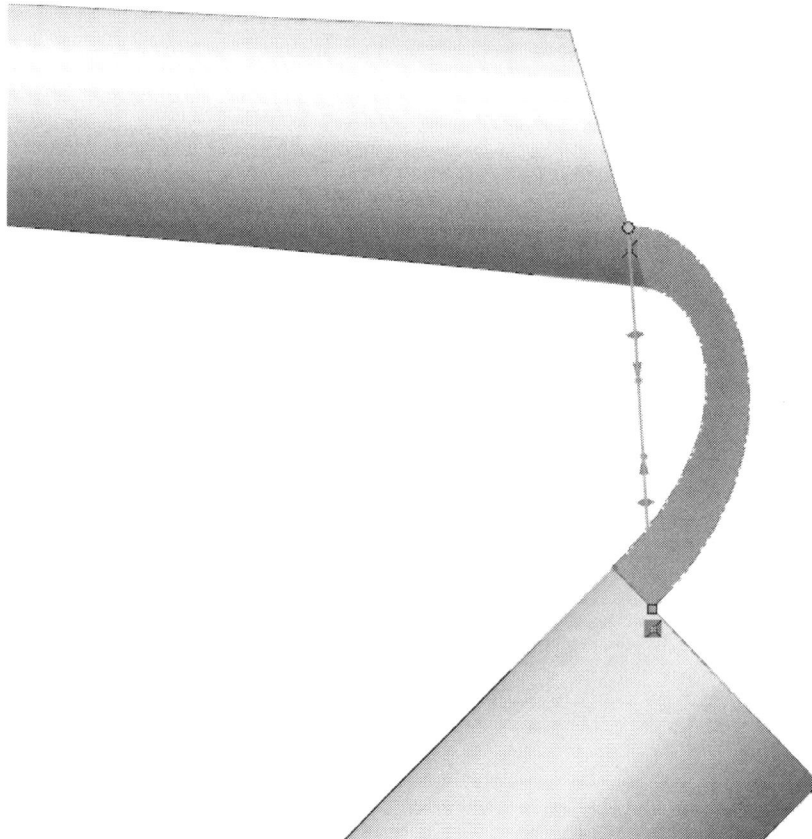

Figure 267 - Model Current Status

Drag the diamond handle of the top Spline Handle to control tangency direction (vector) as shown in the following image.

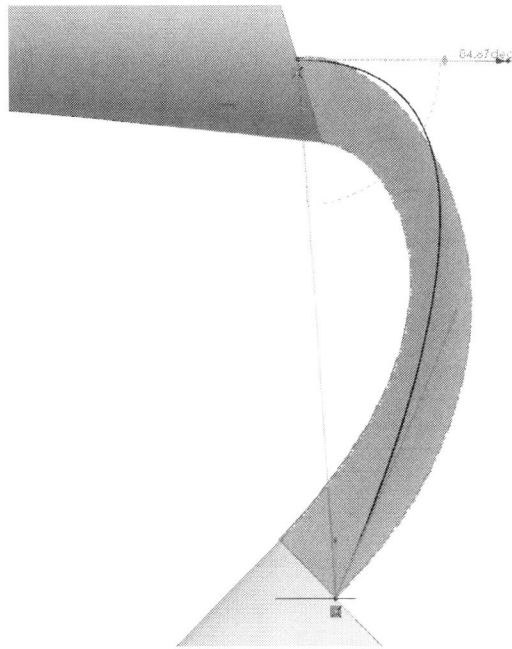

Figure 268 - Editing a Spline

Drag the diamond handle of the bottom Spline Handle to control tangency direction (vector) as shown in the following image.

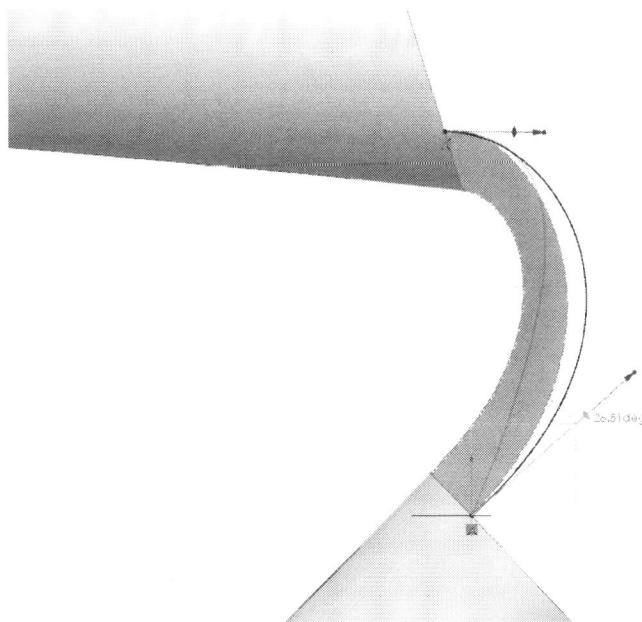

Figure 269 - Editing a Spline

Continue to carefully edit the spline by manipulating the Spline Handles including dragging the circular handles to control both tangent weighting and direction (vector). After you are happy with the spline position you may add fixed relations to the Spline Handles to keep the Spline Fully Defined. Your part should now look as shown in the following image - zoomed to the area

161

of interest - remember to maintain the right view orientation to avoid manipulating the spline handles in 3D Space.

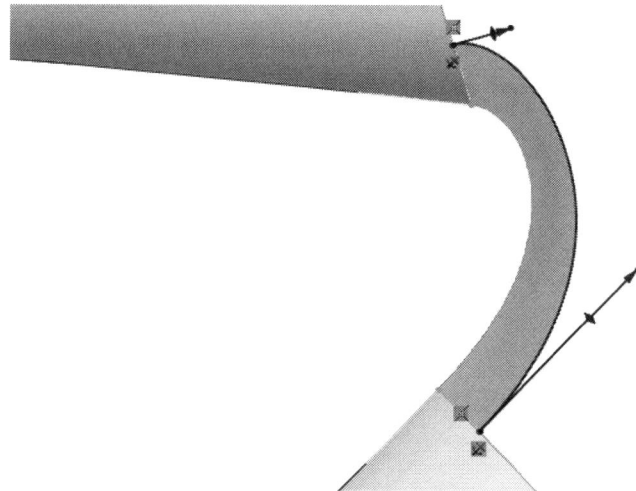

Figure 270 - Tracing a sketch a using a two point spline

ANALYSING THE QUALITY OF A SPLINE USING CURVATURE COMBS

Right-click the two point spline you created and select Show Curvature Combs - curvature combs will appear as shown in the following image - I have changed the Scale to 60 and Density to 280. Curvature combs provide visual enhancement of the slope and curvature of the spline and also help to identify undesirable areas of sudden change in curvature or curvature spikes or flat spots.

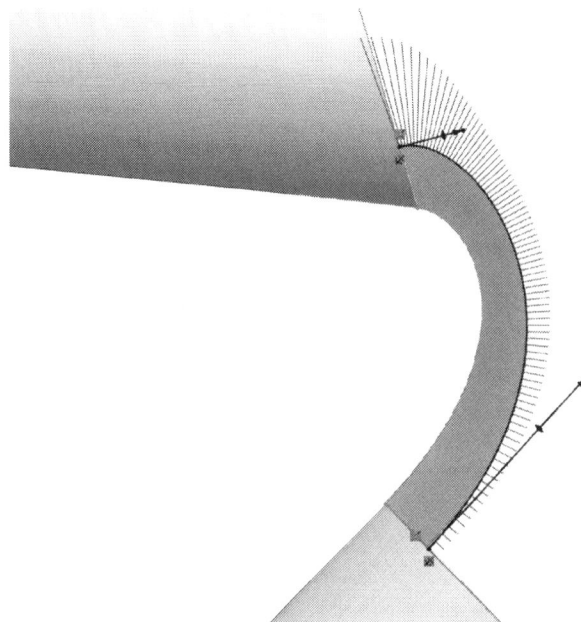

Figure 271 - Curvature Combs with a Scale of 60 and a Density of 280

TIP : You can change the colour of Curvature Combs by clicking Tools > Options > Colors > Temporary Graphics then Edit to choose any color of your preference.

Right-click the two point spline and select Show Curvature Combs to hide the Curvature Combs. So just like Zebra Stripes, Curvature Analysis and Deviation Analysis are Quality Evaluation Tools for surfaces - Curvature Combs are a Quality Evaluation Tool for splines.

Change the view to a 3D View as shown in the following image.

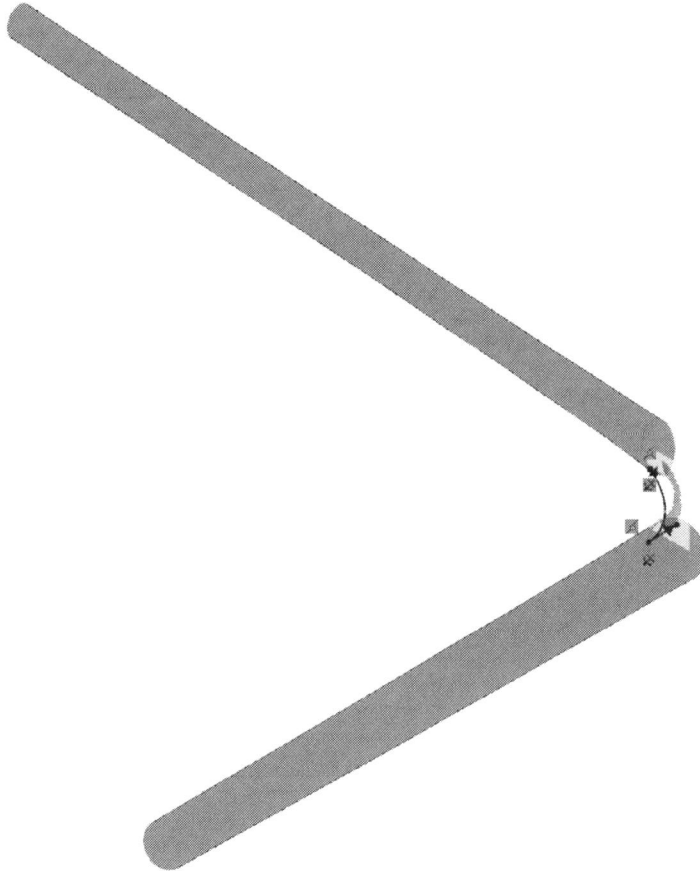

Figure 272 - Part Current Status - 3d View Orientation Using the View Selector

CREATING MIRRORED ENTITIES IN A 3D SKETCH

Instead of trying to recreate the two point spline on the RHS again and struggling to make it match the one we have already created on the LHS, we can use the Mirror Tool by doing the following:

Click Mirror Entities or Tools > Sketch Tools > Mirror.

Select the Two Point Spline we just created for Entities to mirror.

Select the Right Plane or YZ Plane for Mirror about.

Check the Copy Checkbox as shown in the following image. If you don't select the Copy Checkbox the spline will be sort of cut from the LHS and pasted onto the RHS - and the resulting mirrored spline will also be Under Defined.

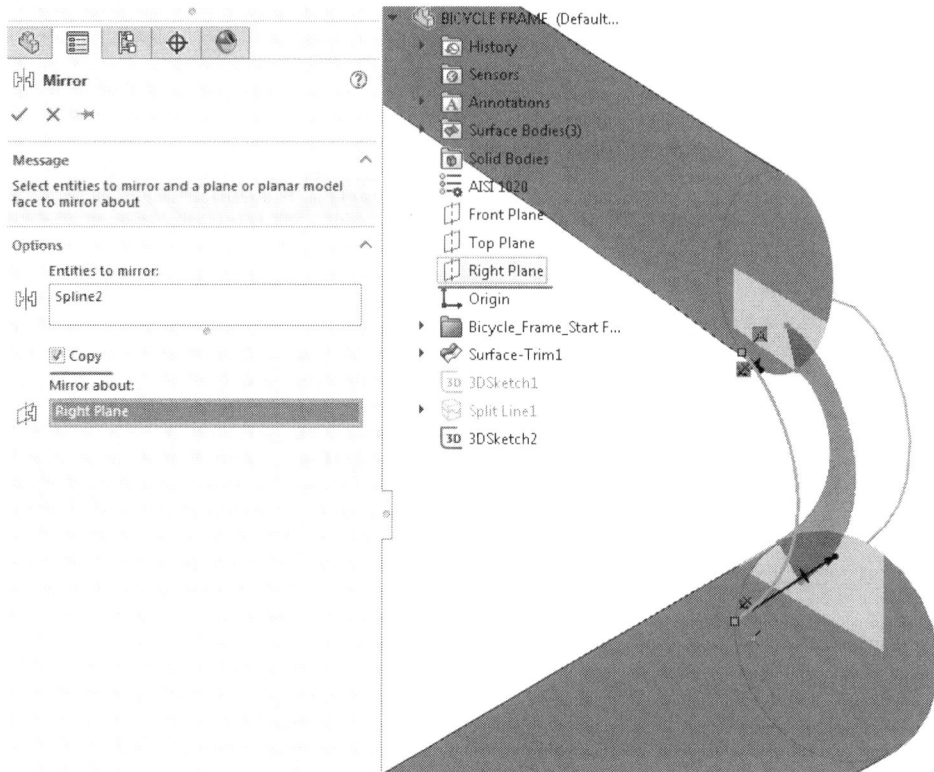

Figure 273 - Creating Mirrored Entities in a 3D Sketch

Click OK.

Your part should now look as shown in the following image - zoomed to the area of interest.

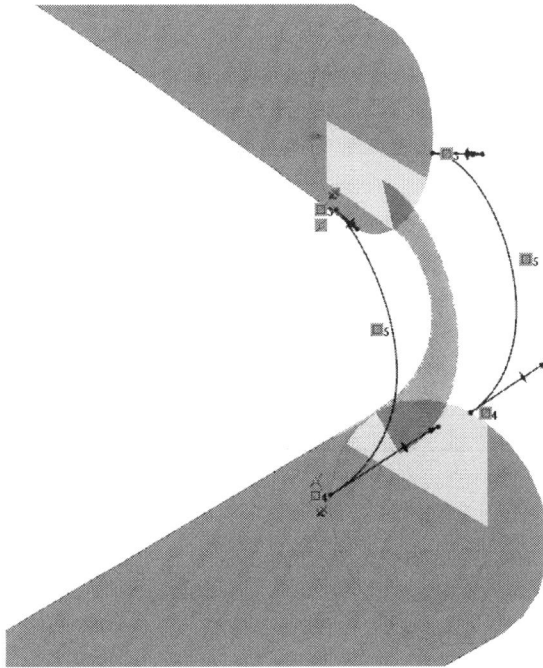

Figure 274 - Part Current status - Mirrored Sketch Entity in a 3D Sketch

USING CONVERT ENTITIES IN A 3D SKETCH

Click Convert Entities (Sketch Toolbar) or Tools > Sketch Tools > Convert Entities. In the Property Manager under Entities to Convert, select the bottom spline in sketch "123" in the Graphics Area. Click OK. Your part should now look as shown in the following image.

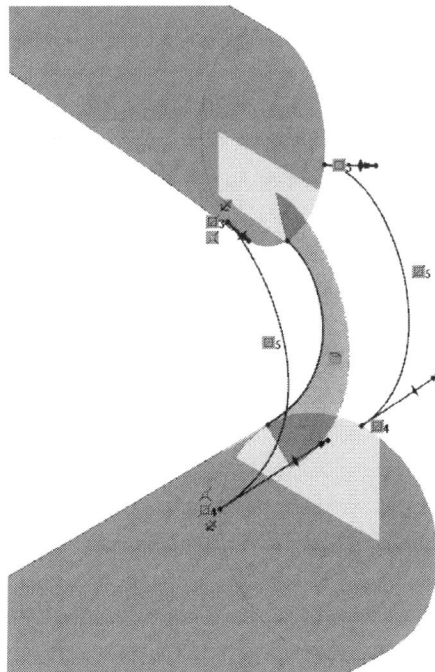

Figure 275 - Part Current status - Using Convert Entities

165

Exit the 3D Sketch. Save your part. Hide Sketch "123" and your part should look as in the following image - in 3D View and normal to the Right Plane.

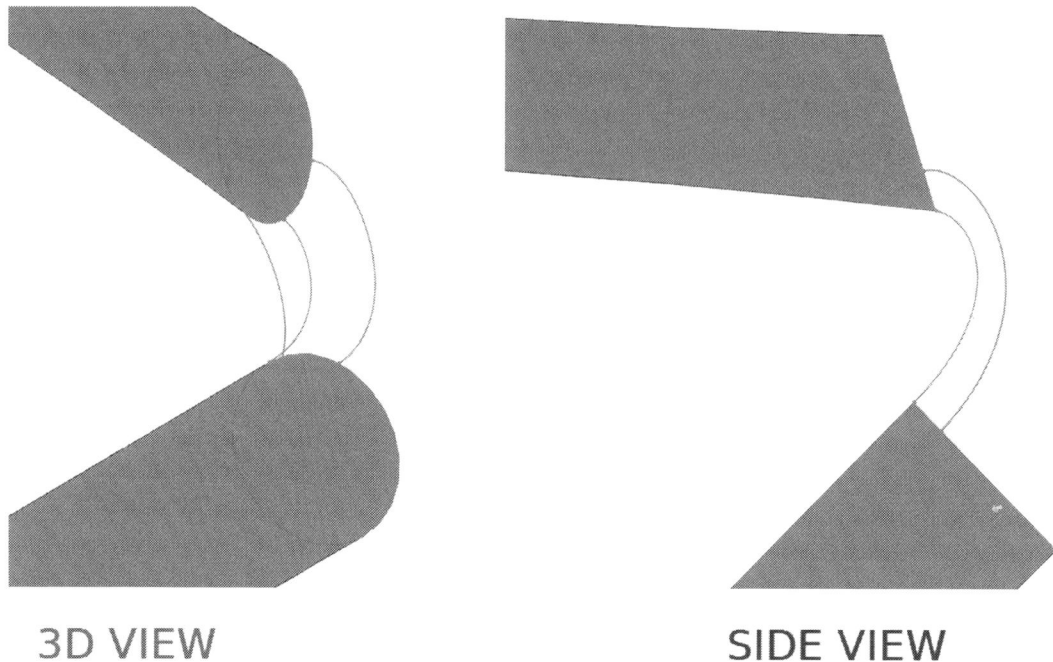

3D VIEW   SIDE VIEW

Figure 276 - Part Current status

Congratulations, you have finished creating Direction 2 Boundaries for the Boundary Surface we are going to create in the following section.

CREATING A BOUNDARY SURFACE

Click Boundary Surface (Surfaces Toolbar) or Insert > Surface > Boundary Surface.

Set Options in the Boundary Surface Property Manager as shown in the following image - **NB:** Use the Selection Manager to select individual splines in the Direction Two curves. Select Mesh Preview under Curvature Display.

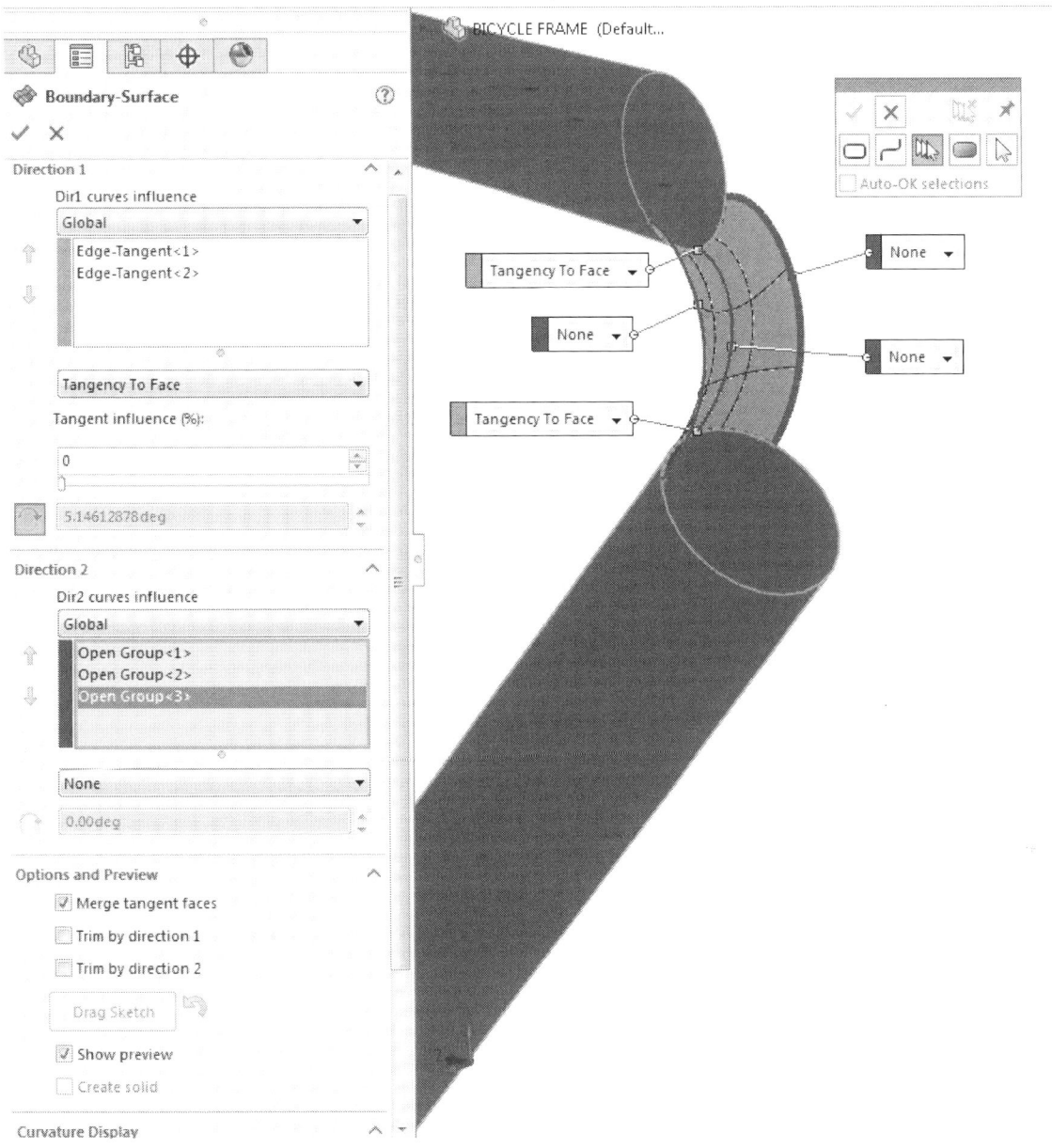

Figure 277 - Part Current status

Click Ok and your part should now look as shown in the following images.

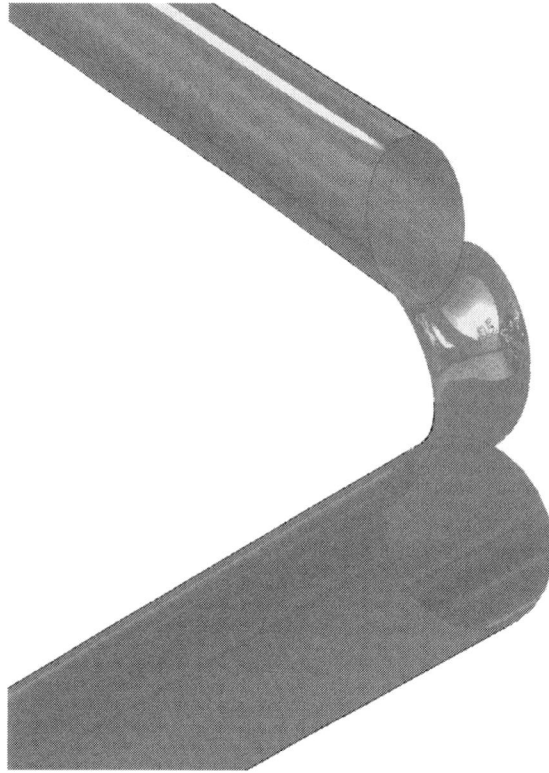

Figure 278 - Part Current status - zoomed to the area of interest- with RealView Graphics Turned On

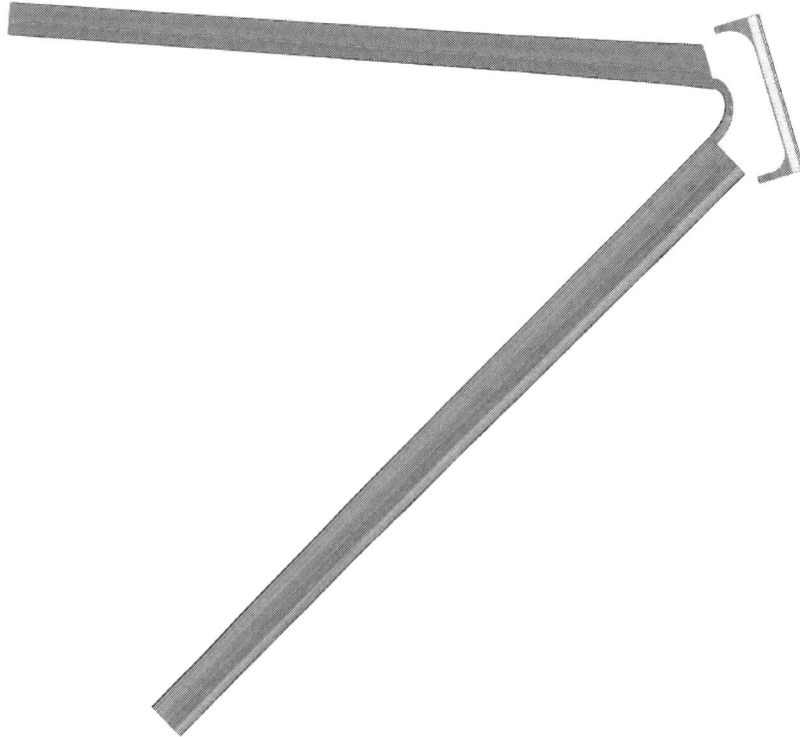

Figure 279 - Part Current status - with RealView Graphics Turned On

You are required to measure the Surface Area created and provide the answer in square millimeters to two Decimal Places.

MEASURING SURFACE AREA

Click Measure (Tools Toolbar) or Tools > Evaluate > Measure.

Select the boundary surface in the Graphics Area as shown in the following image.

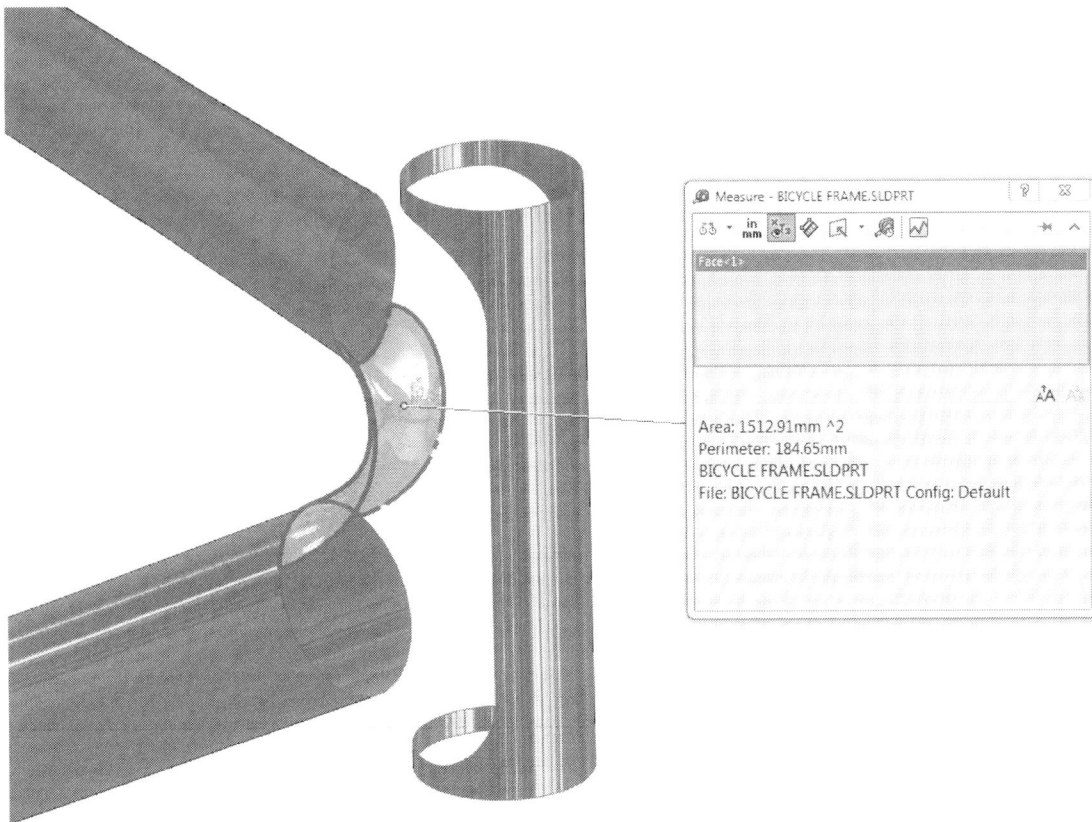

Figure 280 - Using the Measure Tool to measure Surface Area

The answer is 1512.91mm^2. **NB:** Your answer might be slightly different and that's acceptable due to the nature of surface modeling.

CHAPTER 8

In this Chapter we will continue working with the same part from Chapter Seven.

You are required to:

- Create the blend surfaces 4 and 6 as shown in the following images using the Boundary Surface feature. They should follow the dimensions specified in the second image.

Note: The extents of the blend surfaces should be defined by the trimmed tube surfaces.

- The blend surfaces should use the following options.
- Blend Surface 4:

    Tangency to Face 1: Tangent Length = 1.70

    Tangency to Face 2: Tangent Length = 1.70

- Blend Surface 6:

    Tangency to Face 1: Tangent Length = 1.00

    Tangency to Face 3: Tangent Length = 1.80

Note: See surface labels in the following image.

What is the surface area of Blend Surface 6 in square millimeters? Your unit system is still MMGS, Decimal Places: 2 and Material = None.

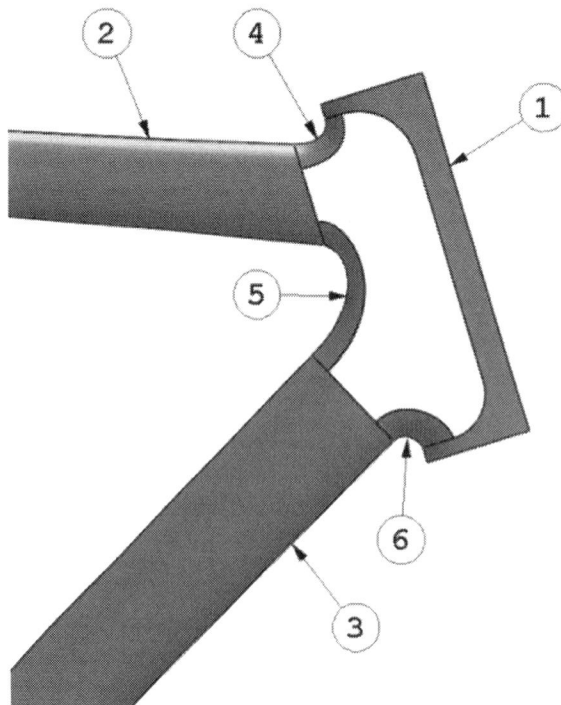

Figure 281 - Surfaces to be created - Blend Surfaces 4 and 6

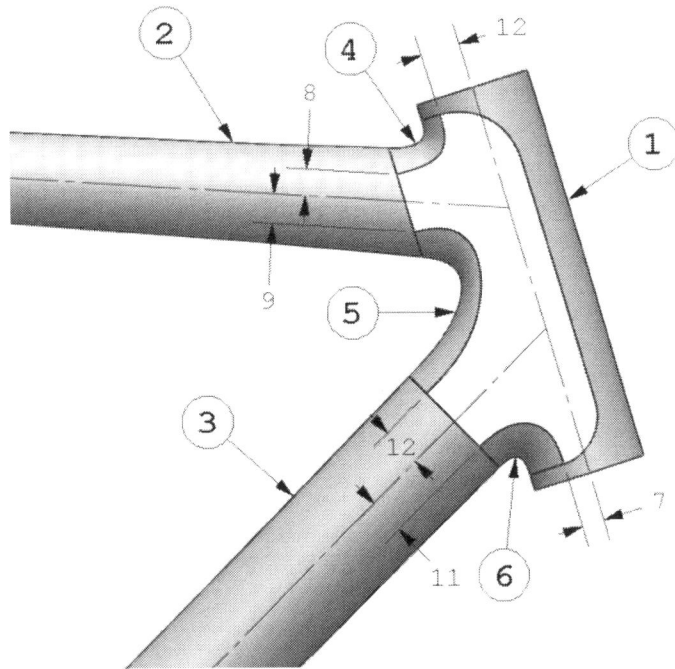

Figure 282 - Surfaces to be created - Blend Surfaces 4 and 6

EDITING AN EXISTING SKETCH

To edit the sketch used to trim the Top Tube, Head Tube and Down Tube, Right Click:

- The sketch in the FeatureManager Design Tree **OR**
- The sketch entity in the graphics area.

Select Edit Sketch.

ADDING REFERENCE DIMENSIONS

Reference Dimensions are Driven Dimensions - thus they show measurements of the model but they do not drive the model and you cannot change their values. However, when you change the model, the reference dimensions update accordingly.

Reference dimensions are usually enclosed in parentheses in Solidworks Drawings - **NB:** Drawings and not sketches. In a sketch, the grey colour of a dimension shows that it is driven. And the black colour shows that a dimension is driving.

To automatically add parentheses around reference dimensions in Solidworks Drawings, check the Add parentheses by default Check Box in Tools > Options > Document Properties > Dimensions as shown in the following image.

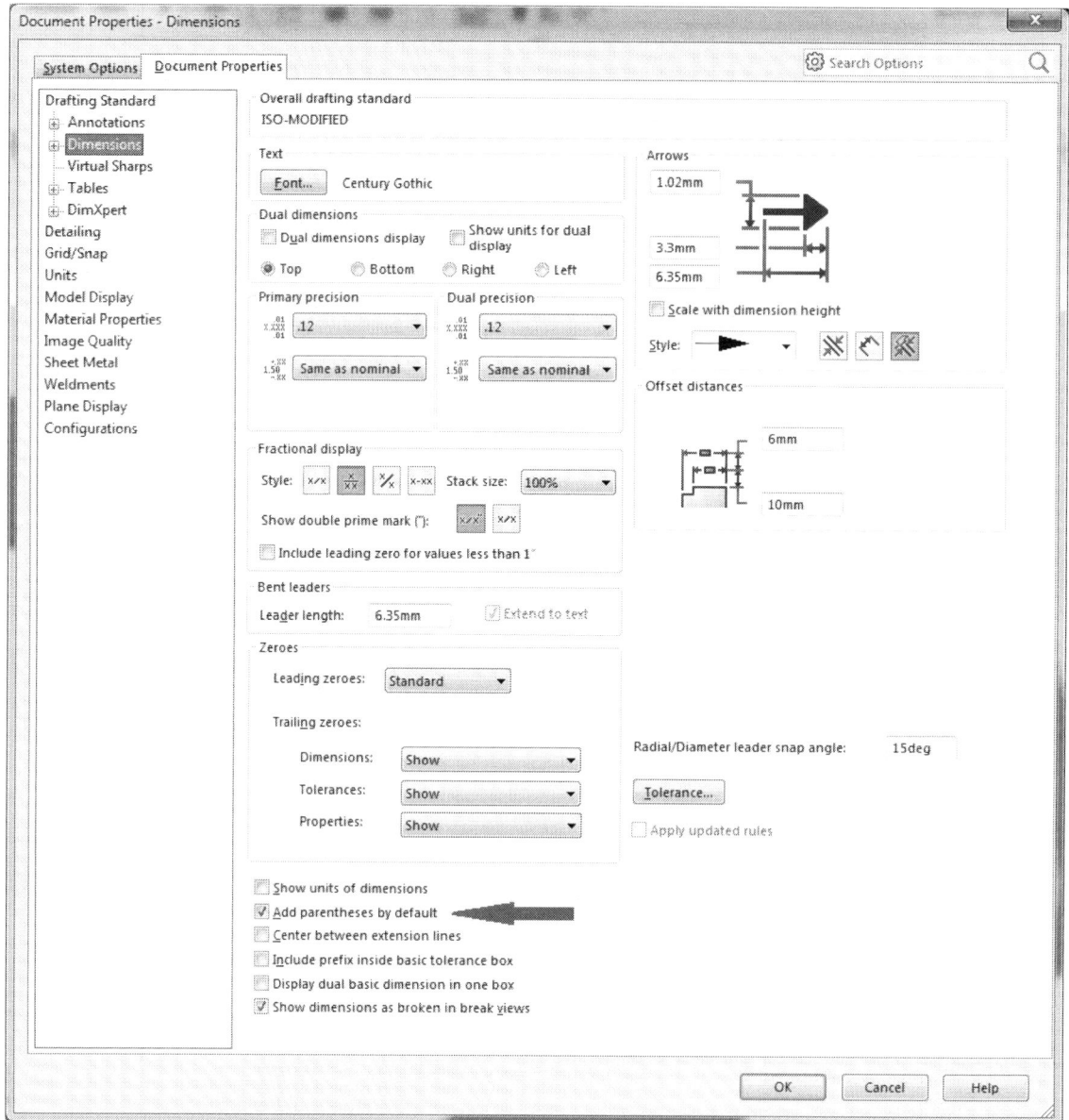

Figure 283 - Automatically adding parenthesis to reference dimensions in Solidworks Drawings

Your part should currently be in sketch edit mode as shown in the following image.

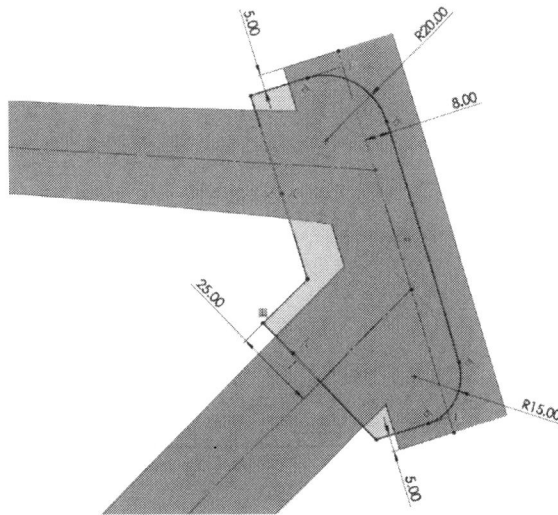

Figure 284 - Editing the Sketch used as the Trim Tool

Now let's add the two driven dimensions - thus the 7mm and 12mm Dimensions indicated with a BLUE Label in the following image. The other two dimensions indicated with a RED Label are already existing in our sketch. To add driven dimensions : - Click Smart Dimension on the Dimensions / Relations toolbar, or click Tools > Dimensions > Smart. Select the items to dimension and add the two dimensions in BLUE or labeled BLUE as shown in the following image. **NB:** Since both dimensions are being created from a Centerline you may need to take precaution with your dimension placement so as not to create a Diametric Dimension - to avoid creating a Diametric Dimension you may also right click your mouse to lock the dimension to a Radial Dimension which makes the position placement of the Dimension relative to the centerline irrelevant in terms of whether the dimension becomes Radial or Diametric.

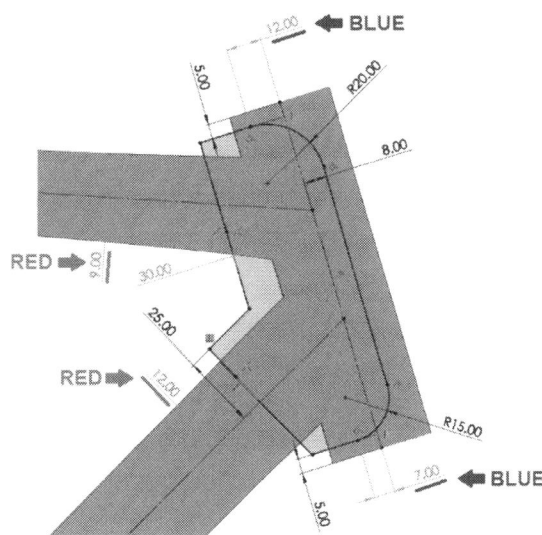

Figure 285 - Adding Driven Dimensions - Radial Dimensions Relative to the Centerline

173

Comparing existing dimensions and driven dimensions we just added above in our Edit Sketch mode to the dimensions required as per this Chapter's question you will notice that we still need to have two more dimensions enclosed in rectangles in the following image - thus the 8mm and 11mm Dimensions.

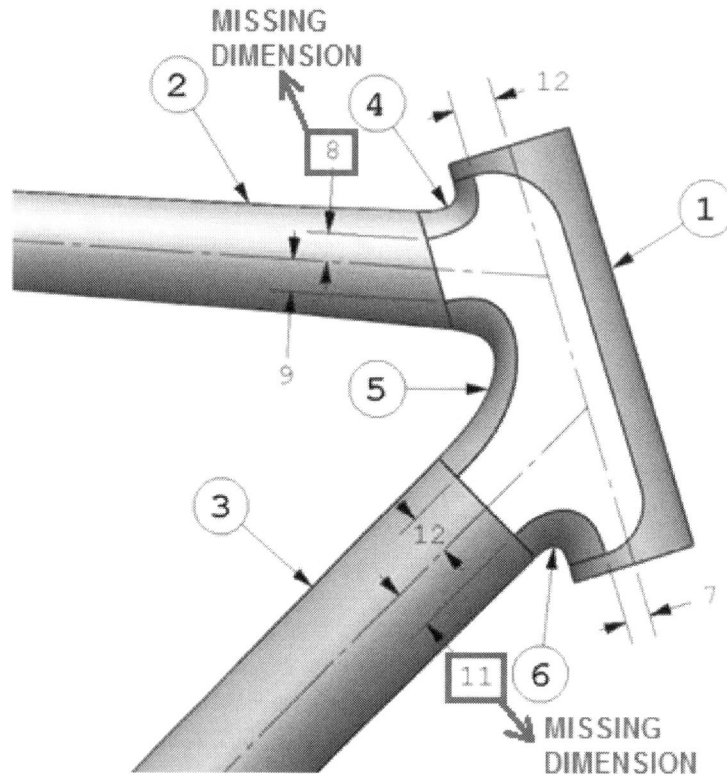

Figure 286 - Missing Dimensions from our Sketch

To add these missing dimensions we may use the Split Entities Tool.

USING SPLIT ENTITIES

Click Split Entities (Sketch toolbar) or Tools > Sketch Tools > Split Entities.

The pointer changes to ⌐o.

Click each line in the sketch at the location where we want the split to occur as indicated in the Missing Dimensions from our Sketch Figure in image above.

The sketch entity splits into two entities, and a split point is added between the two sketch entities as shown in the following image at areas indicated as Split Point 1 and Split Point 2.

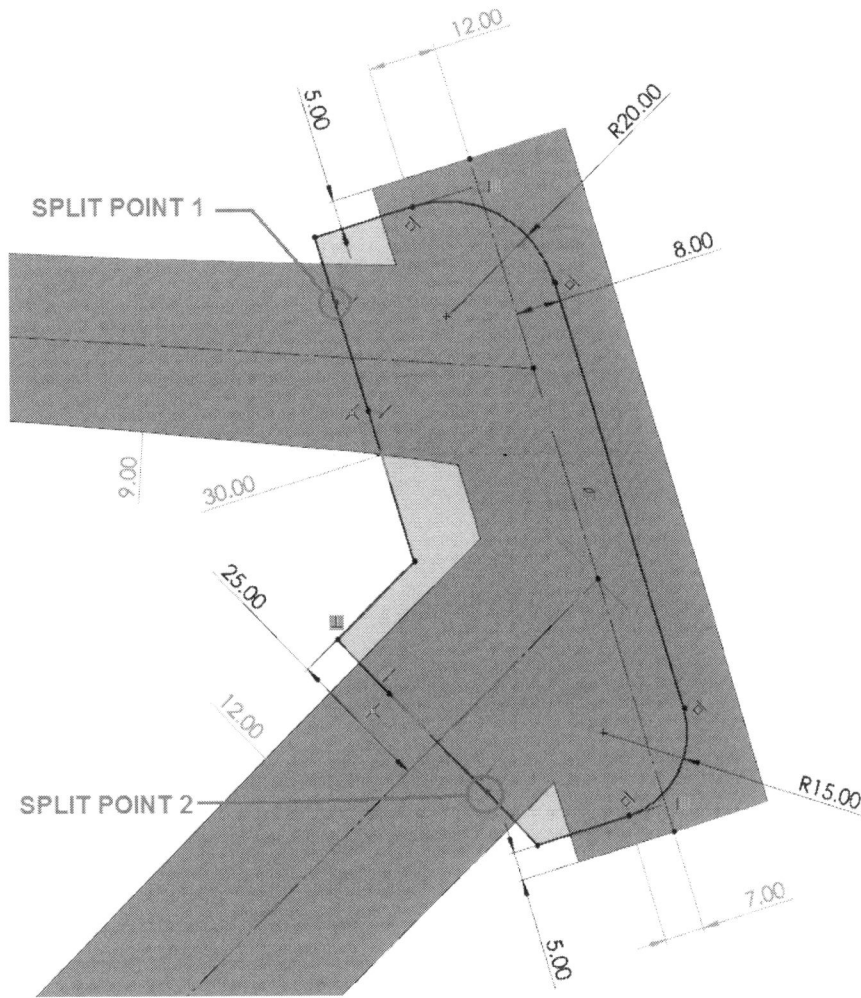

Figure 287 - Adding Split Points

Press the letter D on your keyboard and select OK to accept and finish the Split Entities Command.

ADDING DRIVING DIMENSIONS

Click Smart Dimension on the Dimensions/Relations toolbar, or click Tools > Dimensions > Smart.

Add the 8mm and 11mm dimensions as shown in the following image.

NB: Since both dimensions are being created from a Centerline you may need to take precaution with your dimension placement so as not to create Diametric Dimensions. Place the dimension to be changed to 8mm on top of the centerline and then the dimension to be changed to 11mm below the centerline - both are Radial Dimensions.

175

Your part should now look as shown in the following image.

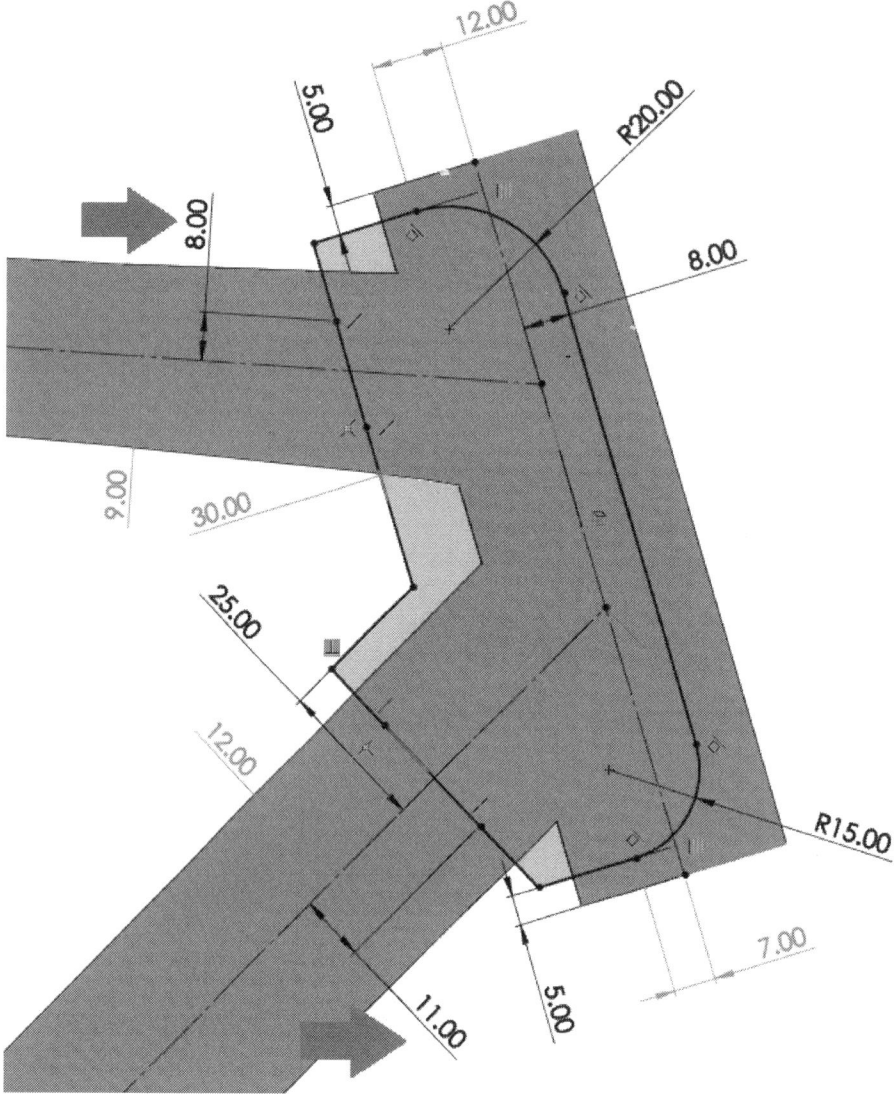

Figure 288 - Adding driving dimensions

Exit the Sketch Mode.

Save your part.

Hide the Sketch we just edited above that is if it's not hidden already.

Your part should now look as shown in the following image, if you move your mouse around the trimmed edges you will notice that the trimmed edges are now broken at the points we have our split points in our trim tool - which is exactly what we want.

Figure 289 - Trimmed Edges broken at Split Points

CREATING BLEND SURFACE 4

You are required to create a Blend Surface - Blend Surface 4 - using the Boundary Surface Feature and options listed below:

- Tangency to Face 1: Tangent Length = 1.70
- Tangency to Face 2: Tangent Length = 1.70

Click Boundary Surface (Surface toolbar) or Insert > Surface > Boundary Surface.

Under Direction 1 - select the two edges as shown in the following image. Right Click on any one of the Selected Edges in the Direction 1 Box and select Flip Connectors if the connectors are misaligned or the surface preview is not aligned.

Click callouts to set the Tangent Type to Tangency To Face.

In the Property Manager, set the Tangent Length to 1.70 on both Edges. Make sure the Apply to all Checkbox is checked in both instances.

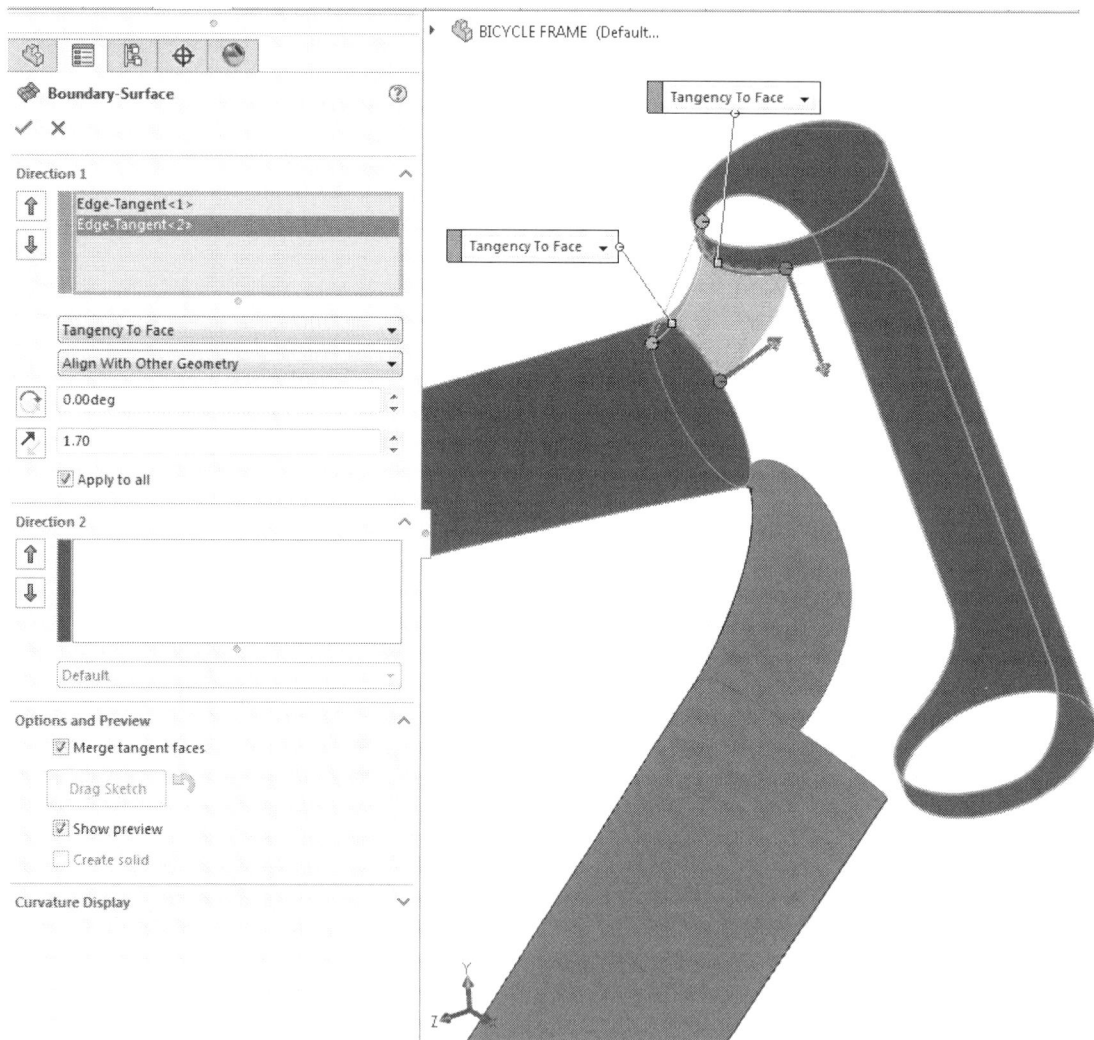

Figure 290 - Creating a Blend Surface

Click OK.

Save your part.

Your part should now look as shown in the following image.

Figure 291 - Blend Surface 4

CREATING BLEND SURFACE 6

You are required to create a Blend Surface - Blend Surface 6 - using the Boundary Surface Feature and options listed below:

- Tangency to Face 1: Tangent Length = 1.00
- Tangency to Face 3: Tangent Length = 1.80

Click Boundary Surface (Surface toolbar) or Insert > Surface > Boundary Surface.

179

Under Direction 1 - select the two edges as shown in the following image. Right Click on any one of the Selected Edges in the Direction 1 Box and select Flip Connectors if the connectors are misaligned or the surface preview is not aligned.

Click callouts to set the Tangent Type to <u>Tangency To Face</u>.

Check the <u>Apply to all</u> Checkbox. Select the first Edge *(Trimmed Edge on the Down Tube)* under Direction1 in the Property Manager *(notice the changes in the Graphics Area as you click on each of the selected edges in the Direction 1 Box which gives you a hint as to which edge is selected)* set the Tangent Length to 1.00 and then select the second edge *(Trimmed Edge on the Head Tube)* and set the Tangent Length to 1.80 *(make sure the <u>Apply to all</u> Checkbox remains checked in both instances)*.

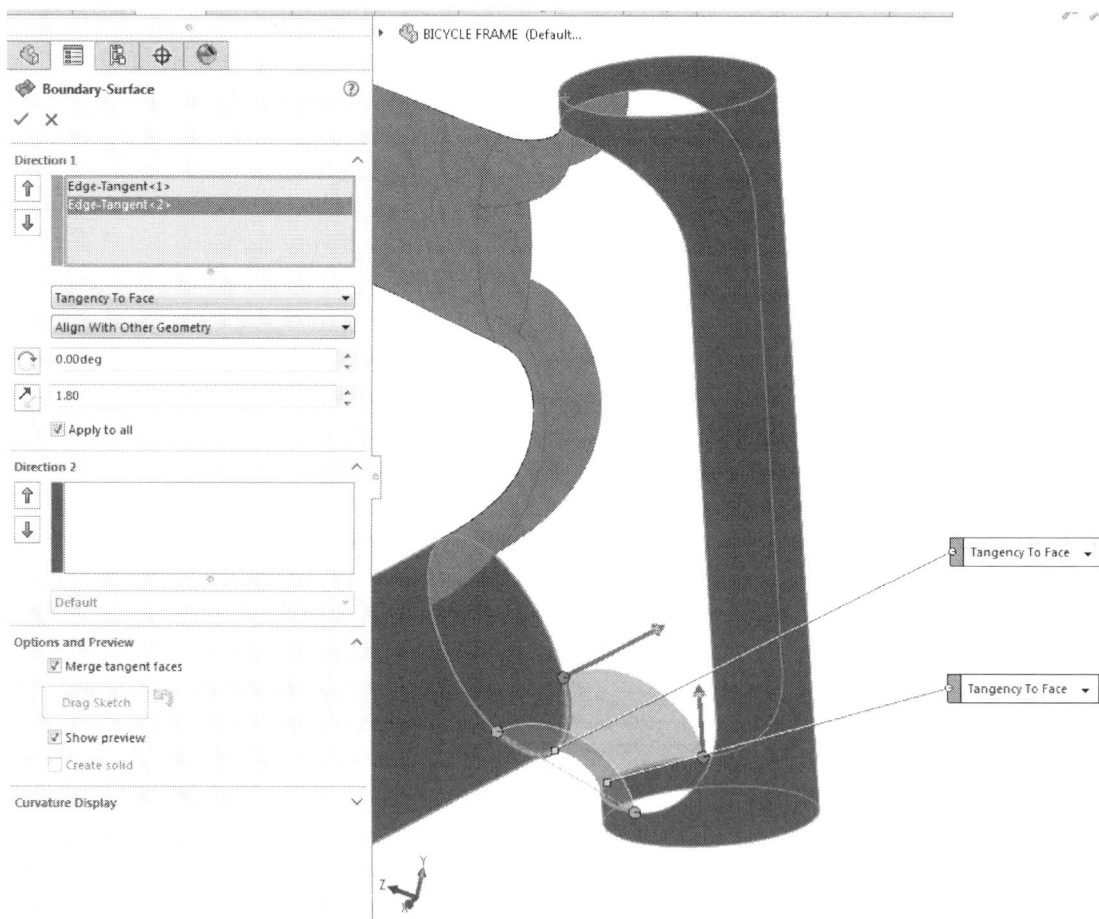

Figure 292 - Blend Surface 6

Click OK.

Save your part.

Your part should now look as shown in the following images.

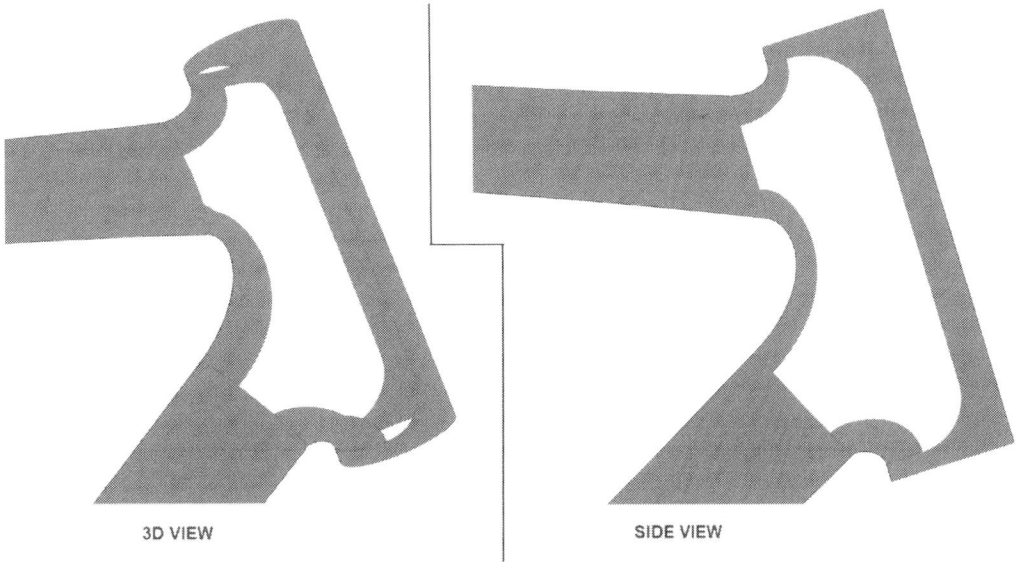

3D VIEW   SIDE VIEW

Figure 293 - Part Current Status

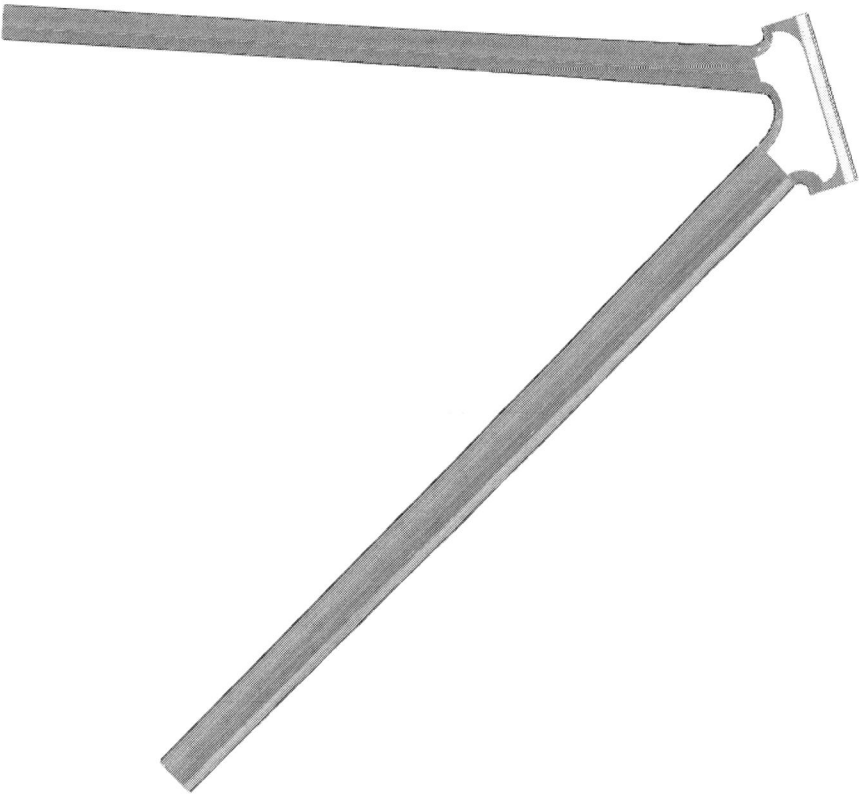

Figure 294 - Part Current Status

You are required to provide the surface area of Blend Surface 6 in square millimeters?

USING THE MEASURE TOOL TO MEASURE SURFACE AREA

Click Measure (Evaluate toolbar) or Tools > Evaluate > Measure.

Select Blend Surface 6 in the Graphics Area as shown in the following image. The answer is therefore 708.92mm^2 - your answer here should not be more than +/- 1mm2 from this given answer.

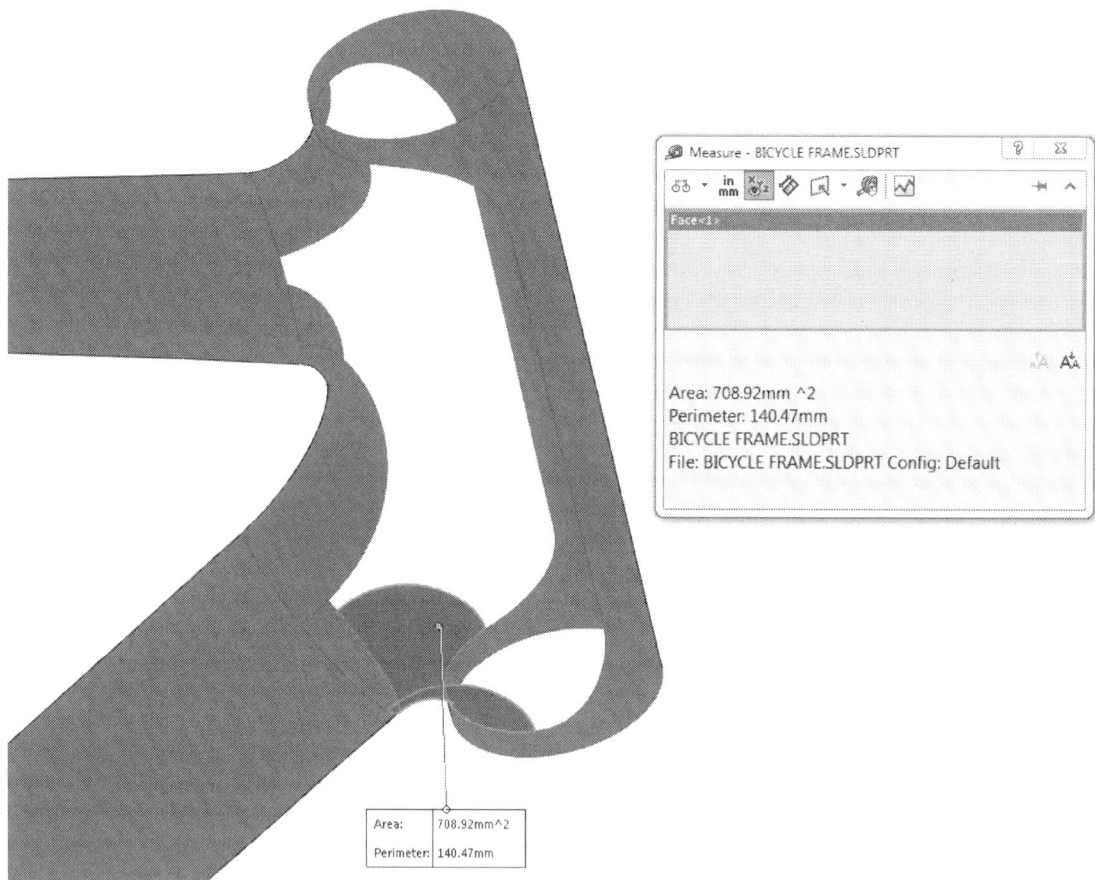

Figure 295 - Using the Measure Tool to measure surface area

CHAPTER 9

In this Chapter we will continue working with the same part from Chapter Eight.

You are required to create the side surfaces 7 and 8 as shown in the following images. These surfaces should use the following options (if applicable):

- Curvature control: Curvature applied to all edges
- Optimize surface: OFF

Measure side surface 7.

What is the surface area of Surface 7 in square millimeters? Your unit system is still MMGS, Decimal Places: 2 and Material = None.

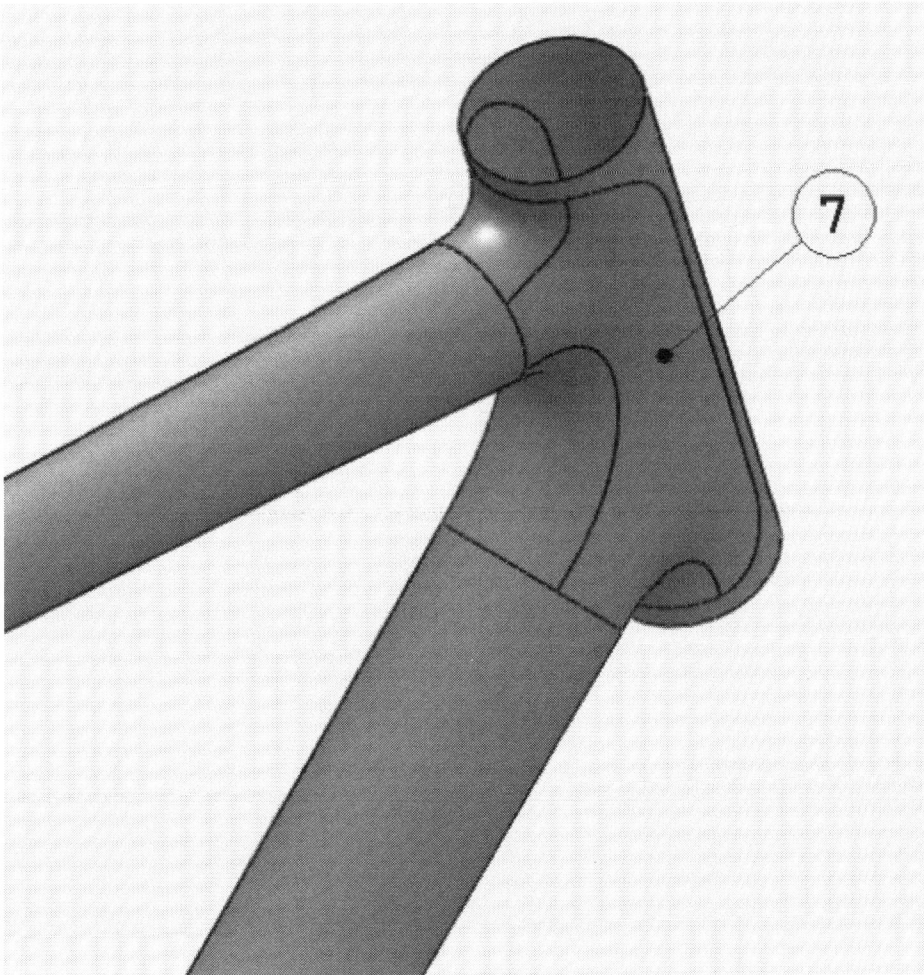

Figure 296 - Side Surface to be created - Side Surface 7

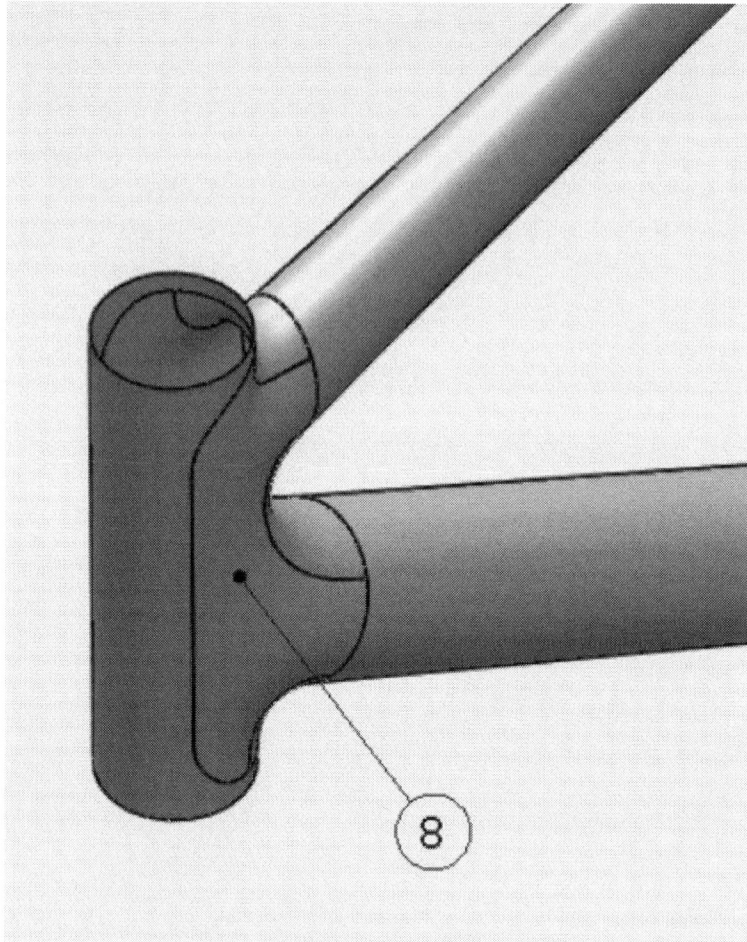

Figure 297 - Side Surface to be created - Side Surface 8

KNITTING SURFACES

Knit the Top Tube, Down Tube, Head Tube and the three blended surfaces.

To knit these surfaces:

Click Knit Surface on the Surfaces toolbar, or click Insert > Surface > Knit.

In the PropertyManager, under Selections:

Select all the surfaces in the Graphics Area or Surfaces Folder in the Feature Manager Design Tree for Surfaces and Faces to Knit as shown in the following image.

Select Merge entities to merge faces with the same underlying geometry.

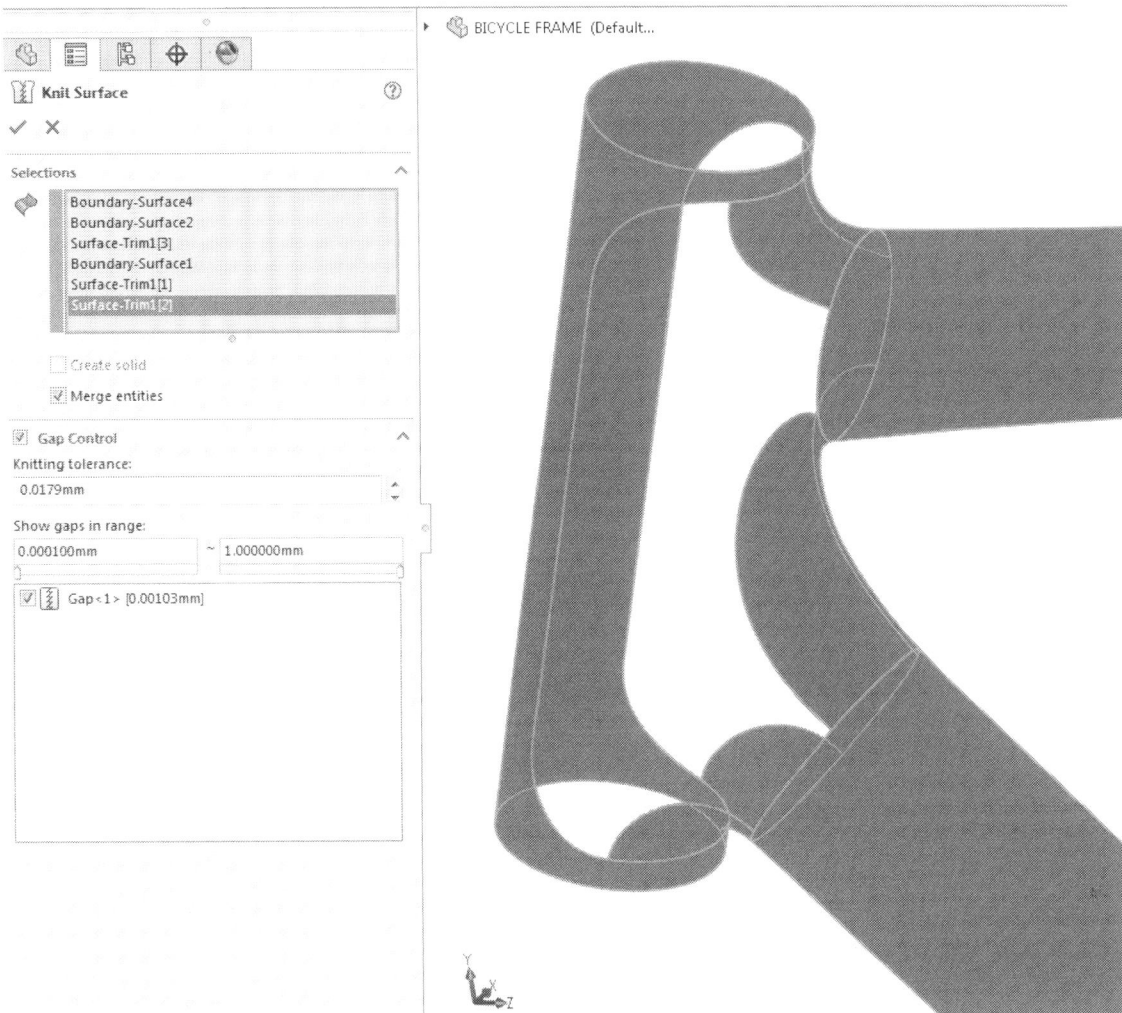

Figure 298 - Knitting Surfaces

Click OK.

USING THE FILLED SURFACE FEATURE

The Filled Surface feature constructs a surface patch with any number of sides, within a boundary defined by existing model edges, sketches, or curves, including composite curves. We can use this feature to construct Surface 7 and then mirror the created surface to create Surface 8.

To create a filled surface, click Filled Surface on the Surfaces toolbar, or click Insert > Surface > Fill, set the PropertyManager options as shown in the following image - make sure that the curvature control is set to Curvature, the Apply to all edges Checkbox is checked and the Optimize Surface checkbox is unchecked.

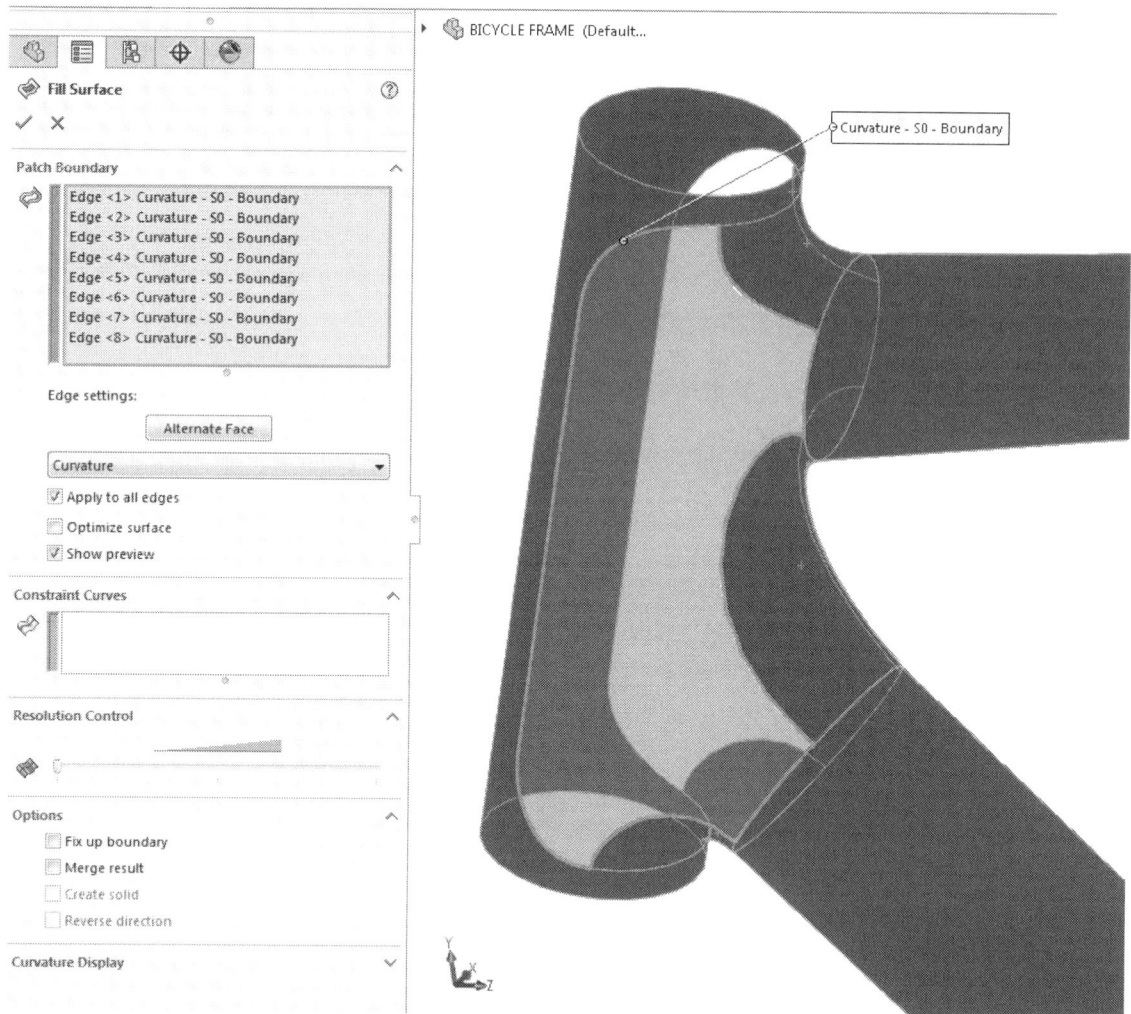

Figure 299 - Using the Filled Surface Feature to create Side Surface 8

Click OK.

Now, you may have noticed that instead of creating the filled surface from the side indicated as Side Surface 7, I have chosen to create the Filled Surface from the opposite side - Side Surface 8. Why? I have done this to avoid the Cumulative Effect of Tolerances because if you remember when we created the Boundary Surface - Surface 5 in Chapter Seven, we started from the Surface 8 Side and then the Surface 7 Side was derived by mirroring a spline in a 3D Sketch - refer to Figure 273 - Creating Mirrored Entities in a 3D Sketch.

Your part should now look as shown in the following image.

Figure 300 - Created Filled Surface on Side Surface 8

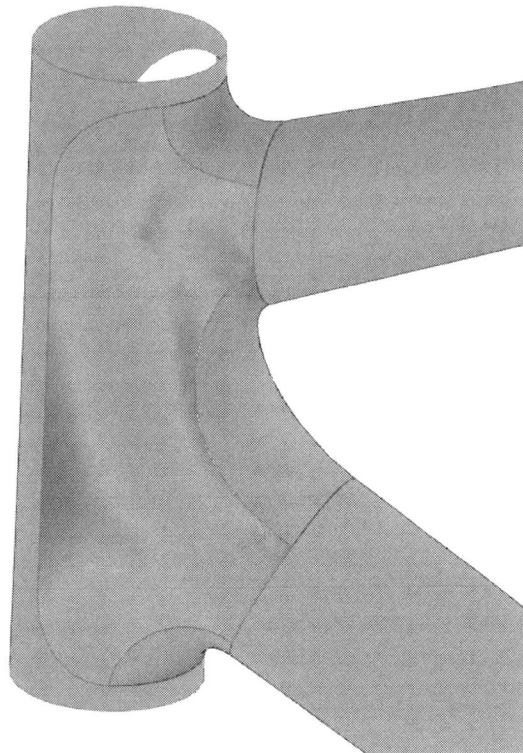

Figure 301 - Created Filled Surface on Side Surface 8 - with Curvature Display On

187

You are required to provide the surface area of Side Surface 7 in square millimeters?

USING THE MEASURE TOOL TO MEASURE SURFACE AREA

Click Measure (Evaluate toolbar) or Tools > Evaluate > Measure.

Select the created Side Surface in the Graphics Area as shown in the following image. The answer is therefore 3760.46mm² - your answer may not be exactly the same as mine due to the nature of surface modeling - especially considering the fact that one of the edges used in creating this filled surface is from a Boundary Surface (Surface 5) which we created from a 3D sketch derived from tracing an image profile.

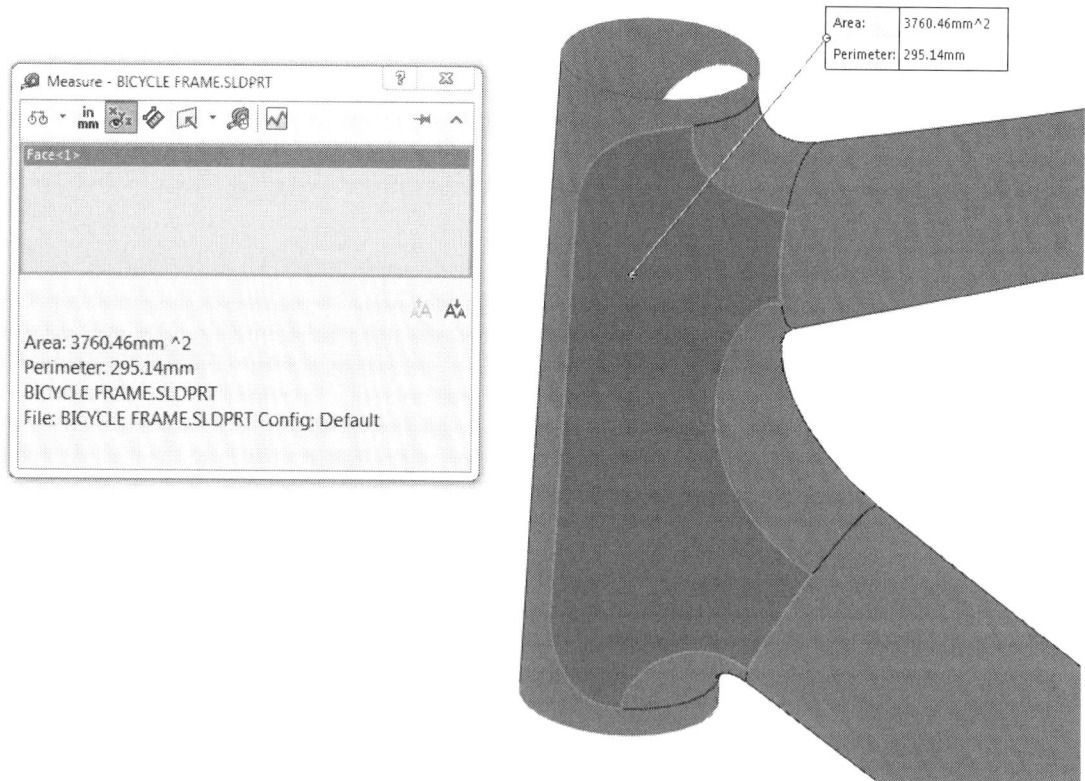

Figure 302 - Surface Area of created Filled Surface

Just for interest's sake, you may mirror the created Filled Surface using the Right Plane as the mirror plane (select Bodies to Mirror). After the mirror operation is done, you may Knit all the surfaces in the Graphics area. Your part should now look as shown in the following image. Now examine your part and if any edge did not knit it means you have to do further investigations with regards to the integrity of your Boundary Surfaces. I would say this is one way to double check your part.

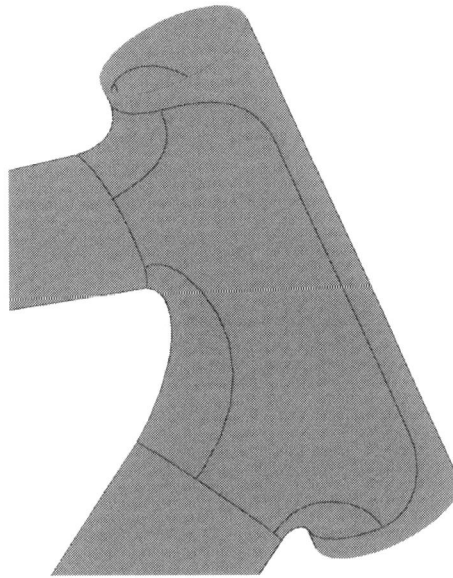

Figure 303 - Part Current Status

Again, just for interest's sake - to further check the integrity of your part try using the Thicken Feature to thicken your part to a 1.6mm material thickness to the inside as shown in the following image - ensure the Merge Result Checkbox is checked. If this operation fails, then again you may need to do further investigation with regards to the integrity of your Boundary Surfaces.

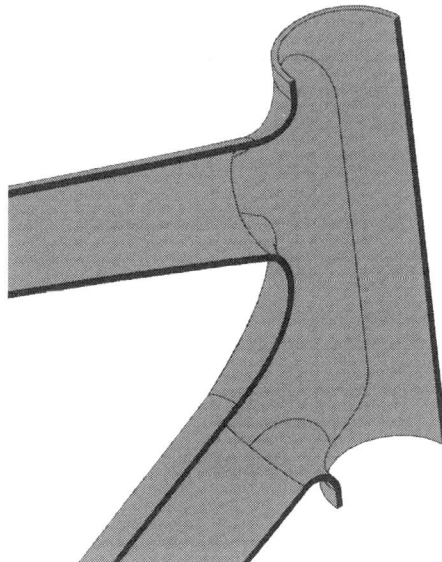

Figure 304 - Sectional View of the Part thickened to a 1.6mm material thickness to the inside

TIP: The best way to use mirroring in advanced surface modeling is to build half (or one quarter) of the part, knit it into a solid, and then mirror the solid body. However, if you decide to mirror surface patches, evaluate and check your model to be sure the results are what you expected.

EXAMPLE - DIMENSIONING AROUND A CENTRELINE (RADIAL AND DIAMETRIC DIMENSIONS)

Open the part named RADIAL AND DIAMETRIC DIMENSIONS.sldprt from the downloaded Chapter 9 Folder. In the Feature Manager Design Tree, right click on Sketch1 and select Edit Sketch. Your part should look as shown in the following image.

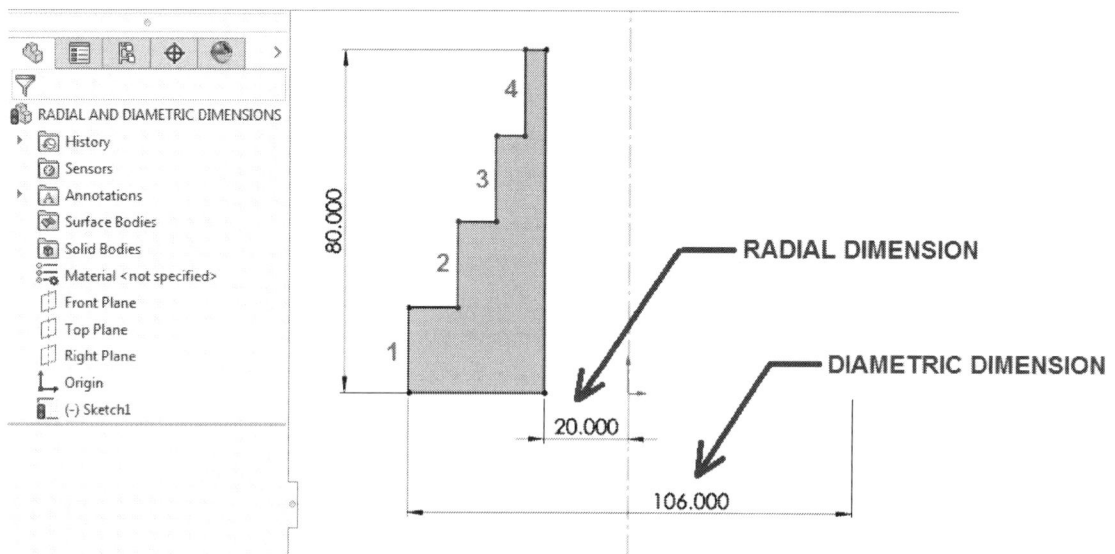

Figure 305 - Practice adding Radial and Diametric Dimensions

Click Smart Dimension (Dimensions/Relations toolbar) or Tools > Dimensions > Smart.

Select the centerline and vertical line number 2. To create a Radial Dimension, move the pointer to the near side or LHS of the centerline. To create a Diametrical Dimension, move the pointer to the far side or RHS of the centerline.

Click to place the dimension. Depending on the version of Solidworks you are using, the pointer changes to ![radial cursor] for radial dimensions and ![diametric cursor] for diametric dimensions.

Select vertical lines 3 and 4 to create additional dimensions without reselecting the centerline - thus, you can create multiple radial or diametric dimensions without selecting the centerline each time.

190

Download the part named INLET CH10 START.SLDPRT in the Chapter 10 Folder from this Google drive location - *http://bit.ly/CSWPA-SU* or Scan the QR Code shown below:

If you experience any problems with downloading any files you may send an email to *cswpasmebook@gmail.com* with the title of the book indicated in your email subject. Open and save the downloaded part to your PC.

Your part should look as shown in the following image.

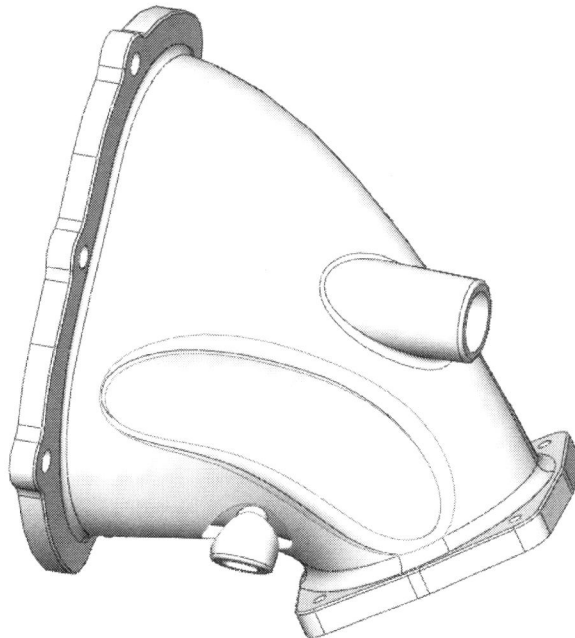

Figure 306 - Downloaded part INLET CH10 START.SLDPRT

This part contains an imported surface model of an inlet. It has some missing faces and in this Chapter you are required to recreate the missing fillet indicated in the following image.

The recreated face should be tangent to the face on which it borders. For reference, the face's radius should be approximately 9.525mm (0.375 inches).

Note: You can recreate this face using any tool.

After recreating the face you are required to measure the surface area of the recreated face and provide an answer in square millimeters to two decimal places.

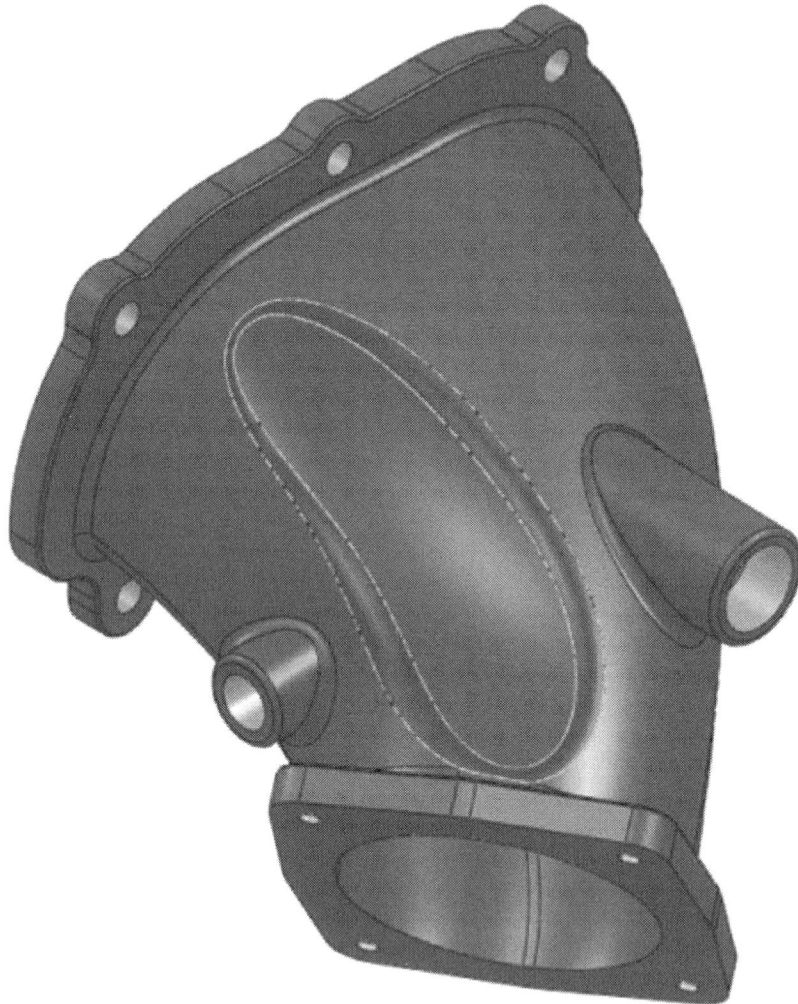

Figure 307 - Downloaded part INLET CH10 START.SLDPRT - Missing Radius to be recreated

CREATING A 3D SKETCH USING CONVERT ENTITIES

Click 3D Sketch (Sketch toolbar) or Insert > 3D Sketch.

Right Click the outer edge on the area where we have the missing fillet and click Select Open Loop as shown in the following image

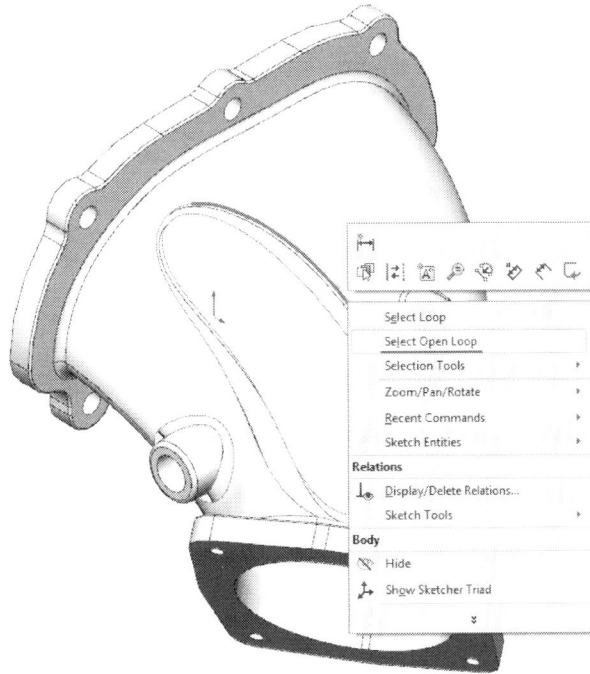

Figure 308 - Using Select Open loop

All edges are selected as shown in the following image.

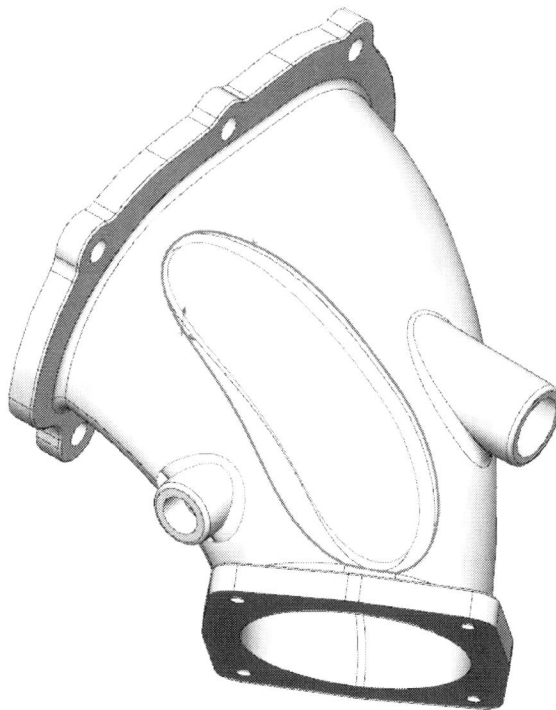

Figure 309 - Selected Edges

Click Convert Entities (Sketch toolbar) or Tools > Sketch Tools > Convert Entities.

193

Your part should now look as shown in the following image.

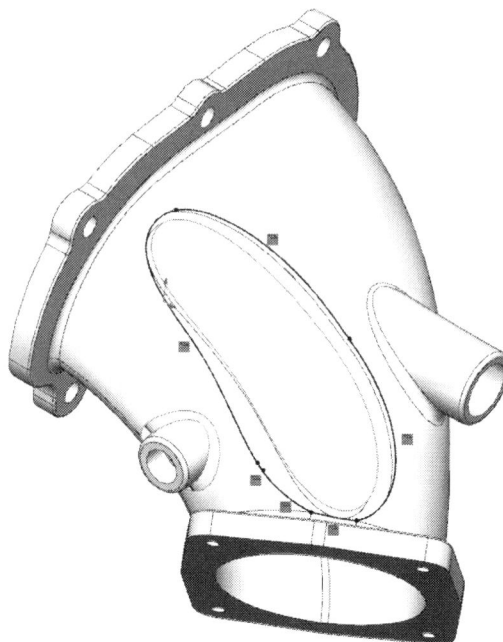

Figure 310 - Using Convert Entities

Repeat the same process for the inner edge. Your part should now look as shown in the following image.

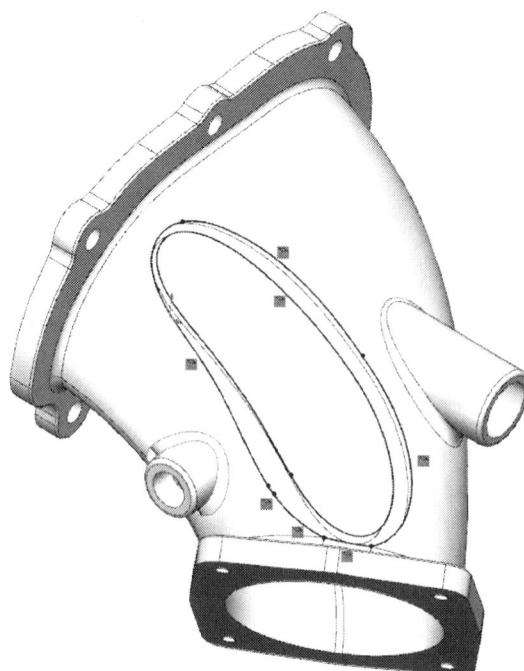

Figure 311 - Using Convert Entities

194

Press the letter "D" on your keyboard and select Exit Sketch - see the following image.

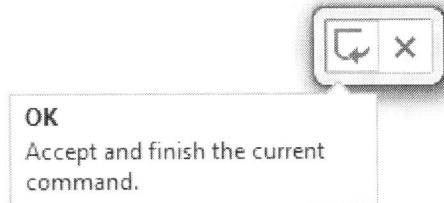

OK
Accept and finish the current
command.

Figure 312 - Using the letter "D" as a shortcut to access the Confirmation Corner

Rename the 3D Sketch we just created to BASELINE and change the colour to magenta *(this is done in the Feature Manager Design Tree)* as shown in the following images.

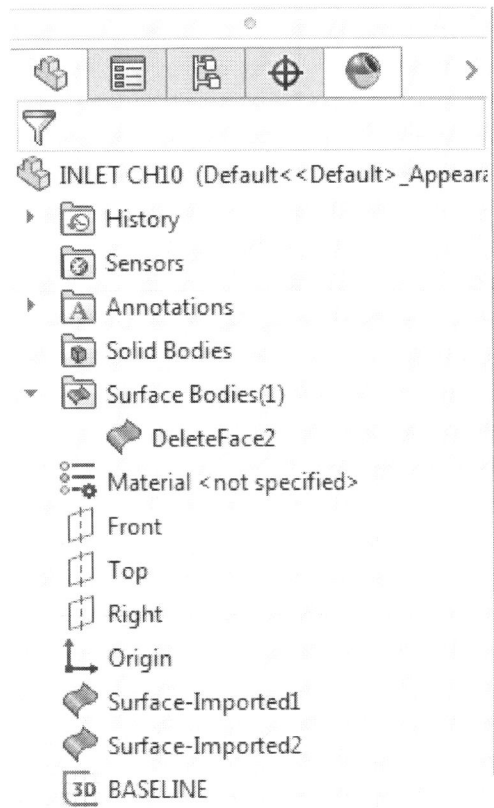

Figure 313 - Renaming a Sketch in the Feature Manager Design Tree

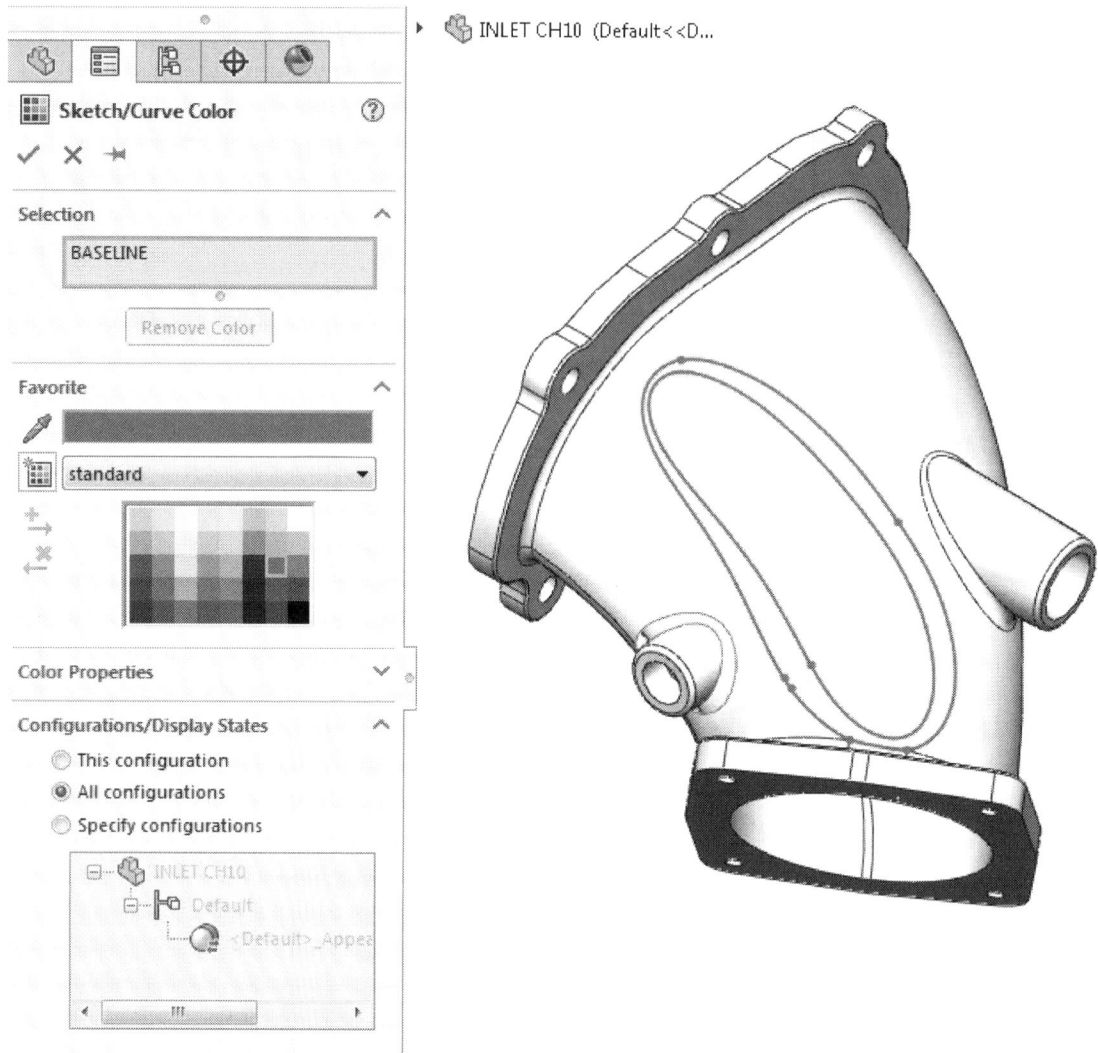

Figure 314 - Changing the color of a Sketch

Click OK.

Ok, so now we have just set our baseline which we shall use to compare the edge of our new fillet to.

Hide the BASELINE 3D sketch we just created.

EXTEND SURFACE

Click Extend Surface (Surfaces toolbar) or Insert > Surface > Extend.

In the PropertyManager:

Under Edges/Faces to Extend, select the edge in the graphics area as shown in the following image.

Select the Distance End Condition and set a value of 4mm.

Select Same Surface under Extension Type.

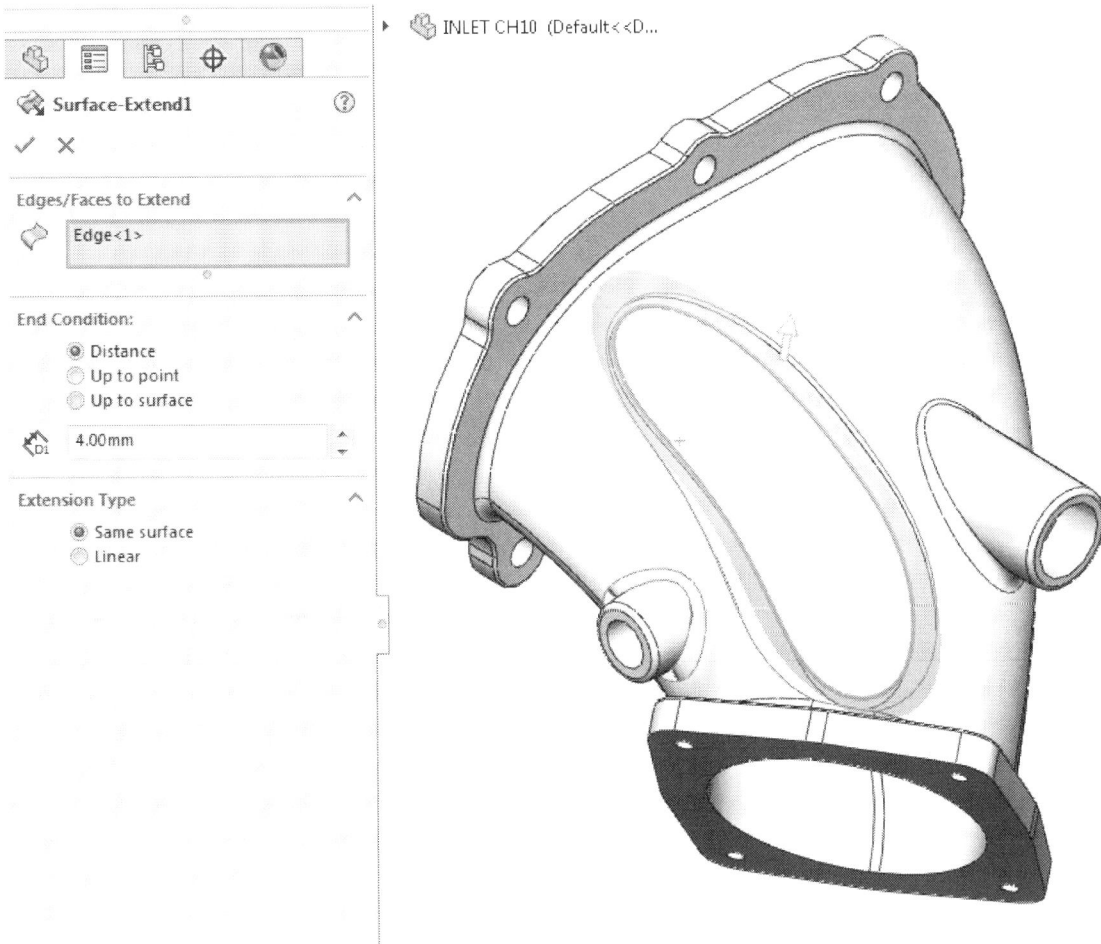

Figure 315 - Using Extend Surface

Click OK.

Save your part.

Hide the surface body we just extended for easy access to the inner edge which we would like to extend as well.

To hide the surface body, click the surface body in the Graphics Area and select Hide or click the surface body in the Surface Bodies Folder in the Feature Manager Design Tree and select Hide.

Click Extend Surface (Surfaces toolbar) or Insert > Surface > Extend.

In the PropertyManager:

Under Edges/Faces to Extend, select the edge in the graphics area as shown in the following image.

Select the Distance End Condition and set a value of 6mm.

Select Same Surface under Extension Type.

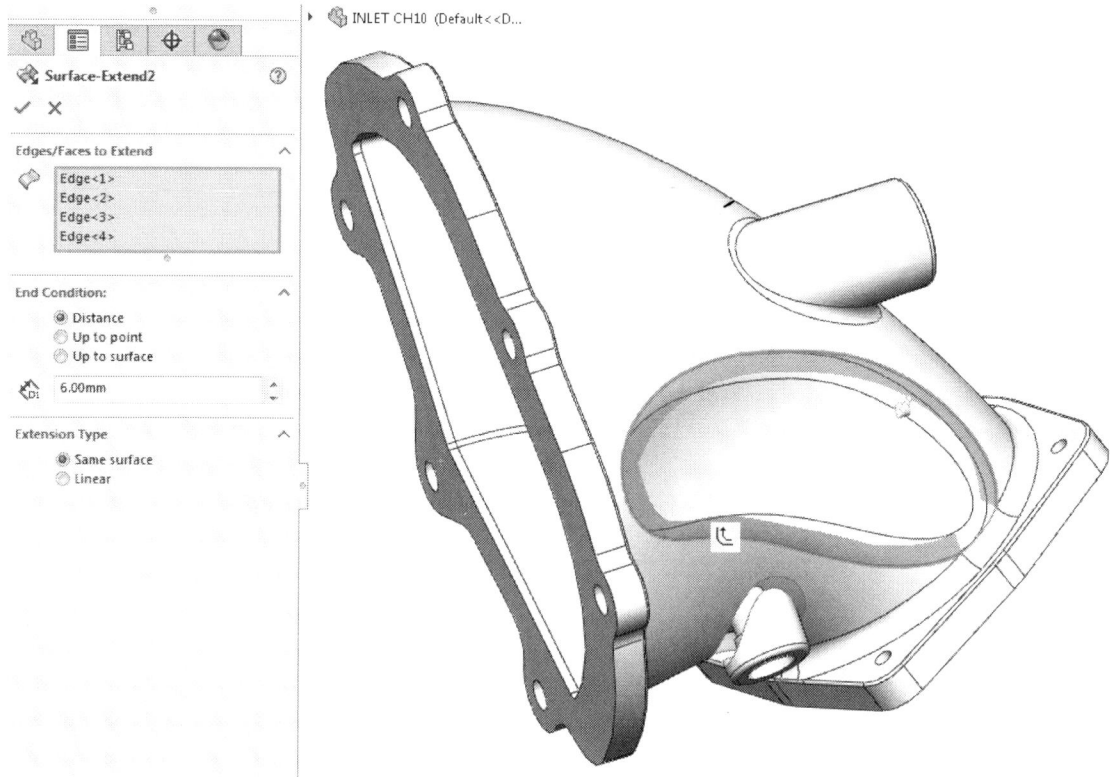

Figure 316 - Using Extend Surface

Click OK.

Save your part.

CREATING A BOUNDARY SURFACE

Click Boundary Surface (Surface toolbar) or Insert > Surface > Boundary Surface.

Select the edges under Direction 1 and Direction 2 as shown in the following image.

Set the PropertyManager options as shown in the following image.

Take note of the Tangent Type set on each edge.

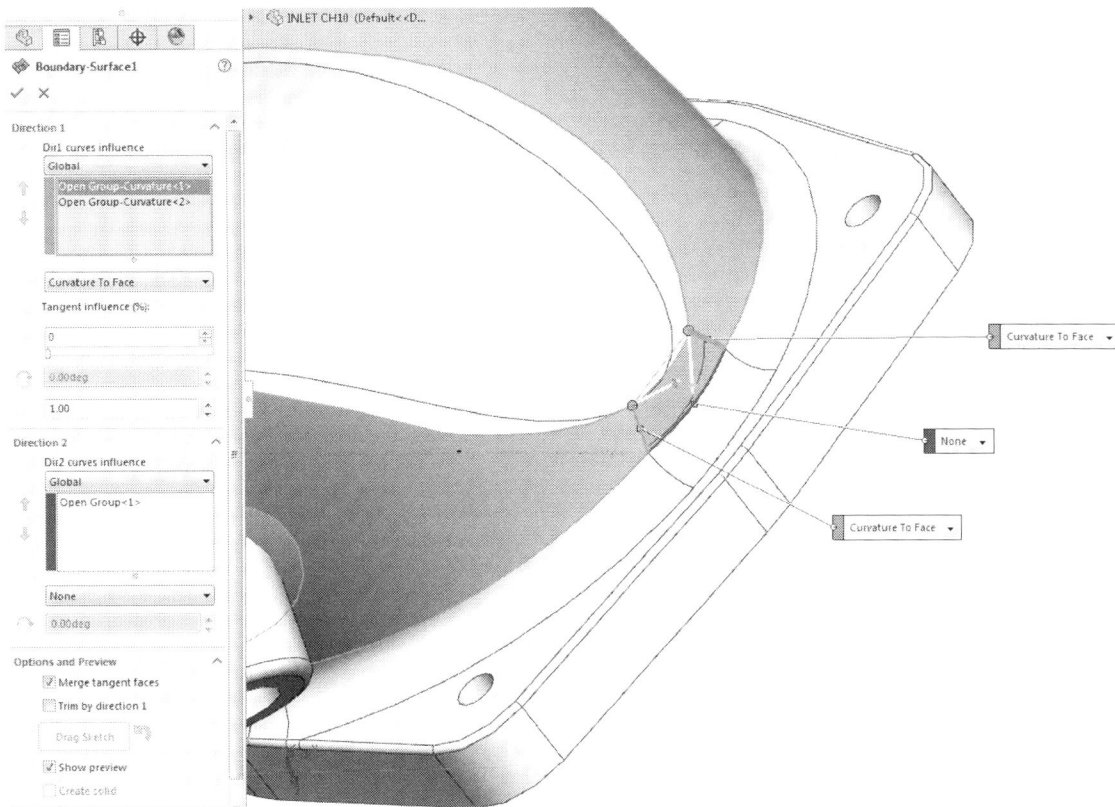

Figure 317 - Creating a Boundary Surface

Click Ok.

Save your part.

KNIT SURFACE

Click Knit Surface on the Surfaces toolbar, or click Insert > Surface > Knit.

In the PropertyManager, under Selections:

Select the Boundary Surface we just created and the main surface body of the Inlet as shown in the following image for Surfaces and Faces to Knit.

Select Merge entities to merge faces with the same underlying geometry.

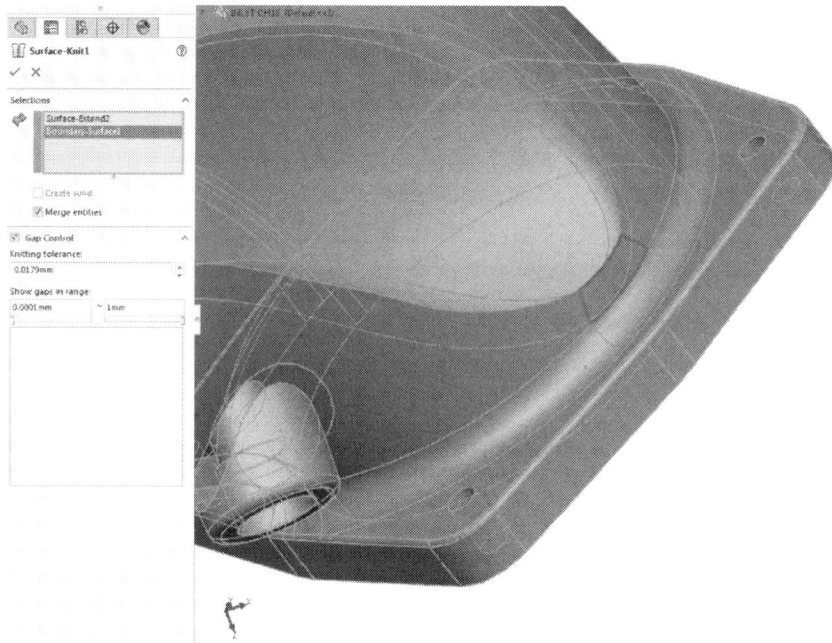

Figure 318 - Knitting Surfaces

Click Ok. Save your part. Your part should now look as shown in the following image.

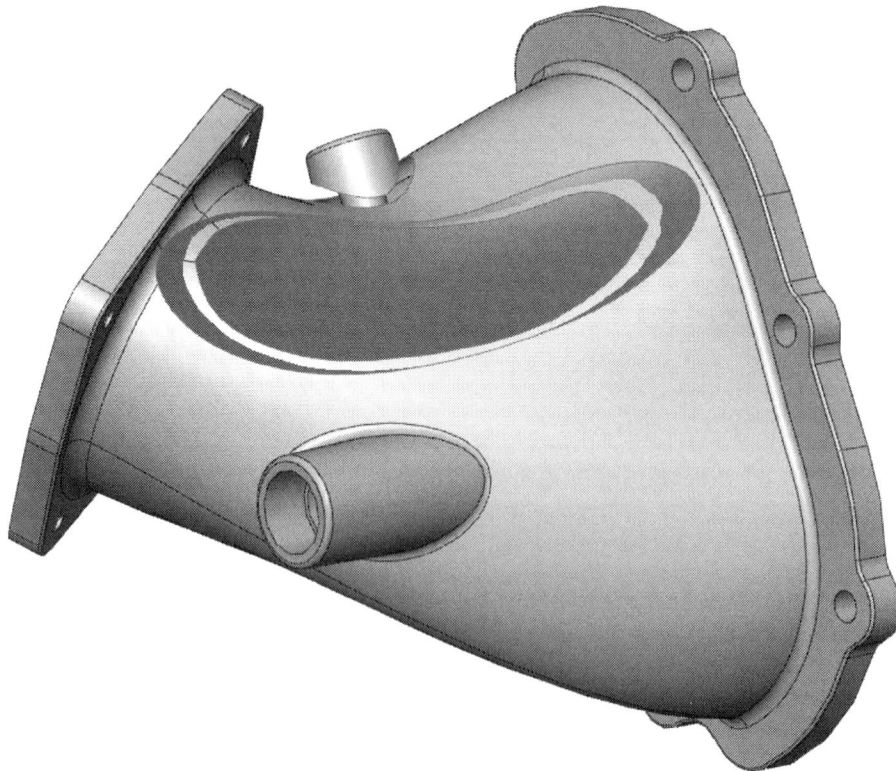

Figure 319 - Part Current Status

200

TRIM SURFACE

Click Trim Surface on the Surfaces toolbar, or click Insert > Surface > Trim.

In the PropertyManager, under Trim Type, select Mutual.

Select the two surface bodies in the graphics area for Trimming Surfaces to use to trim themselves.

Set the other options as shown in the following image.

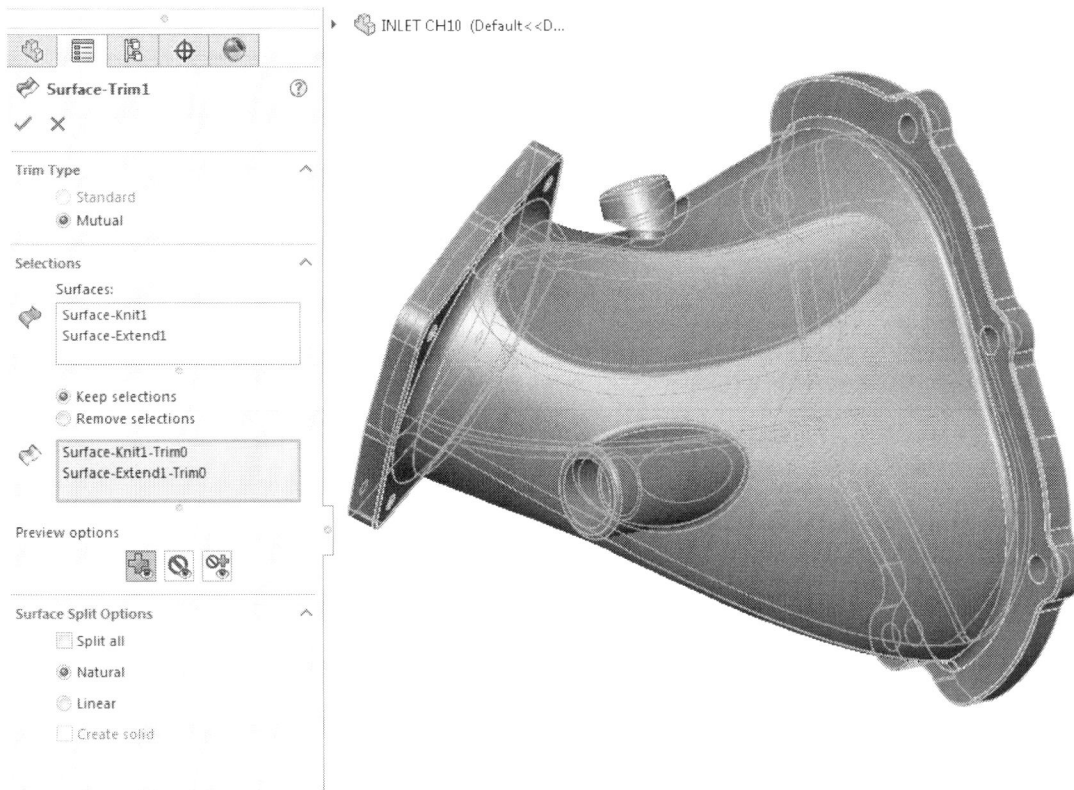

Figure 320 - Question 21 of 19 - Cut List Folders

Click OK.

Save your Part.

Your part should now look as shown in the following image.

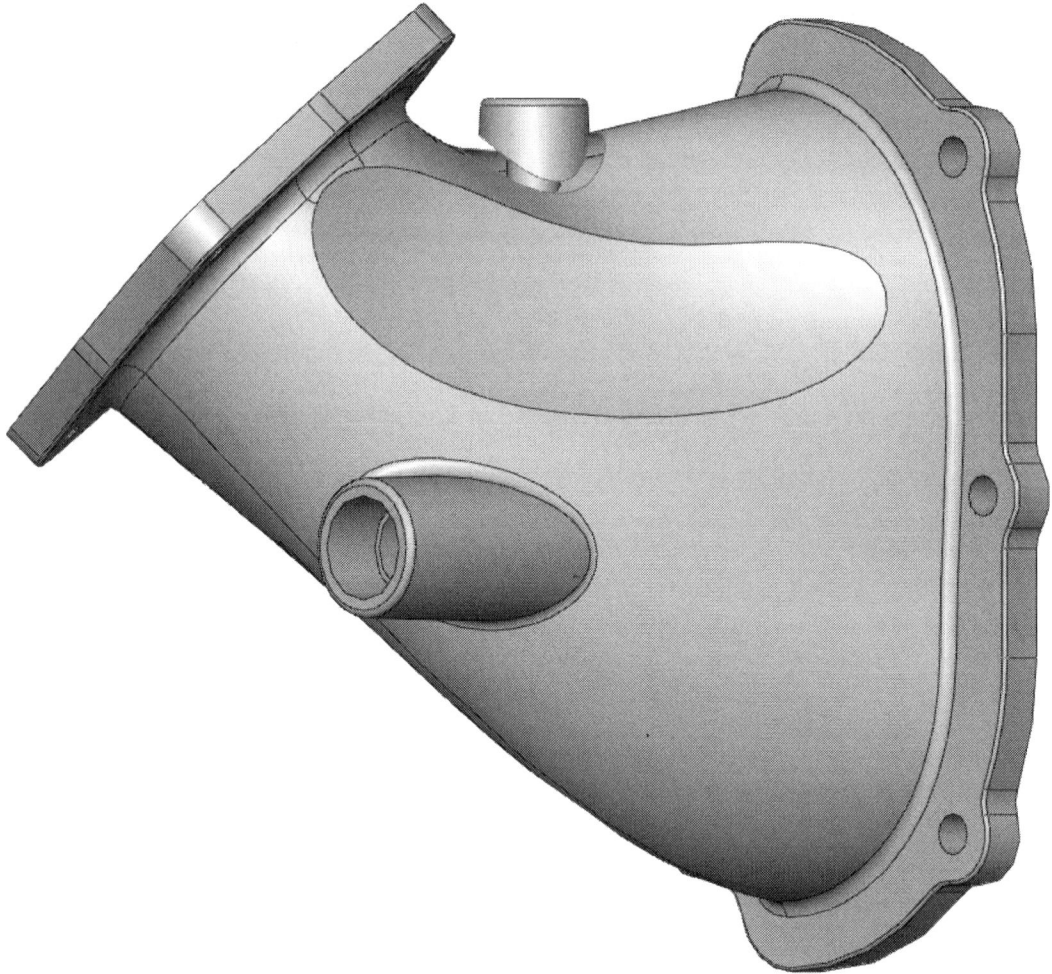

Figure 321 - Part Current Status

FILLET SURFACE

Click Fillet (Features toolbar) or Insert > Surface > Fillet/RoundInsert > Surface > Fillet/Round

Select edges as shown in the following image.

Set the Fillet Property Manager Options as shown in the following image.

Enter a radius value is 9.525mm.

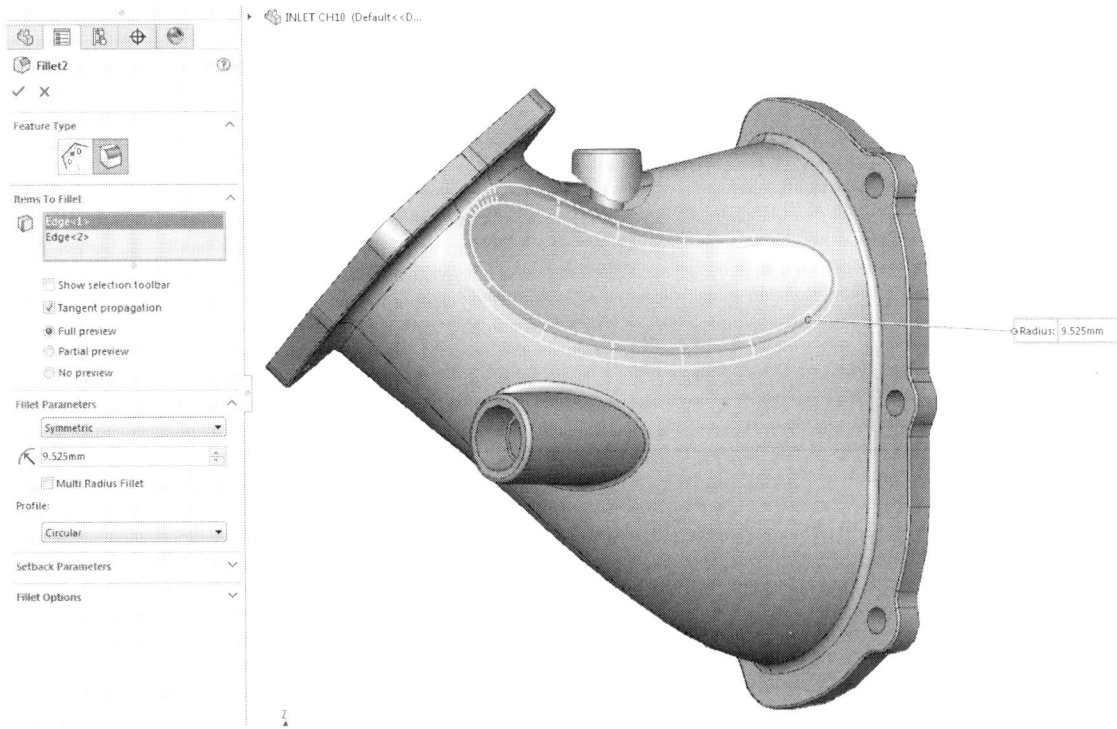

Figure 322 - Adding a Fillet

Click Ok. Save your part. Your part should now look as shown in the following image.

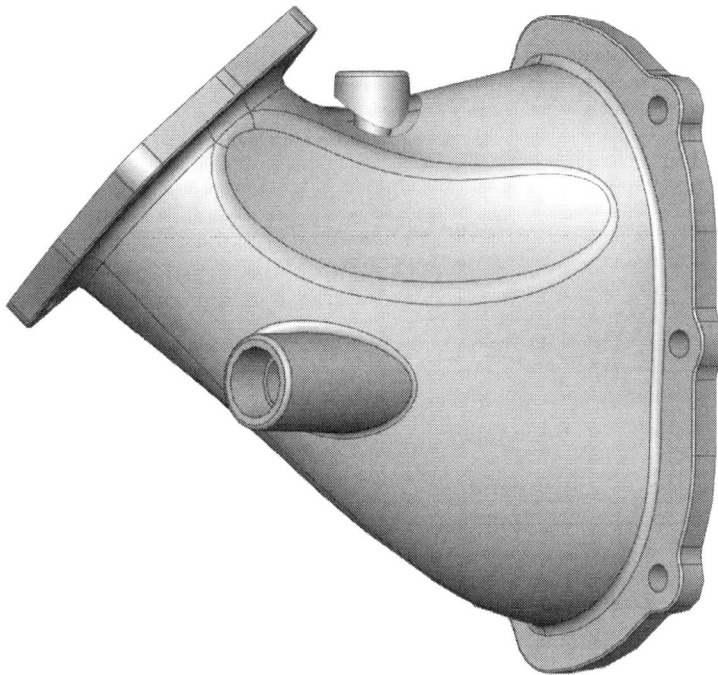

Figure 323 -Part Current Status

CHECKING THE FILLET RESULT

Show the BASELINE sketch we created at the beginning of this chapter and compare its location versus the edge of the fillet we just created - see the following image. You will notice that the two are a match which is what we are looking for as a guide to see if we have not erred in the process of recreating the missing fillet as required.

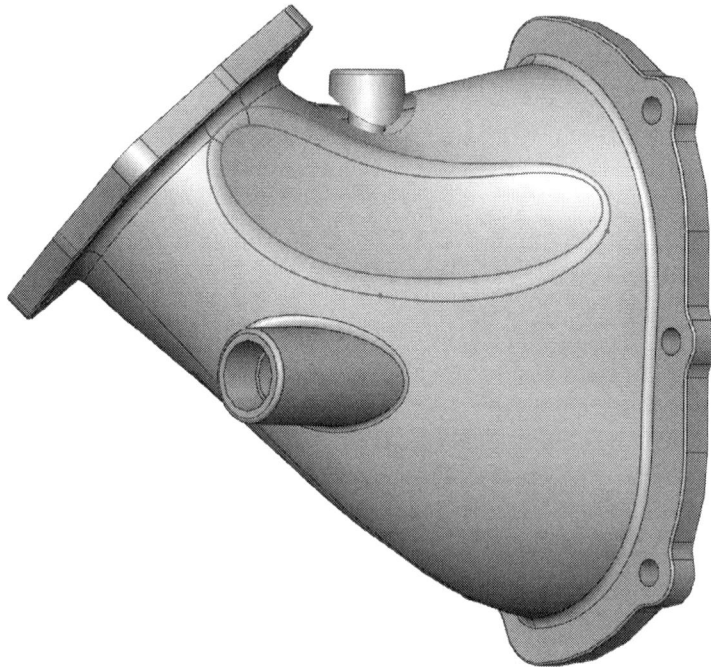

Figure 324 -Part Current Status

Hide the Baseline Sketch.

Save your part.

USING THE MEASURE TOOL TO MEASURE SURFACE AREA

Click Measure (Tools toolbar) or Tools > Evaluate > Measure.

Select the two faces making up the fillet we just created.

Your answer is thus 1753.05 square millimeters as shown in the following image.

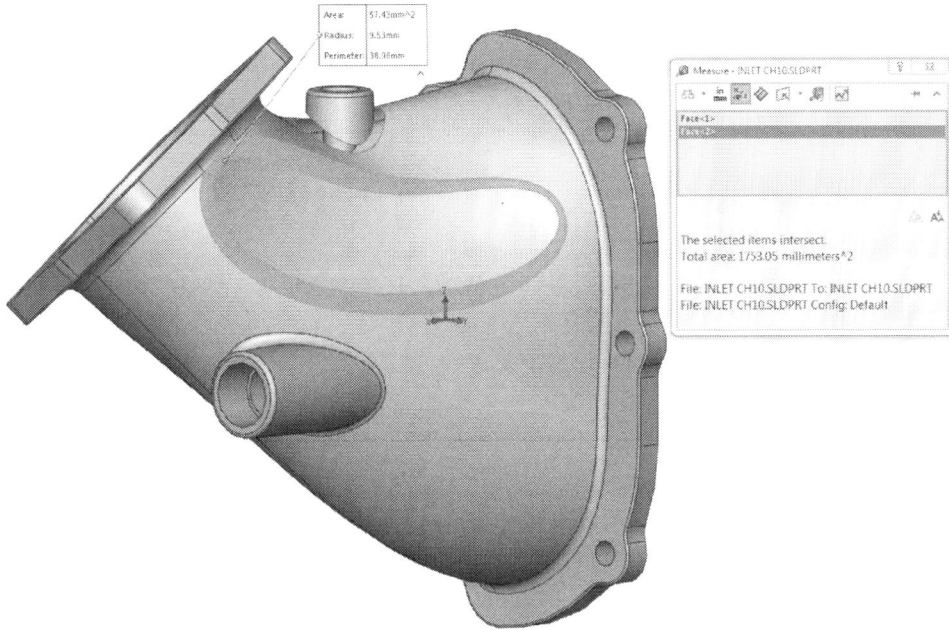

Figure 325 -Using the Measure Tool to measure Surface Area

CURVATURE ANALYSIS

Click Curvature (Evaluate toolbar), or click View > Display > Curvature.

You will notice that there is some inconsistency in the curvature of the fillet we just created as reveled by the Black Colour around the area shown in the following image.

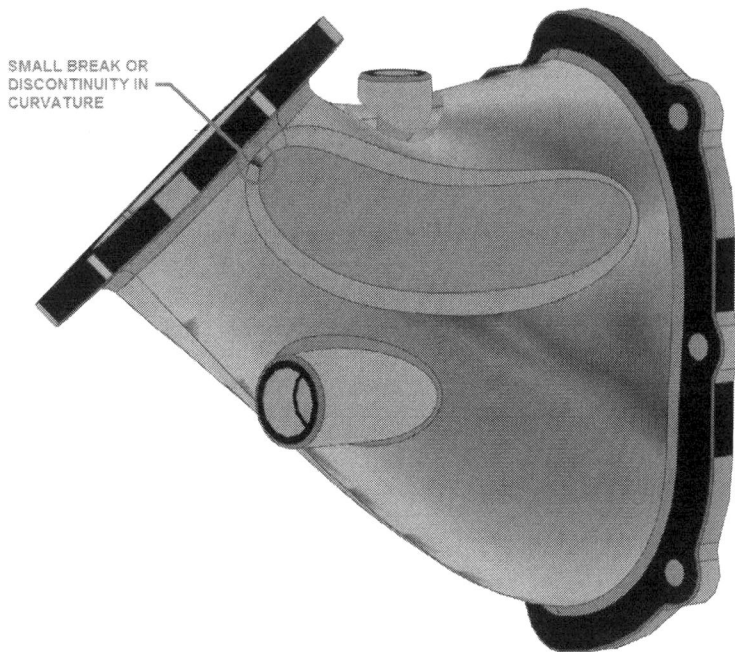

SMALL BREAK OR
DISCONTINUITY IN
CURVATURE

Figure 326 - Curvature Display

Click View > Display > Curvature to remove the Curvature Colour Display.

DELETING AND PATCHING FACES

We may use the Delete and Patch function remove the joining face in the fillet face and have the adjoining faces extend to form an unbroken surface.

Click Delete Face on the Surfaces toolbar, or Insert > Face > Delete. The Delete Face PropertyManager appears.

In the graphics area, click the joining face that we want to delete as shown in the following image.

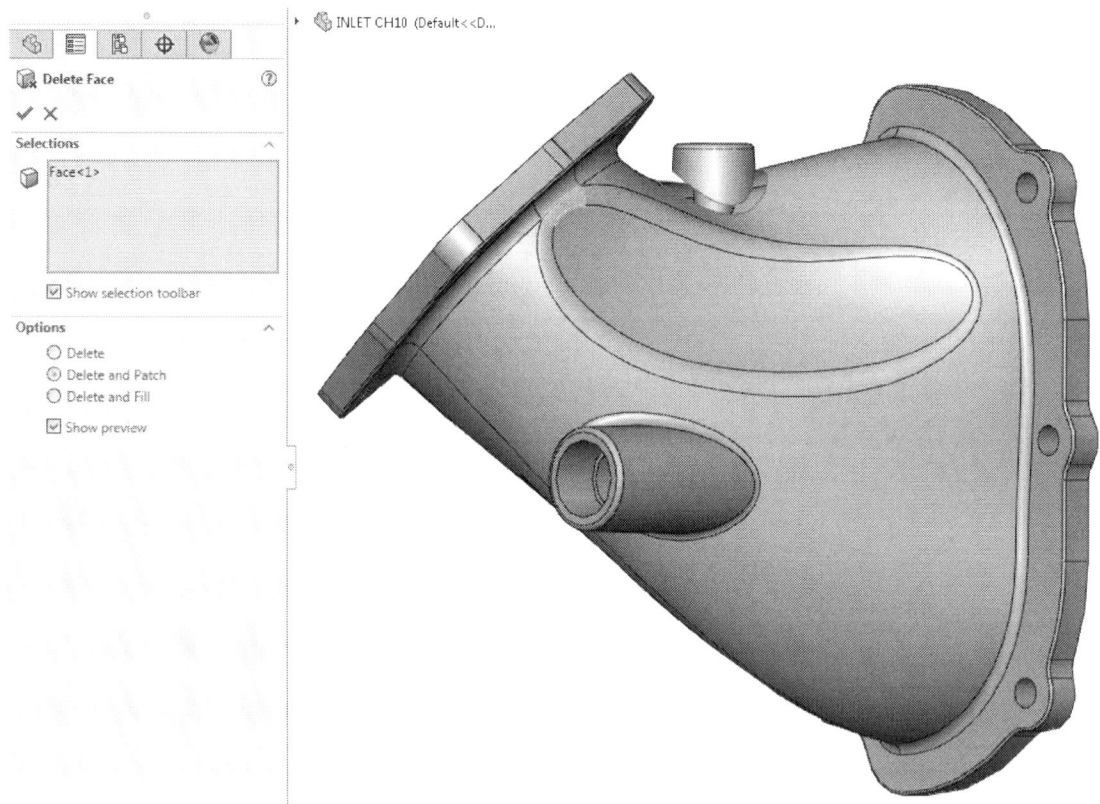

Figure 327 - Deleting and patching faces

Under Options, click Delete and Patch.

Click OK.

Save your Part.

Your part should now appear as shown in the following image.

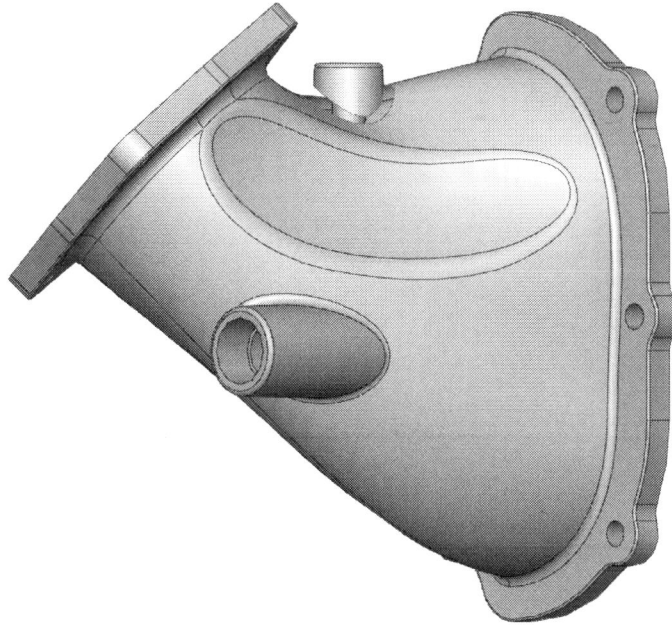

Figure 328 - Deleting and patching faces - Part Current Status

As you can see, we now have an unbroken surface. Now you must double check if it is still a fillet because, eeh , let me just say sometimes unpredictable things do happen with the Delete and Patch Command. To do so, use the Measure Tool you should get a reading of the radius of 9.525mm as shown in the following image.

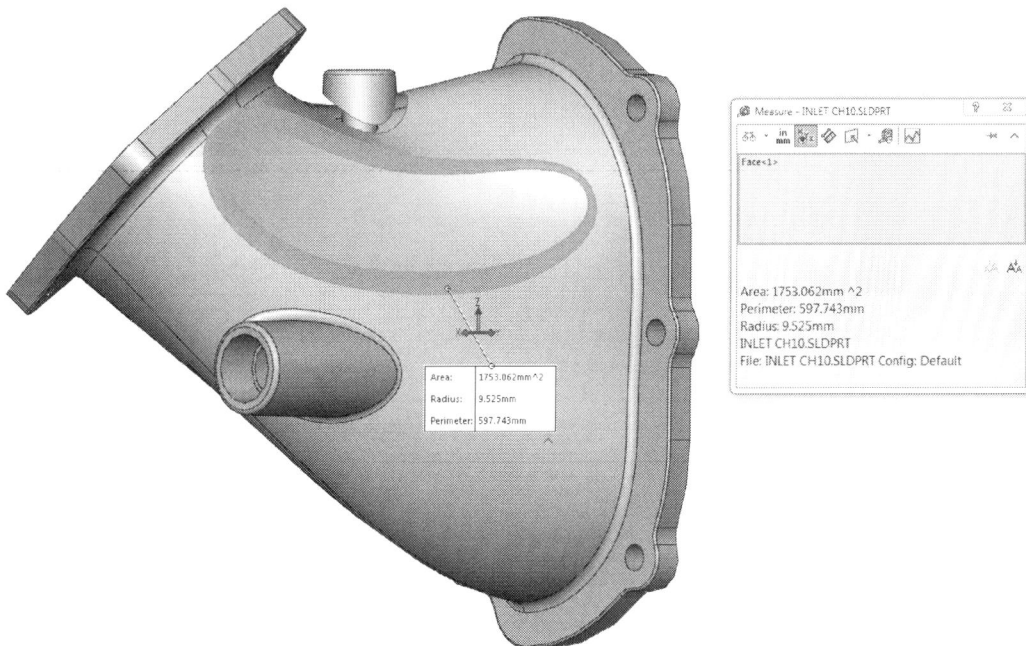

Figure 329 - Checking the result of Delete and Patch

207

CURVATURE ANALYSIS

Click Curvature (Evaluate toolbar), or click View > Display > Curvature.

You will notice that there is consistency now in the curvature of the fillet we created as shown in the following image.

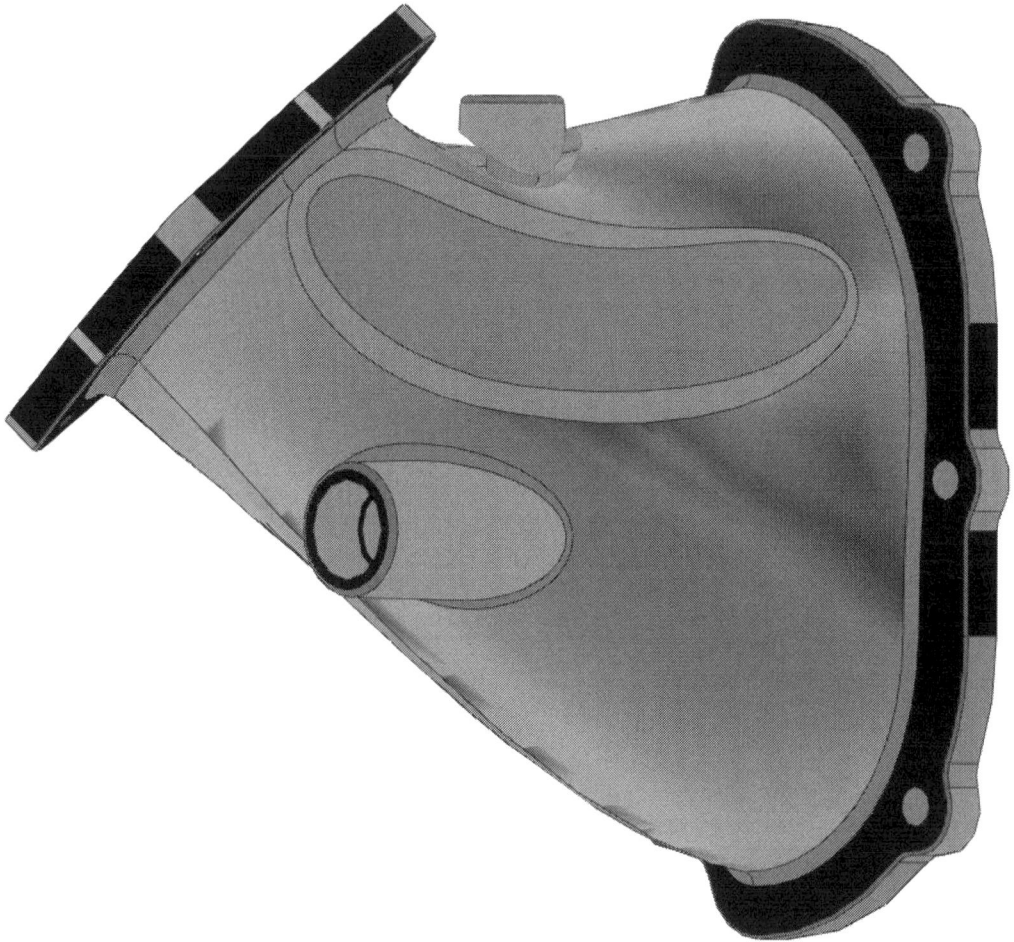

Figure 330 - Curvature Display

Click View > Display > Curvature to remove the Curvature Colour Display.

Save your part.

On top of checking if the resulting surface after running the Delete and Patch Command is still a radius of 9.525mm, I would also advise comparing the Baseline Sketch to the edge of the fillet again after running the Delete and Patch Command.

Now let's check the Surface Area again:

USING THE MEASURE TOOL TO MEASURE SURFACE AREA

Click Measure (Tools toolbar) or Tools > Evaluate > Measure.

Select the fillet face we just created.

Your answer is thus 1753.06 square millimeters as shown in the following image.

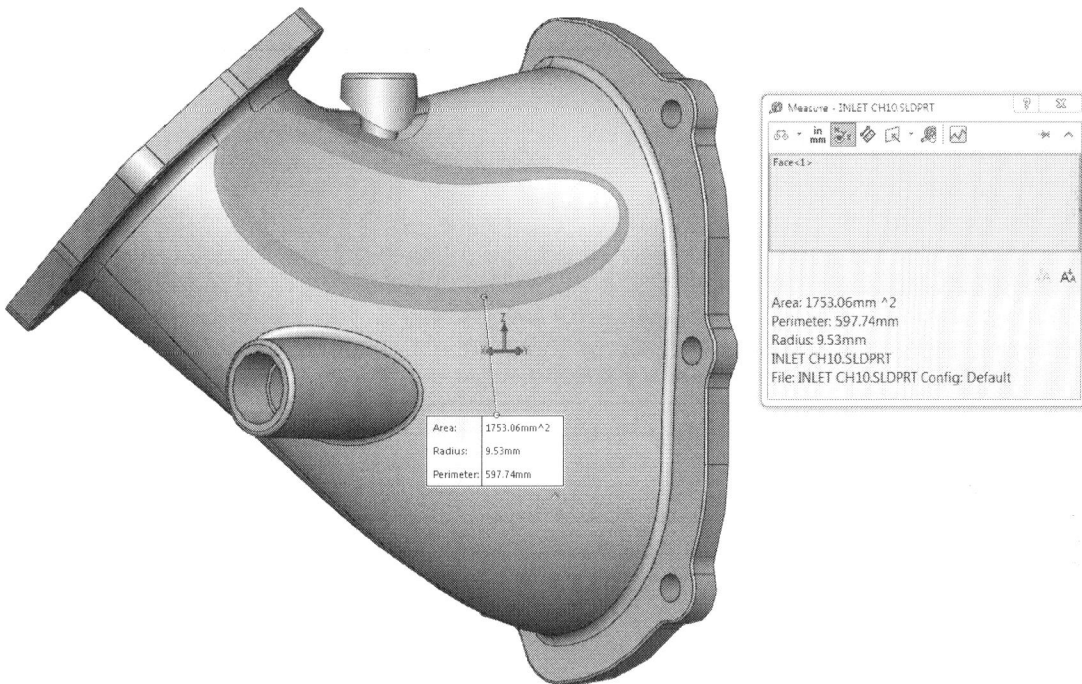

So in essence, we still have the same surface area but the main difference is that we now have a much better Surface and consequently will have a better quality part or product.

Save your part and move on to Chapter 11.

This Chapter is a continuation from Chapter 10.

You are required to modify the surface model by removing the port indicated in the following image.

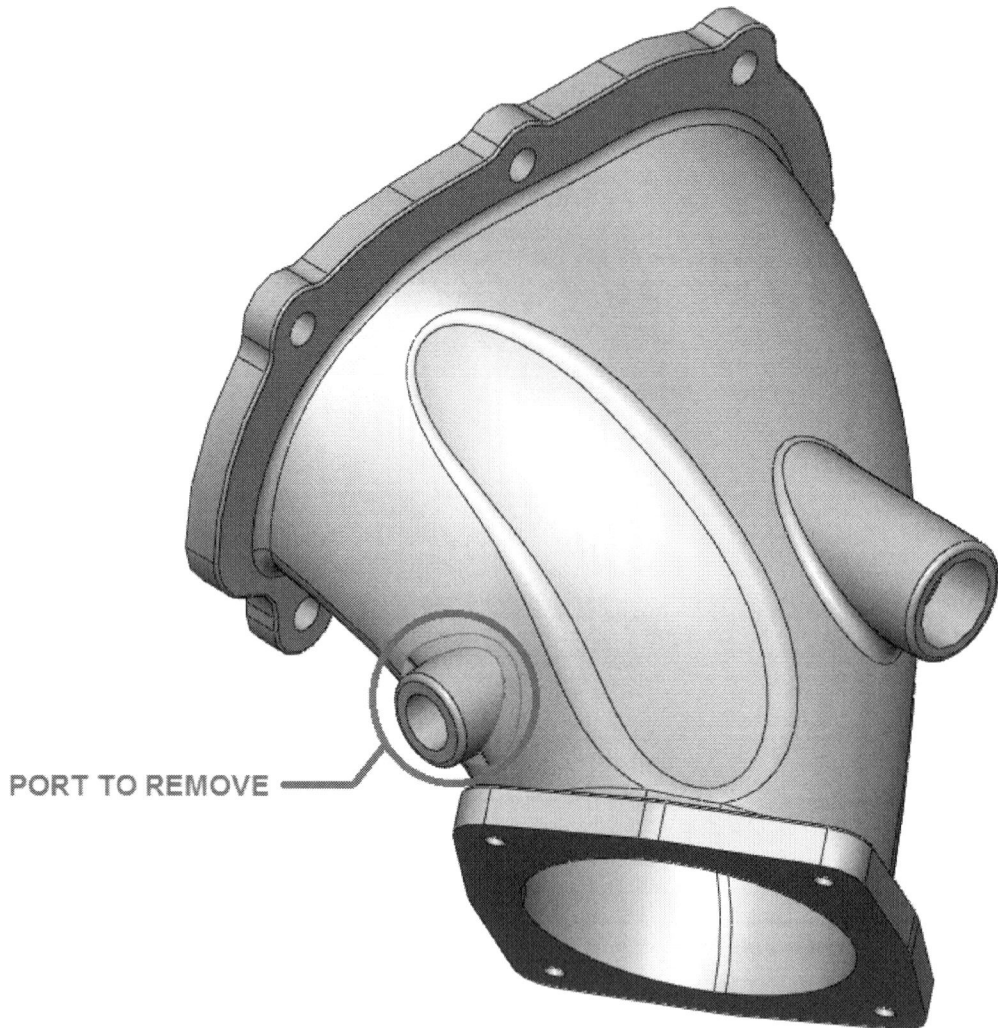

Figure 332 - Port to remove

After removing the port you should modify the surface model so that no trace of the removed port is evident as shown the following images.

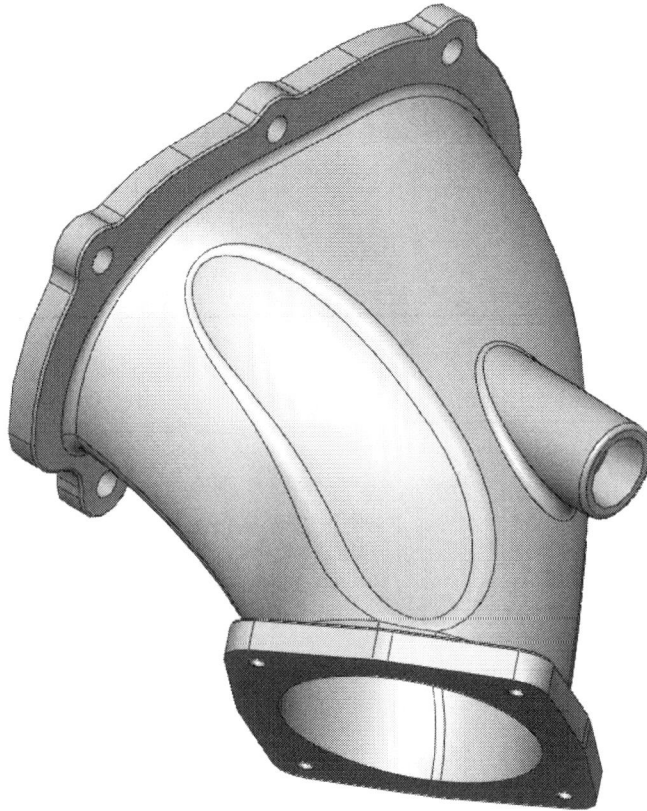

Figure 333 - No trace of removed port on outer surface

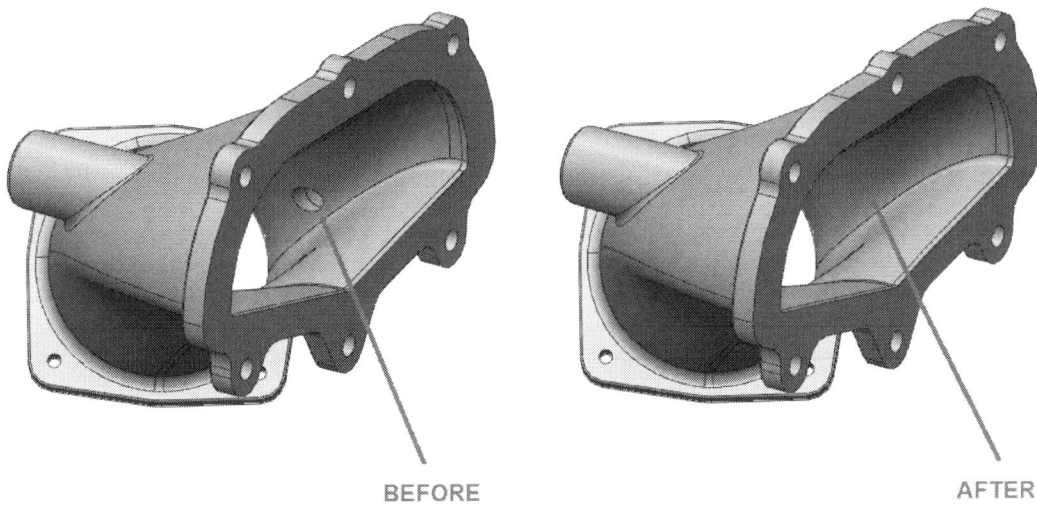

BEFORE

AFTER

Figure 334 - No trace of removed port on inner surface - Before and After

211

You are then required to measure and give the surface area of face 1 shown in the following image.

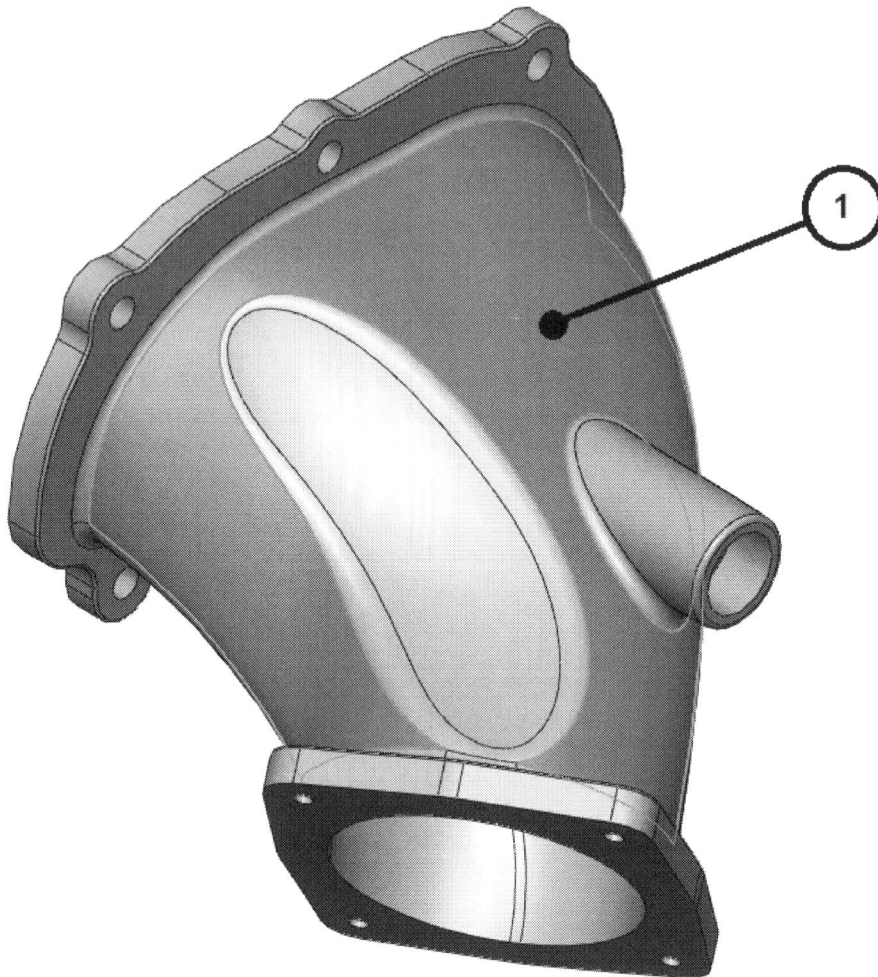

Figure 335 - Face whose area is to be measured

DELETING FACES

Click Delete Face on the Surfaces toolbar, or Insert > Face > Delete.

The Delete Face PropertyManager appears.

In the graphics area, click the six faces forming the port we want to delete. The names of the faces appear under Selections to delete.

Under Options, click Delete as shown in the following image.

212

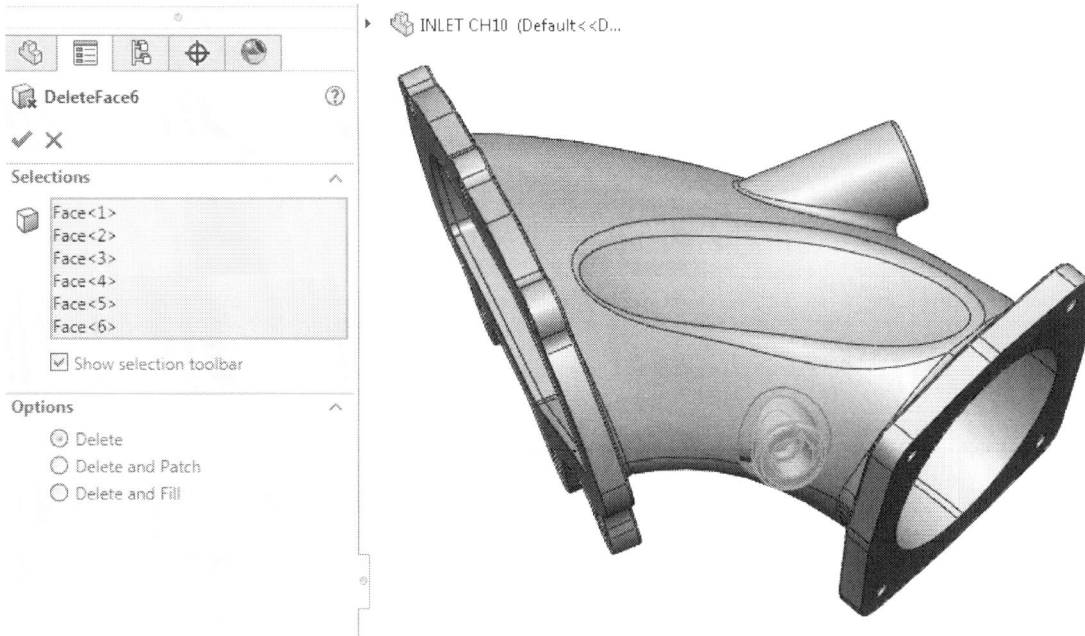

Figure 336 - Deleting faces using the Delete Face feature

Click OK.

Your part should now look as shown in the following image.

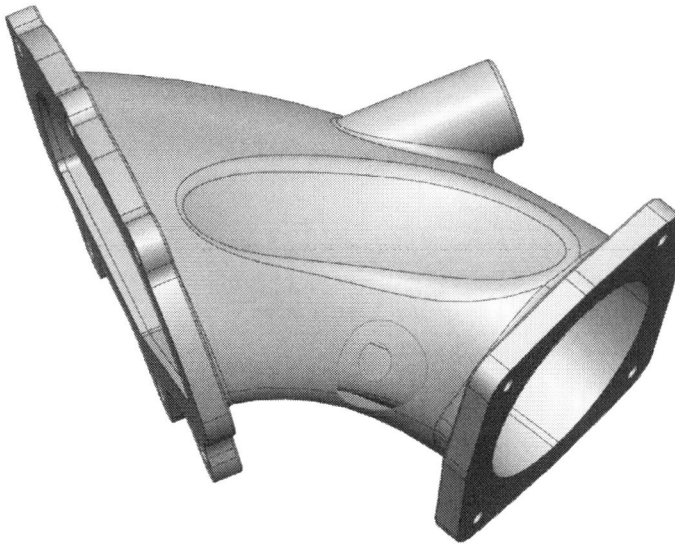

Figure 337 - Deleting faces using the Delete Face feature

DELETING HOLES FROM A SURFACE

To delete a hole or holes from a surface:

213

Press the Ctrl Key on your keyboard and select the two closed profiles or holes on the surface body as shown in the following image.

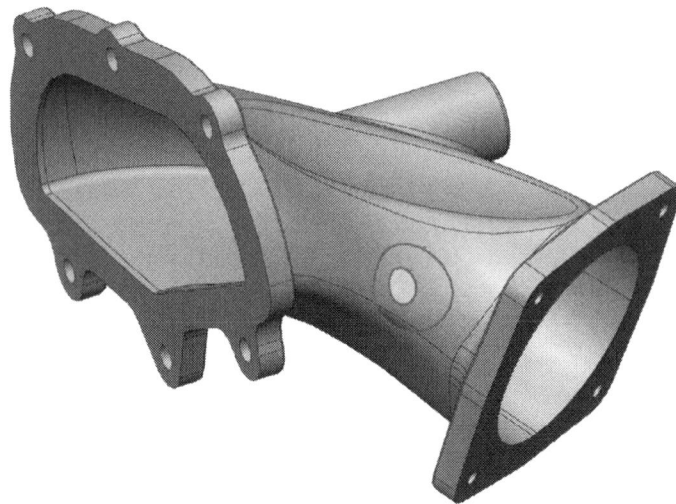

Figure 338 - Selected holes or closed profiles

Press and release the Delete key on your Keyboard - the Choose Option Dialog Box appears.

Select the Delete Holes(s) Radio Button in the Choose Option Dialog Box as shown in the following image.

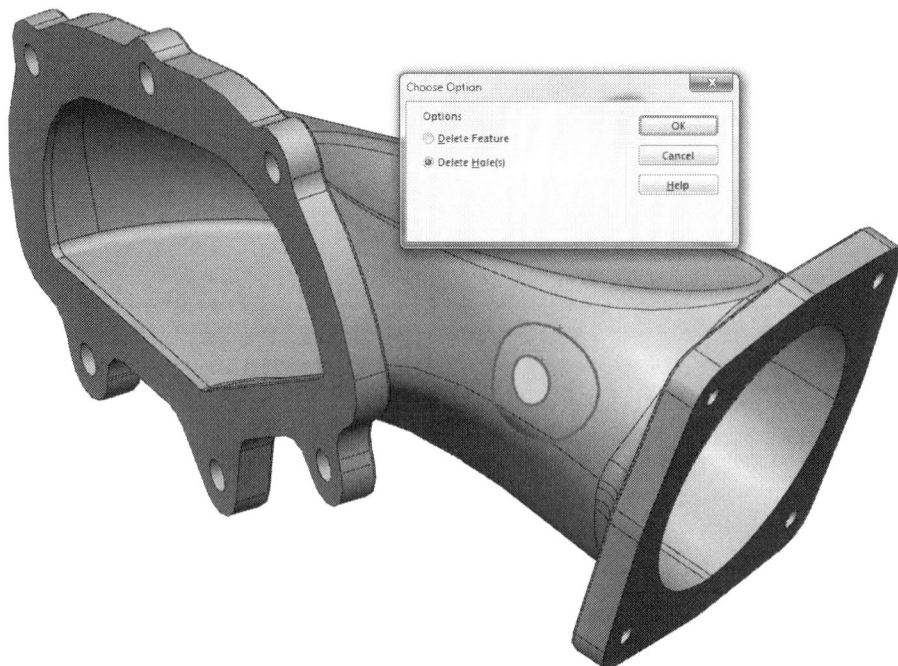

Figure 339 - Choose Option Dialog Box

Click the OK command button on the Choose Option Dialog Box.

Your part should now look as shown in the following image - from inside and outside.

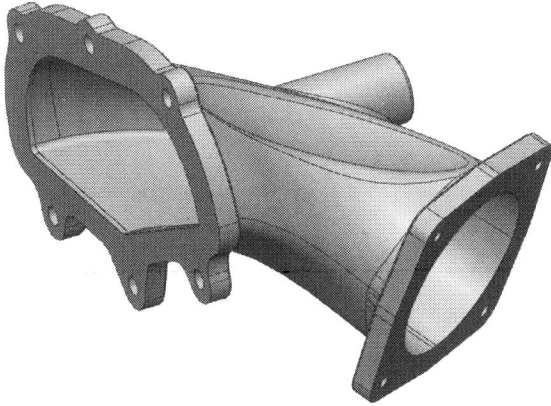

PART CURRENT STATUS - EXTERIOR   PART CURRENT STATUS - INTERIOR

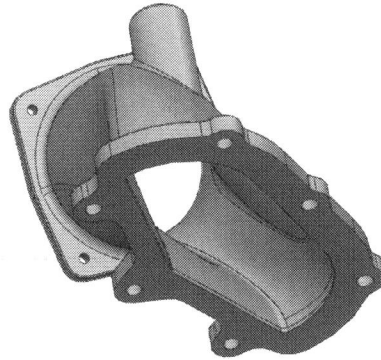

Figure 340 - Part current status

USING THE MEASURE TOOL TO MEASURE SURFACE AREA

Click Measure (Tools toolbar) or Tools > Evaluate > Measure.

Select the face labeled as face 1. Your answer is thus 16149.42 square millimeters as shown in the following image.

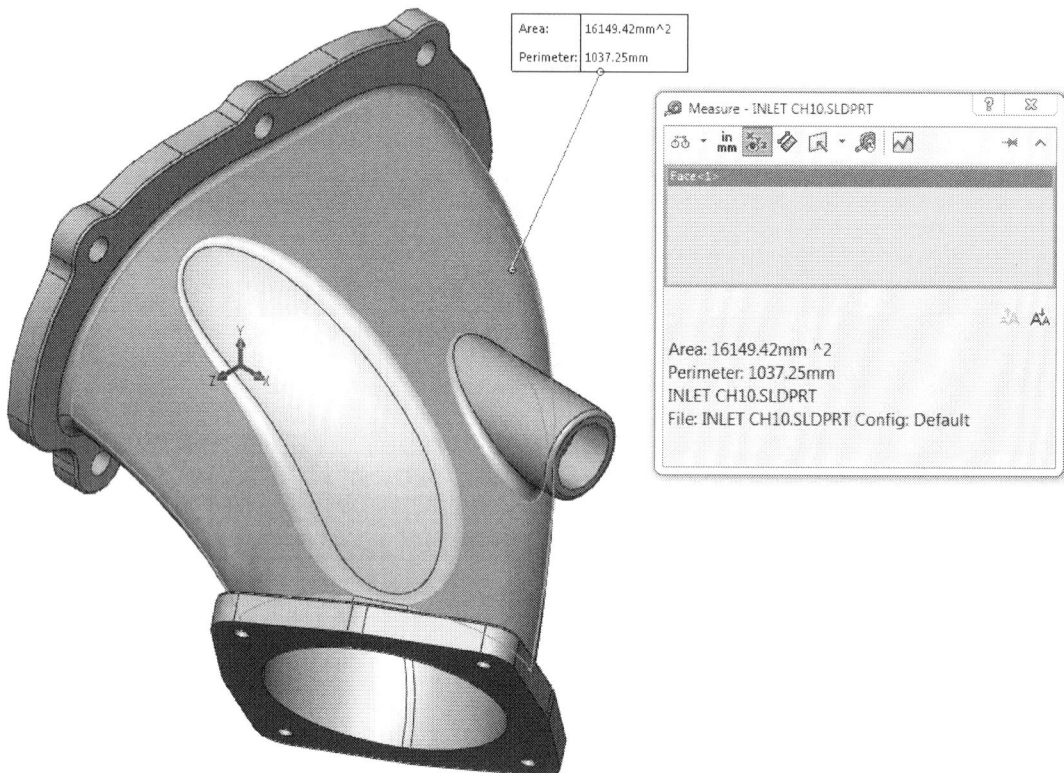

Figure 341 - Surface Area of Face 1

Close the Measure Tool Dialog box.

Save your part.

CHAPTER 12

This Chapter is a continuation from Chapter 11.

In this Chapter you are required to covert the model into a solid and then apply material - Aluminium 201.0-T43 (Density = 0.0028 g/mm^3 to the solid.

After creating the solid, you are required to measure its mass and give your answer in grams to two decimal places.

CREATING A SOLID FROM AN ENCLOSED VOLUME

Click Thicken on the Surfaces toolbar, or click Insert > Boss/Base > Thicken.

In the Thicken Property Manager under Surface to Thicken, select the surface body in the Graphics Area.

Check the Create solid from enclosed volume Checkbox and the Merge Result Checkbox as shown in the following image.

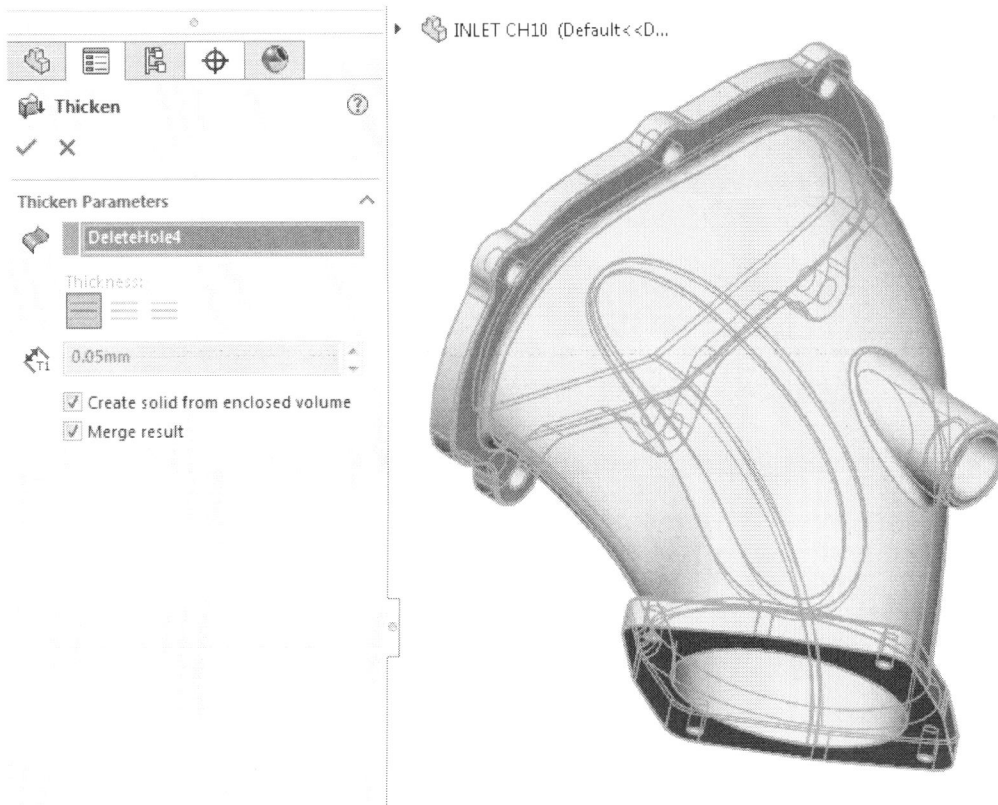

Figure 342 - Creating a solid from an enclosed volume

Click OK.

Apply the specified material which you should find under Solidworks Materials > Aluminium Alloys as shown in the following image.

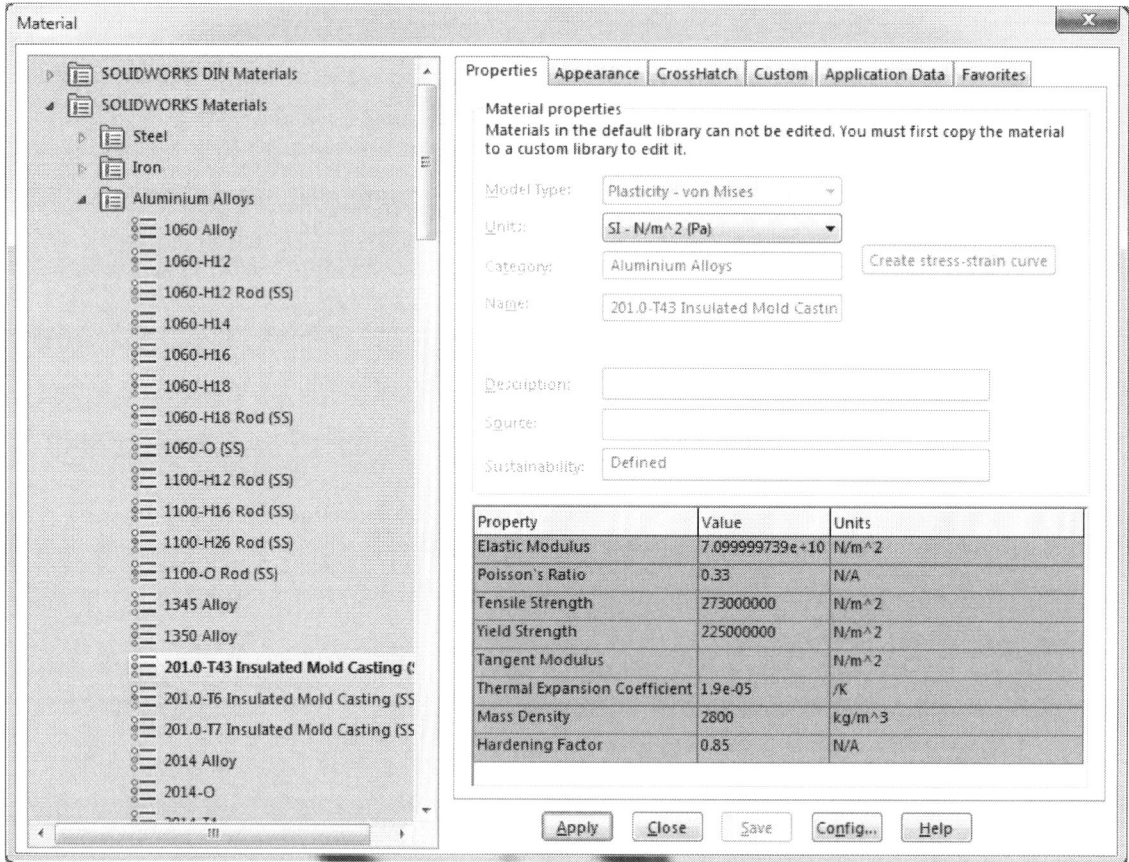

Figure 343 - Solidworks Materials - Aluminium Alloys

Save your part.

Your part should now look as shown in the following image.

INLET CH10 (Default<<Default>_Appearar
- ▸ History
- Sensors
- ▸ Annotations
- ▸ Solid Bodies(1)
- Surface Bodies
- 201.0-T43 Insulated Mold Casting (SS)
- Front
- Top
- Right
- Origin
- Surface-Imported1
- Surface-Imported2
- 3D BASELINE
- Surface-Extend1
- Surface-Extend2
- Boundary-Surface1
- Surface-Knit1
- Surface-Trim1
- Fillet2
- DeleteFace4
- DeleteFace6
- DeleteHole4
- Thicken1

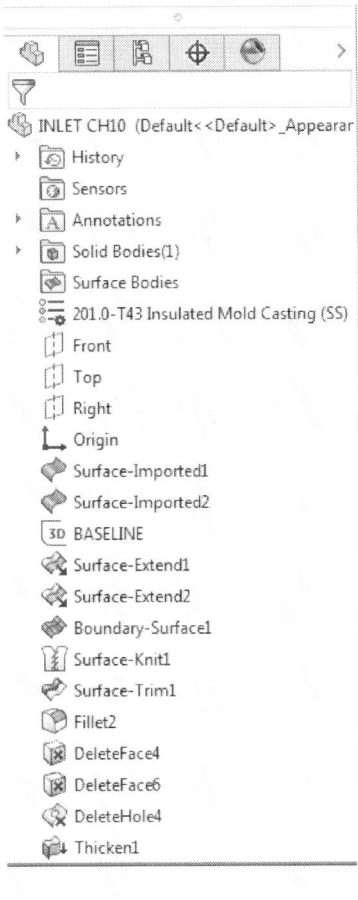

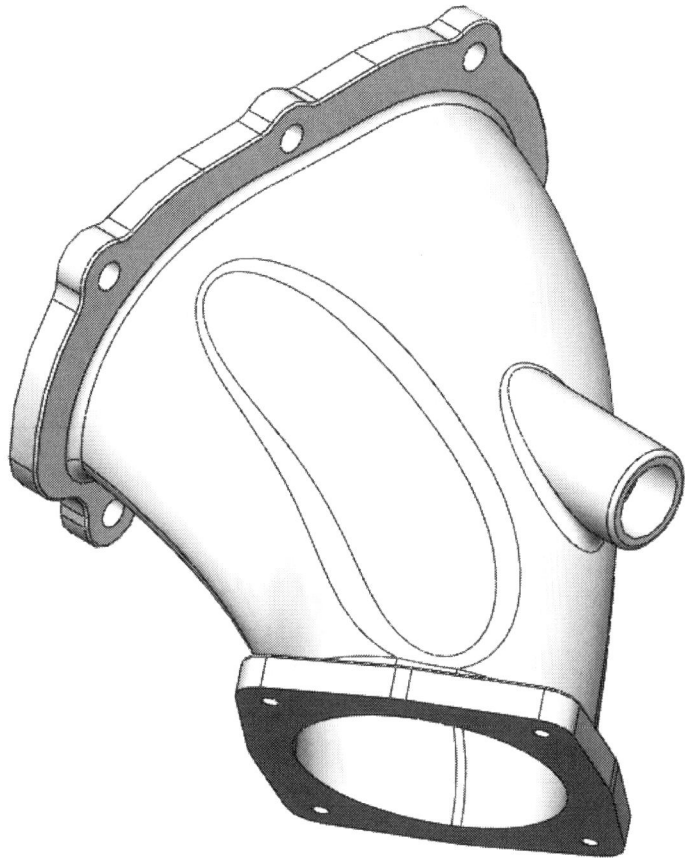

Figure 344 - Part current status

DISPLAYING MASS PROPERTIES OF A PART

Click Mass Properties (Evaluate toolbar) or Tools > Evaluate > Mass Properties.

The calculated mass properties appear in the dialog box as shown in the following image - thus the mass of the part is 928.96 grams.

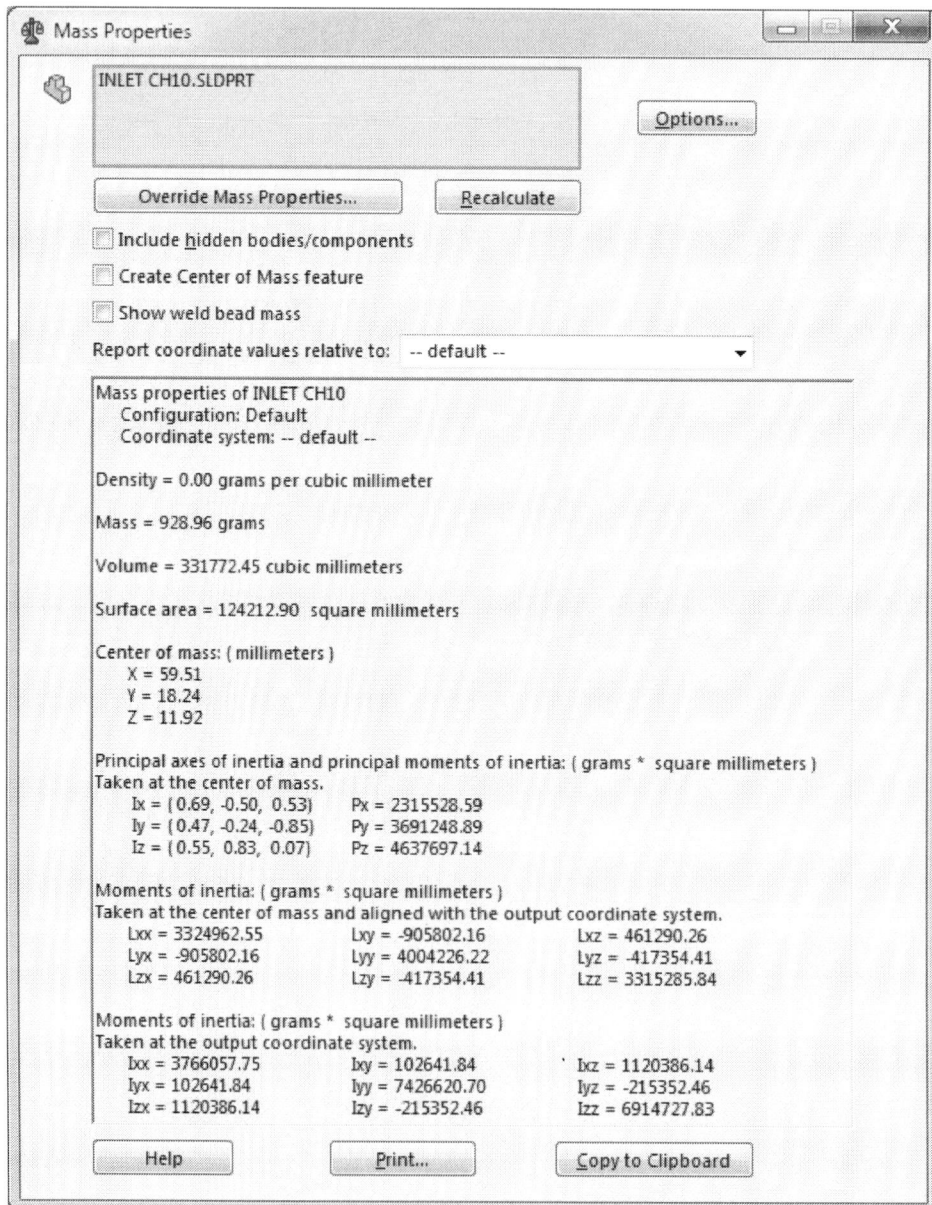

Figure 345 - Displaying Mass Properties of a Part

In this Chapter you are provided with an imported model consisting of a solid body and two surface bodies. You are then required to use the surfaces to create a parting line and a 1.20mm reveal in the solid model sphere - which is actually a molded plastic ball.

Download the part named MOLDED PLASTIC BALL START from the Chapter 13 Download folder from this Google Drive Location - *http://bit.ly/CSWPA-SU* or Scan the QR Code shown below:

If you experience any problems with downloading any files you may send an email to *cswpasmebook@gmail.com* with the title of the book indicated in your email subject. Open and save the downloaded part to your PC.

The part should look as shown in the following image.

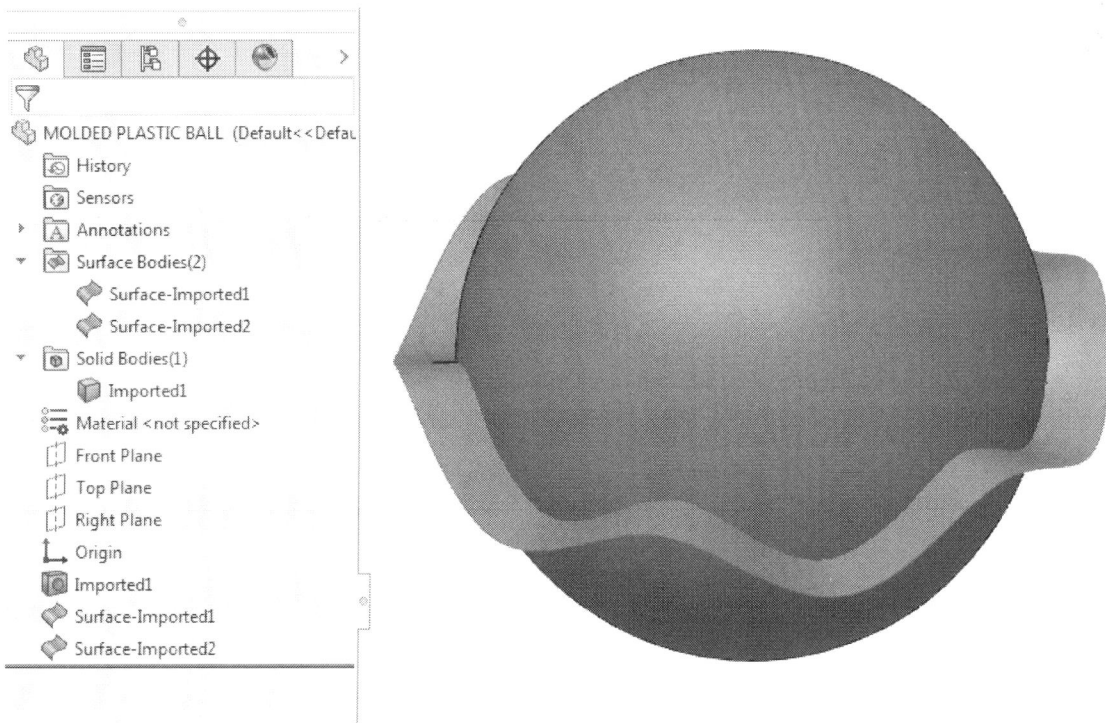

Figure 346 - Molded Plastic Ball

You are required to grow the exterior edge (edge towards the outside of the sphere) of surface 1 (burnt orange surface) outward in order to trim it off with the sphere.

Note: The interior edge of surface 1 must remain unchanged.

Trim surface 1 and the spherical surface to create a parting line surface set to be used to create a plastic ball solid.

After extending and trimming the surfaces, measure surface 1's surface area and give your answer in square millimeters to two decimal places.

No material should be applied.

See the following image with the solid body hidden showing the two surface - surface 1 and the spherical surface.

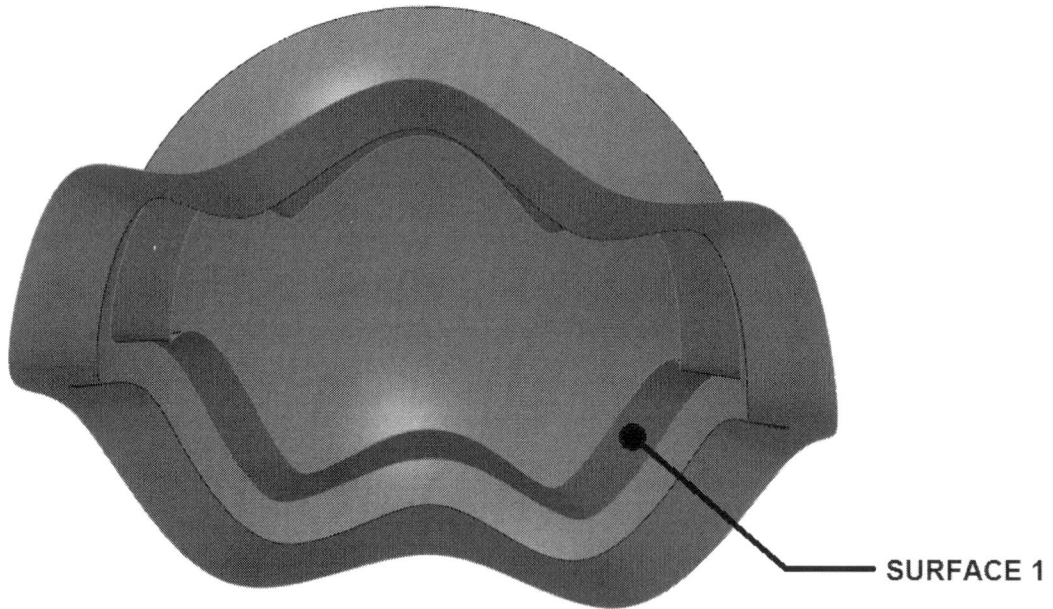

Figure 347 - Surface Bodies with the Solid Body Hidden

EXTEND SURFACE

To extend the exterior edge of Surface 1, we may use the Extend Surface feature:

With the solid body in the model hidden, Click Extend Surface (Surfaces Toolbar) or Insert > Surface > Extend.

Select the outer edge of surface 1 in the graphics area under selected faces/ edges in the Extend Surface Property Manager.

Under End Condition, select the Distance radio button and enter a distance of 5mm and then select Same Surface under Extension Type as shown in the following image.

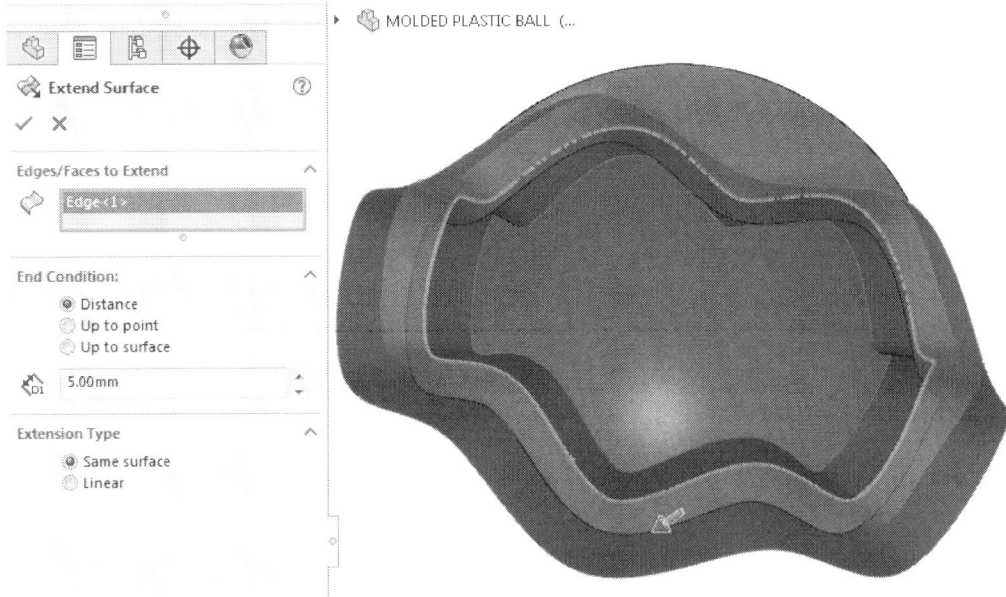

Figure 348 - Extending the outer edge of Surface 1

Click Ok.

Your part should now look as shown in the following image.

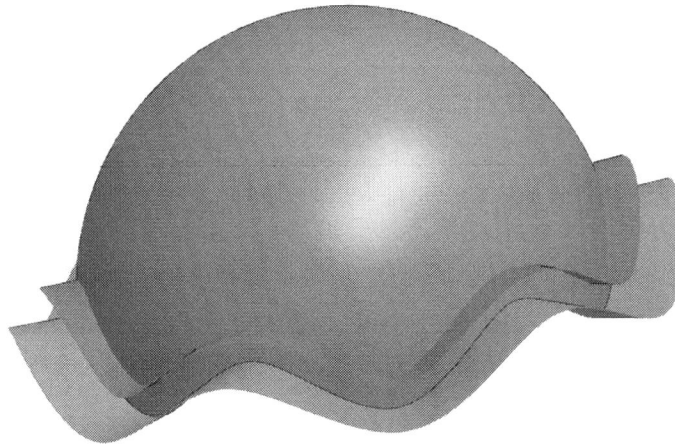

Figure 349 - Part current status

TRIM SURFACE

Click Trim Surface on the Surfaces toolbar, or click Insert > Surface > Trim.

In the PropertyManager, under Trim Type, select Mutual.

Under Selections, Trimming Surfaces - select the spherical surface and surface 1 in the graphics area.

Select Keep selections.

Under Surface Split Options, select Natural as shown in the following image.

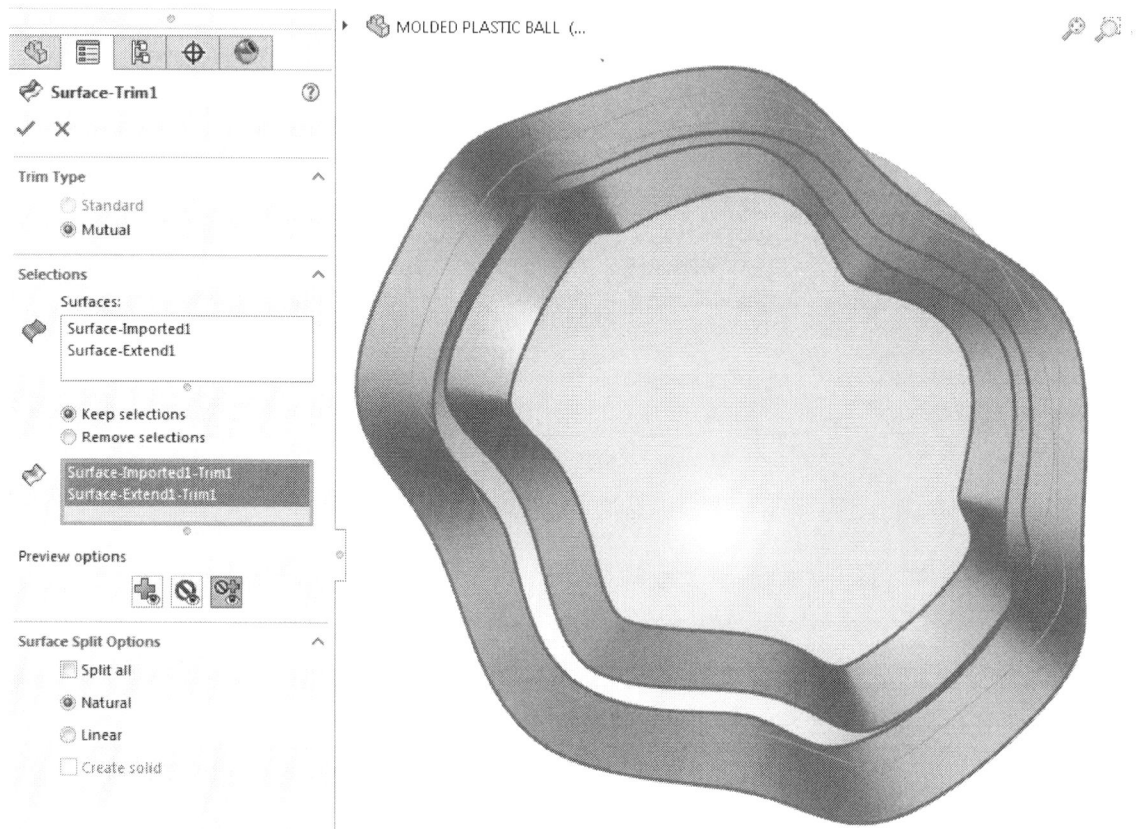

Figure 350 - Surface Mutual Trim

Click OK.

Your part should now look as shown in the following image - with the solid body still hidden.

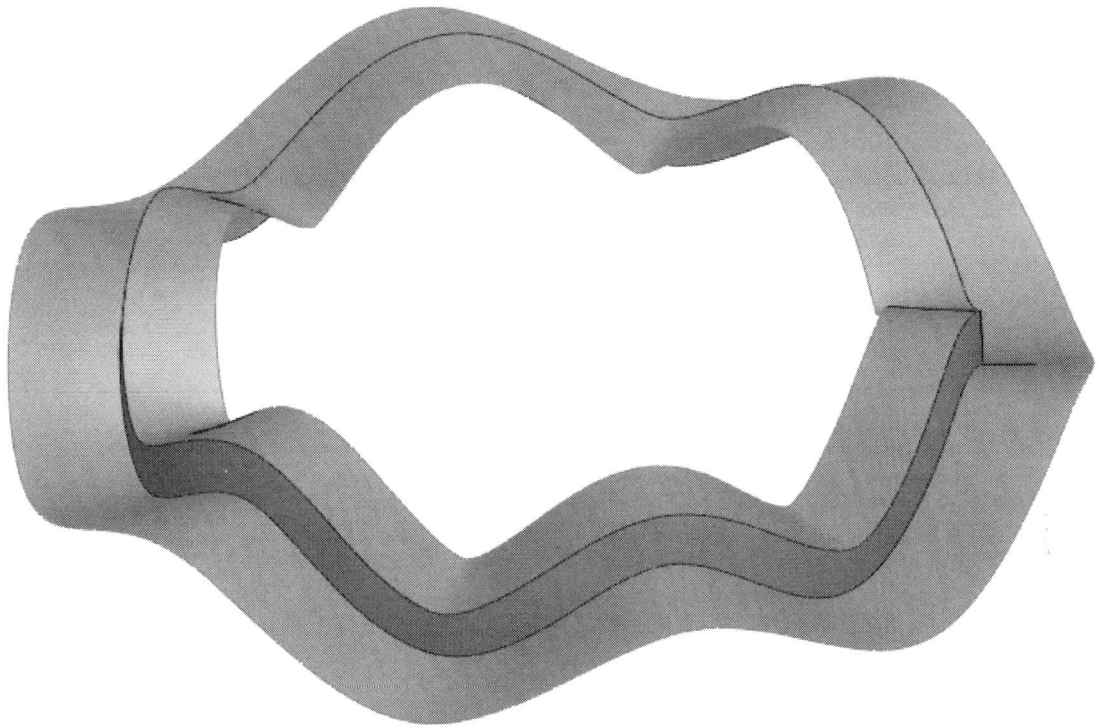

Figure 351 - Part Current Status after the Mutual Trim

MEASURING THE SURFACE AREA OF A SURFACE BODY

Click Measure (Evaluate toolbar) or Tools > Evaluate > Measure.

Select Surface 1 in the Graphics Area as shown in the following image - the answer is thus 1268.29 square millimeters.

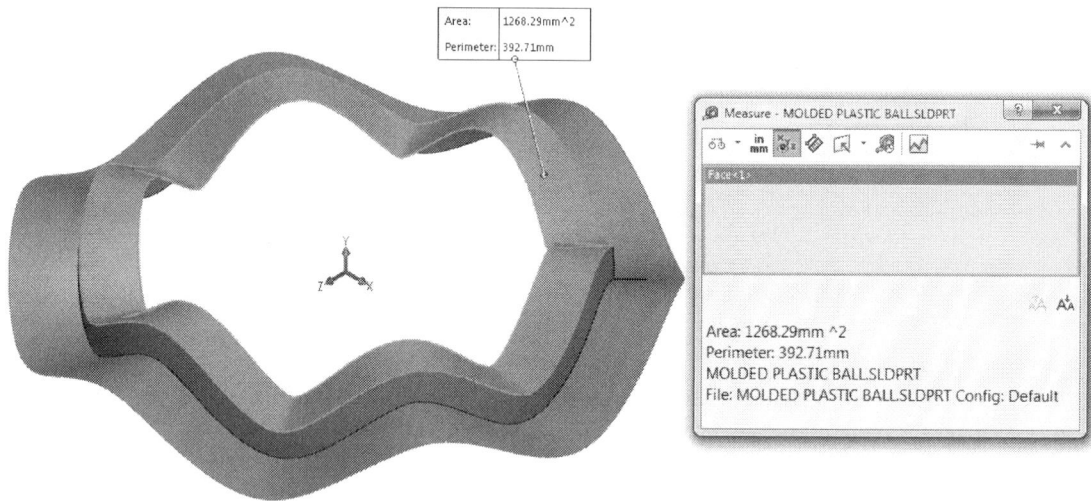

Figure 352 - Measuring the surface area of Surface 1

Close the Measure Tool.

Save your part.

CHAPTER 14

This Chapter is a continuation from Chapter 13.

Show the Solid Body in the model.

Use the surfaces trimmed in the previous chapter to separate the sphere into two pieces.

Measure the surface area of the top part of the ball (the part colored in green in the following image) and give the answer in square millimeters to two decimal places.

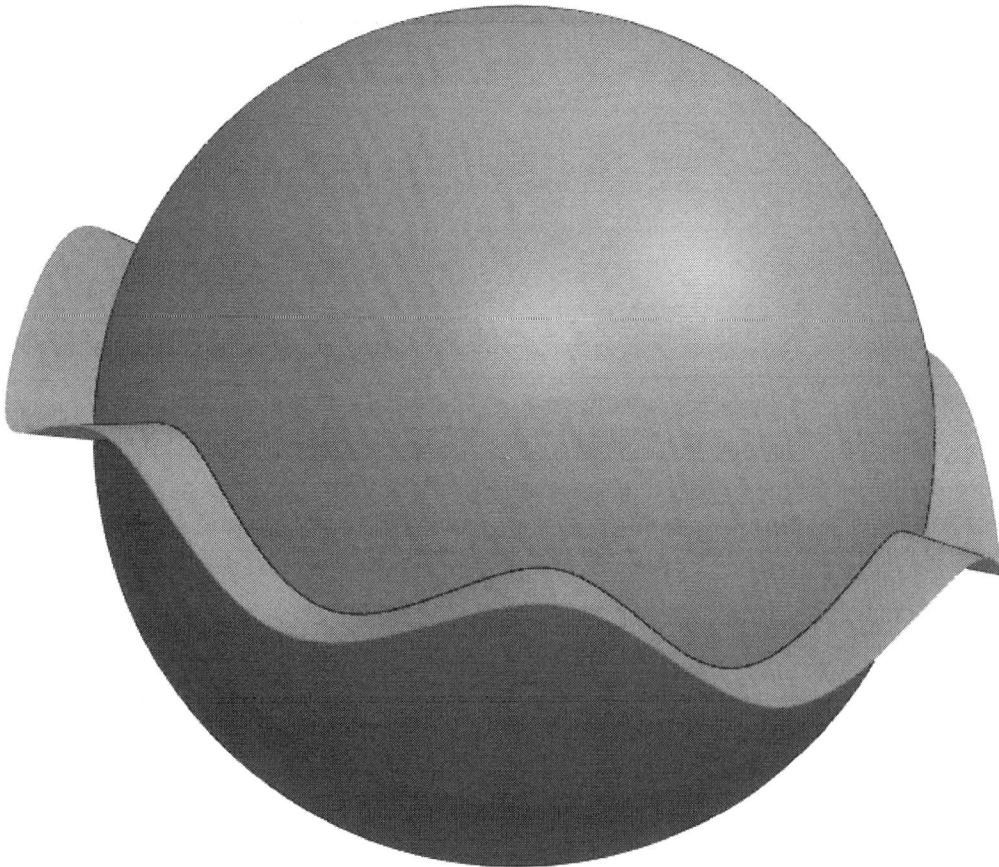

Figure 353 - Surface Area to measure

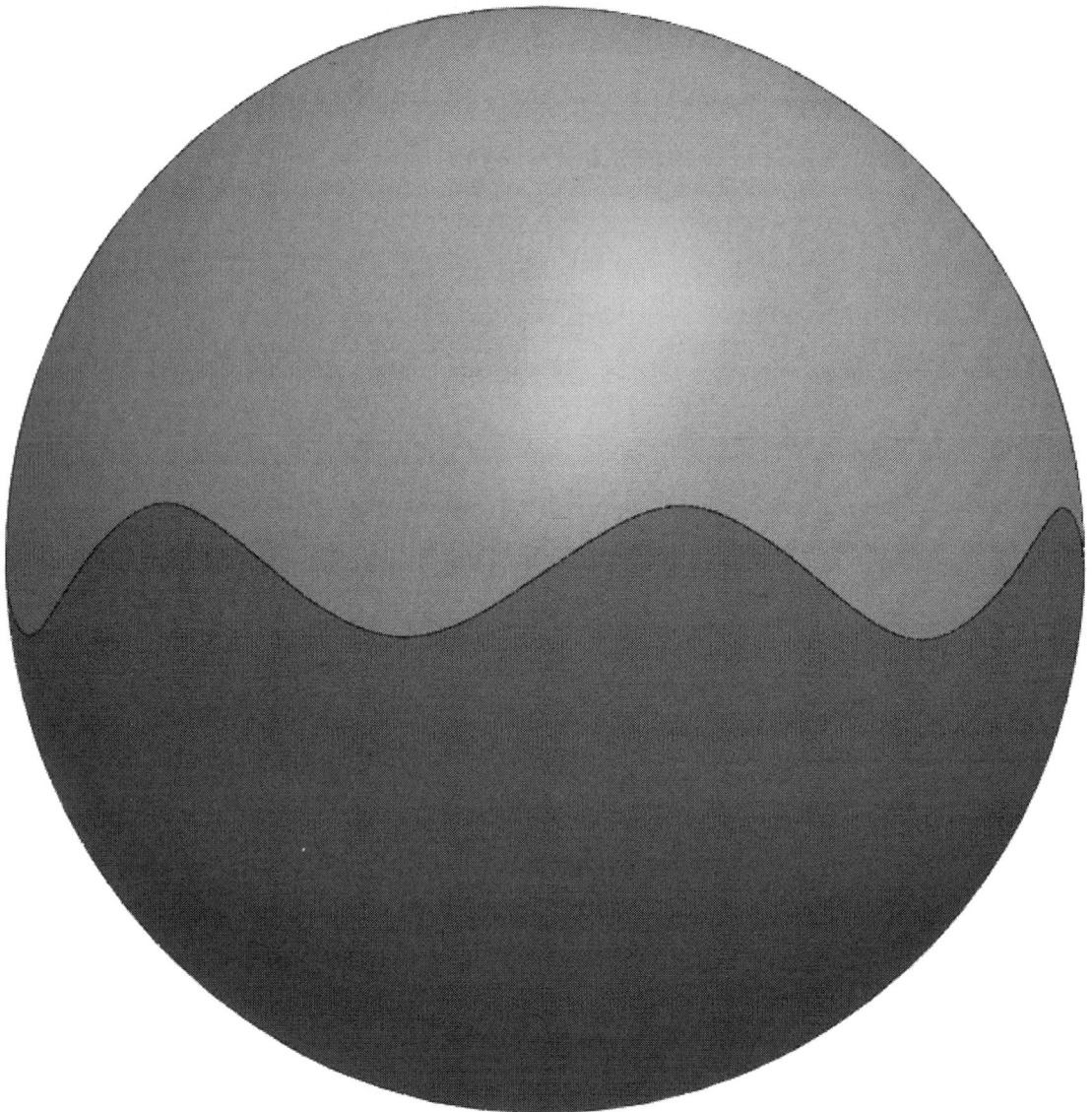

Figure 354 - Surface Area to measure

SPLITTING A PART

You can use Split Feature to divide a solid body or part into multiple bodies. You can keep the bodies within the part or save them into separate part files. You can save them during the creation of the Split feature, or use Save Bodies to save them after the split is complete.

Click Split (Features toolbar) or Insert > Features > Split.

Select a Part Template from the New Solidworks Document Dialog Box and click OK.

Select an Assembly Template from the New Solidworks Document Dialog Box and click OK.

In the Split PropertyManager, select the Surface Body in the Graphics Area under Trim Tools.

Select the All bodies radio button under Target Bodies.

Click the Cut Bodies command button.

Resulting bodies lists the split bodies in the part after you click Cut Bodies - select all as shown in the following image.

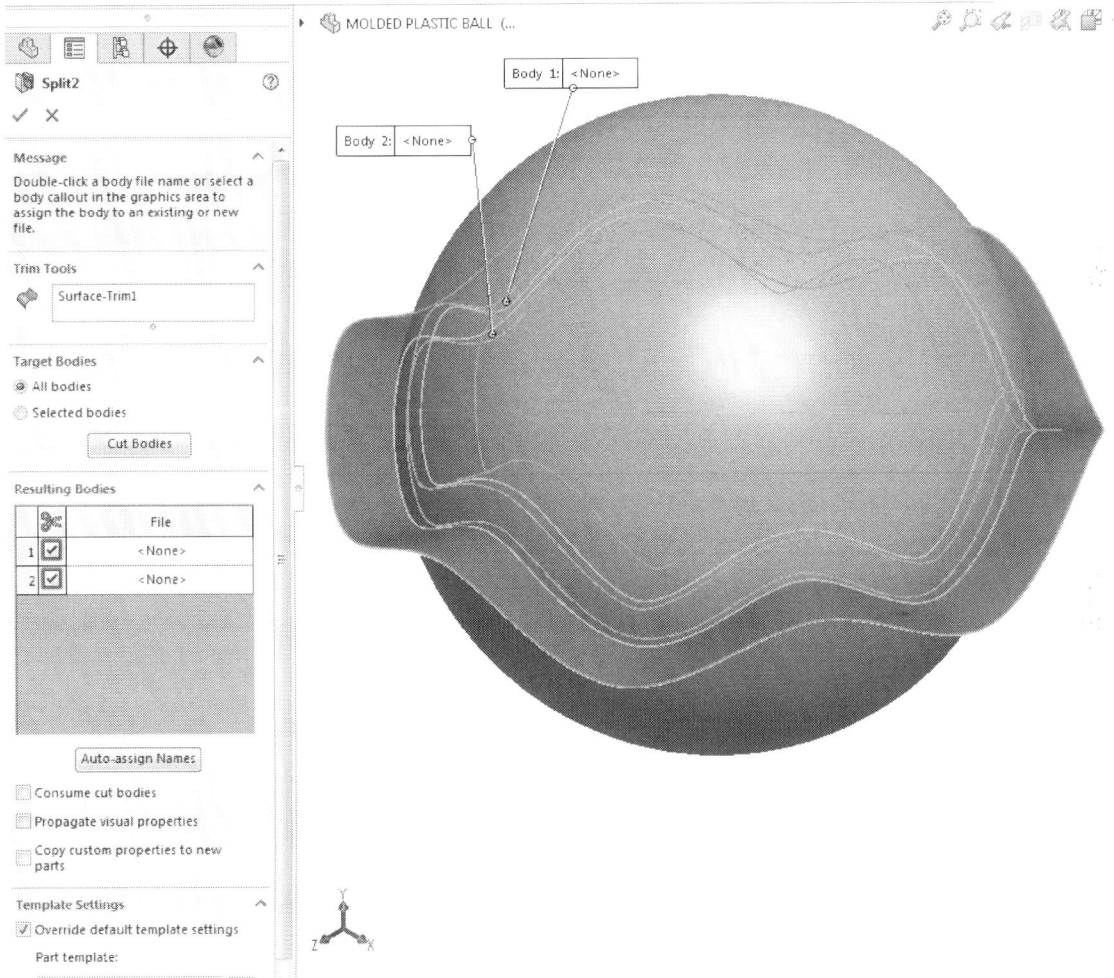

Figure 355 - Split Feature

Click OK.

You will notice that two solid bodies appear in the Feature Manager Design Tree as shown in the following before and after image.

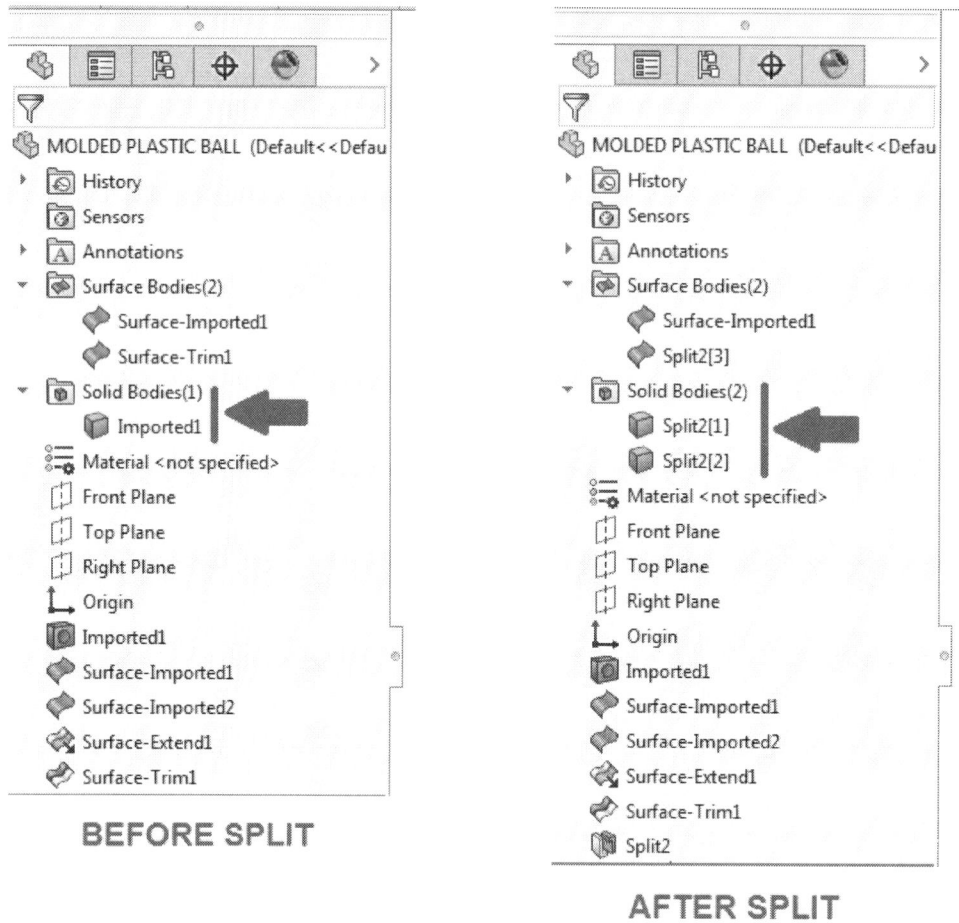

BEFORE SPLIT

AFTER SPLIT

Figure 356 - Change in the number of Solid Bodies in the Solid Bodies Folder in the Feature Manager Design Tree

HIDING ALL SURFACE BODIES

To hide all Surface bodies, right click the Surface Bodies Folder in the Feature Manager Design Tree and select Hide as shown in the following image.

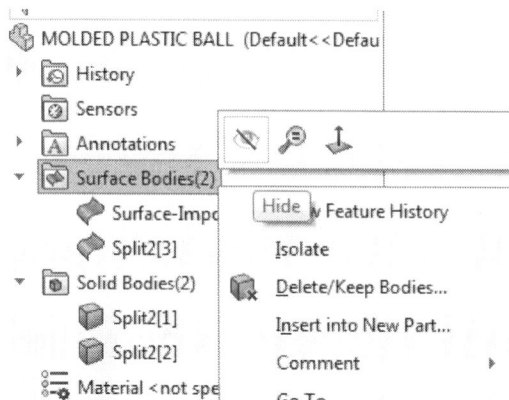

Figure 357 - Hiding all Surface Bodies in the Part

230

Your part should now look as shown in the following image.

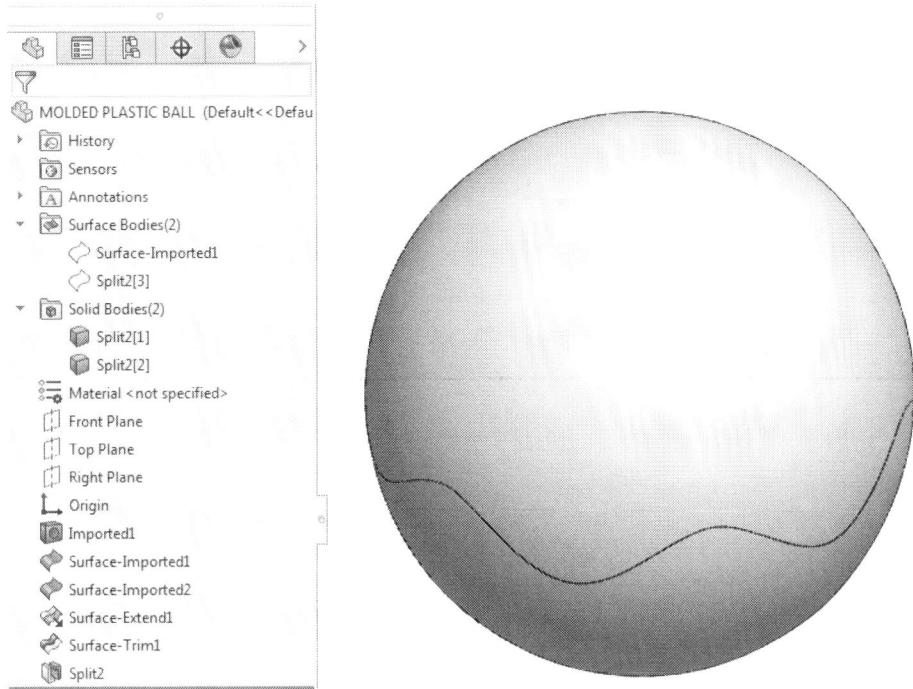

Figure 358 -Part Current Status

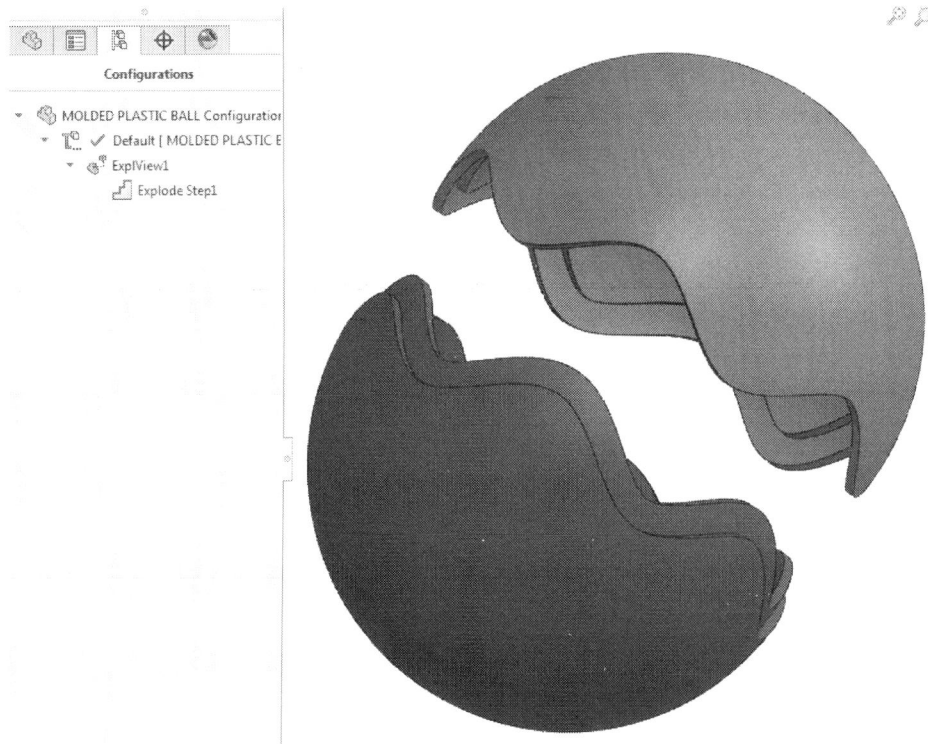

Figure 359 -Part Current Status in Exploded View

231

MEASURING SURFACE AREA

Click Measure (Evaluate toolbar) or Tools > Evaluate > Measure.

Select the top surface in the Graphics Area as shown in the following image - the answer is thus 6475.08 square millimeters.

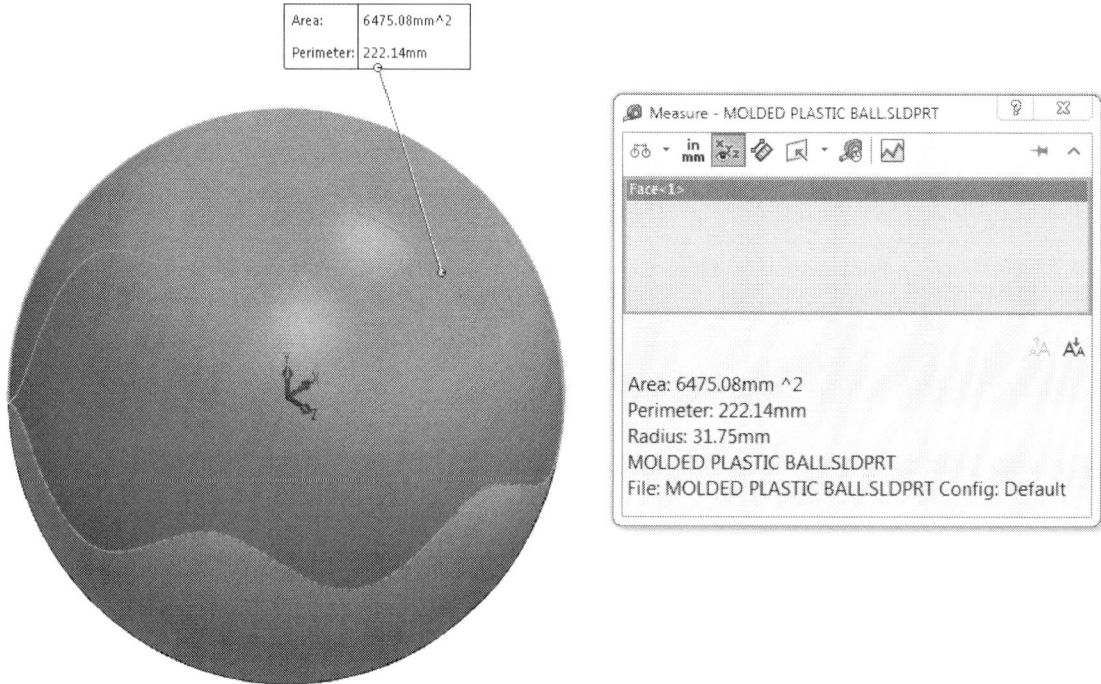

Figure 360 - Measuring surface area

Close the Measure Tool.

Save your part.

CHAPTER 15

This Chapter is a continuation from Chapter 14.

You are required to create the reveal - that is the gap between the green (top part) and red (bottom part) parts of the ball.

The following conditions must be met:

There must be a uniform gap of 1.20mm all throughout the split between the green and red parts of the ball.

The green (top) part of the ball is the part that will be affected by the change. The red (bottom) part of the ball will remains unchanged - see the following two images.

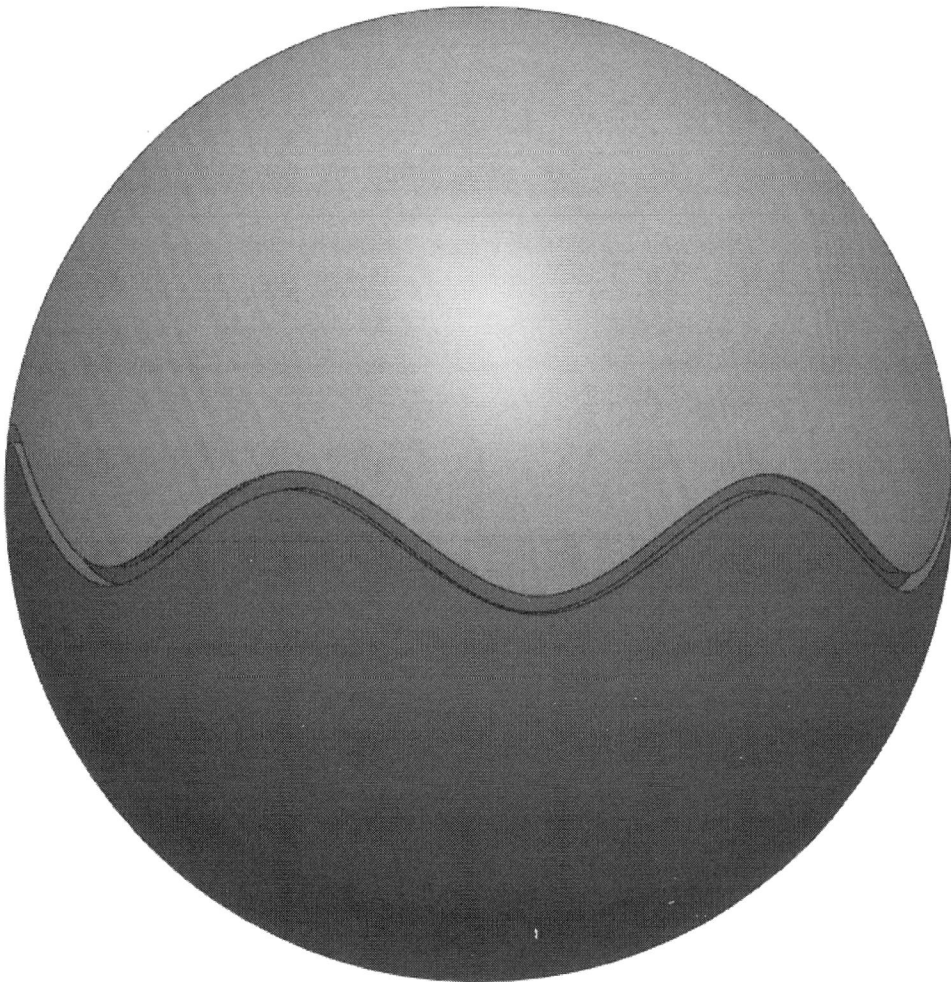

Figure 361 - 1.2mm reveal to be created

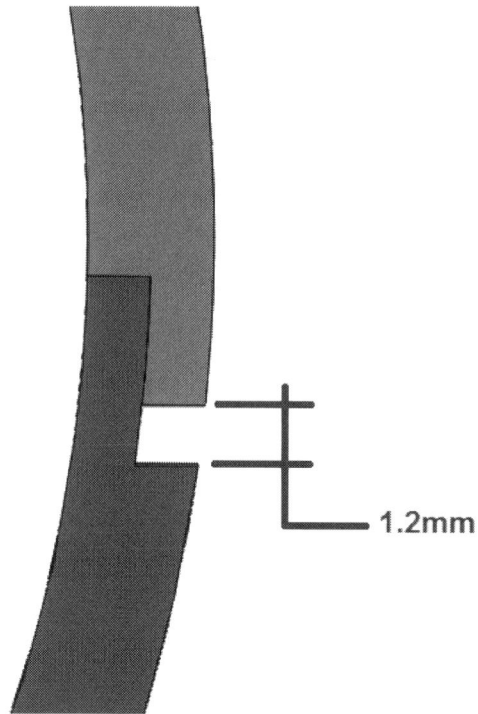

Figure 362 - 1.2mm reveal to be created - Cross Sectional View

You are then required to measure the surface area of the top part (green) of the ball and give your answer in square millimeters to the nearest two decimal places.

HIDE ALL SOLID BODIES

Click on the Solid Bodies Folder in the Feature Manager Design Tree and select hide to hide all Solid Bodies in the part as shown in the following image.

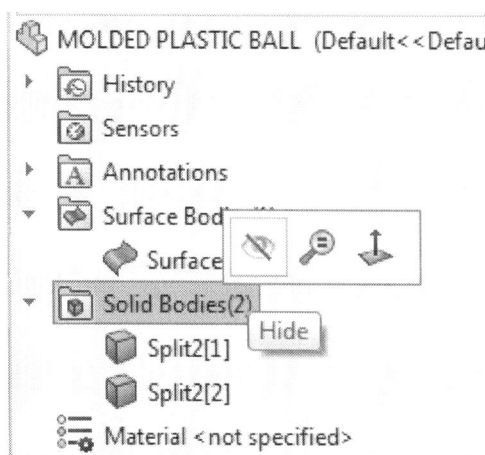

Figure 363 - Hide Solid Bodies in the Part

OFFSET SURFACE

Click Offset Surface (Surfaces toolbar) or Insert > Surface > Offset.

In the PropertyManager:

Select the face in the graphics area for Surface or Faces to Offset as shown in the following image.

Set a value for Offset Distance of 1.20mm.

If necessary, select Flip Offset Direction ↗ to change the direction of the offset.

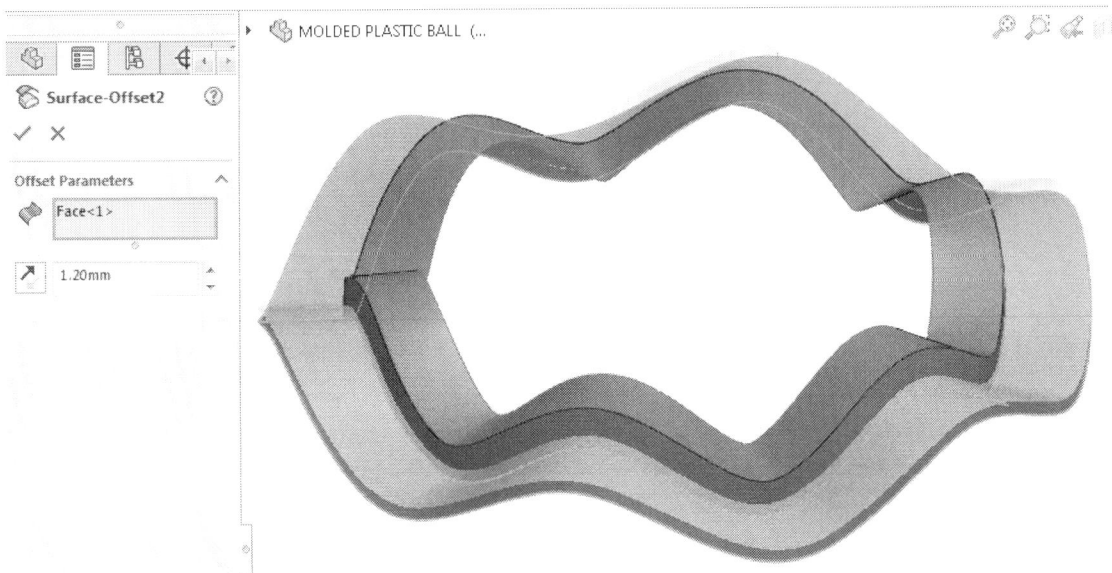

Figure 364 - Creating an offset surface

Click OK .

In the Graphics Area, click on the surface face as shown in the following image and select hide to hide the selected surface body.

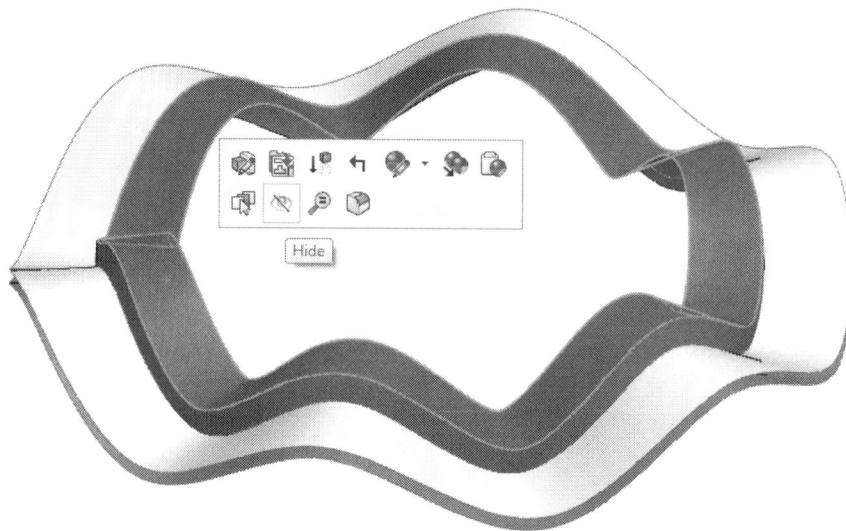

Figure 365 - Hiding a selected surface body

EXTEND SURFACE

To extend the interior edge of the Offset Surface, we may use the Extend Surface feature:

With the solid bodies and the other surface body in the model hidden, Click Extend Surface (Surfaces Toolbar) or Insert > Surface > Extend.

Select the inner edge of the offset surface in the graphics area under selected faces/ edges in the Extend Surface Property Manager.

Under End Condition, select the Distance radio button and enter a distance of 5mm and then select Same Surface under Extension Type as shown in the following image.

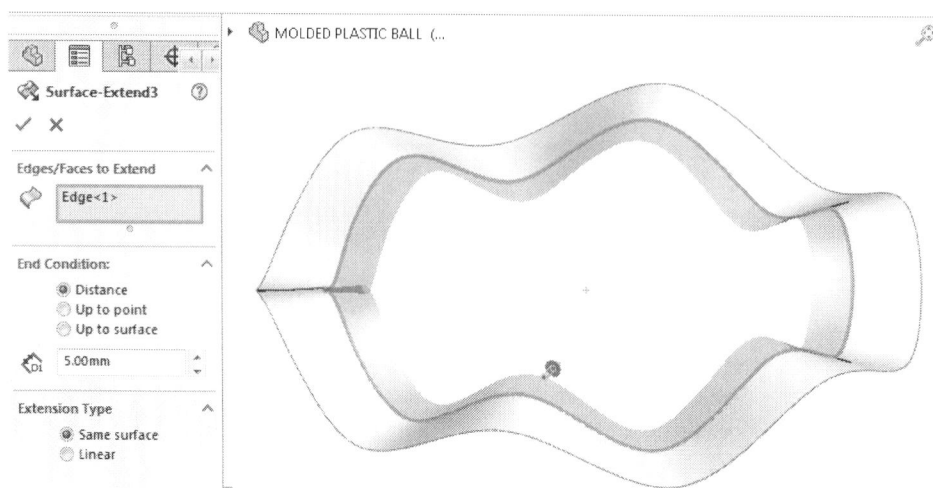

Figure 366 - Extending the inner edge of the Offset Surface

236

Click Ok.

Unhide the top solid part (green part) as shown in the following image.

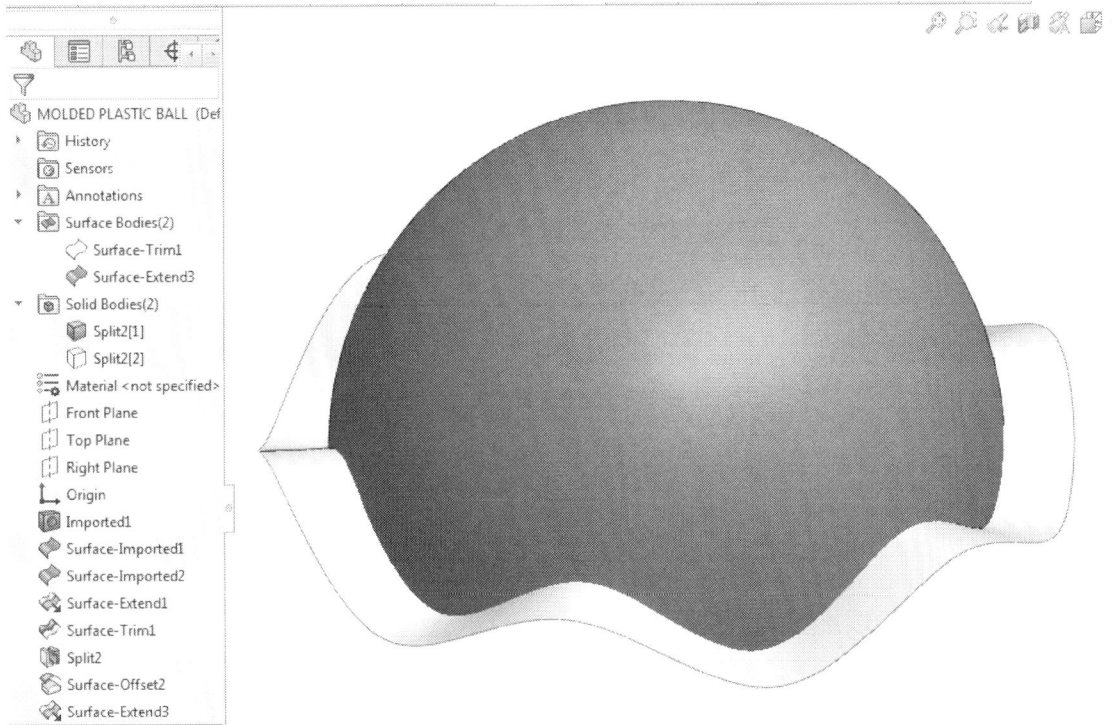

Figure 367 - Part current status

CUT WITH SURFACE

Click Cut With Surface on the Features toolbar, or click Insert > Cut > With Surface.

In the PropertyManager, under Surface Cut Parameters, select the surface to use to cut the solid bodies as shown in the following image.

Examine the preview. If necessary, click Flip cut to reverse the direction of the cut.

The arrow points in the direction of the solid to discard. Since we have a Multibody part, under Feature Scope, uncheck the Auto-select Check Box *(automatically selects all relevant intersecting bodies)* and select the Top Part *(Green Part)* in the Graphics Area. The surface cuts only the bodies you select. If you add new bodies to the model that are intersected by the cutting surface, right-click, select Edit Feature, and select those bodies to add them to the list of selected bodies. If you do not add the new bodies to the list of selected bodies, they remain intact.

The solid bodies you select are highlighted in the graphics area, and listed under Feature Scope.

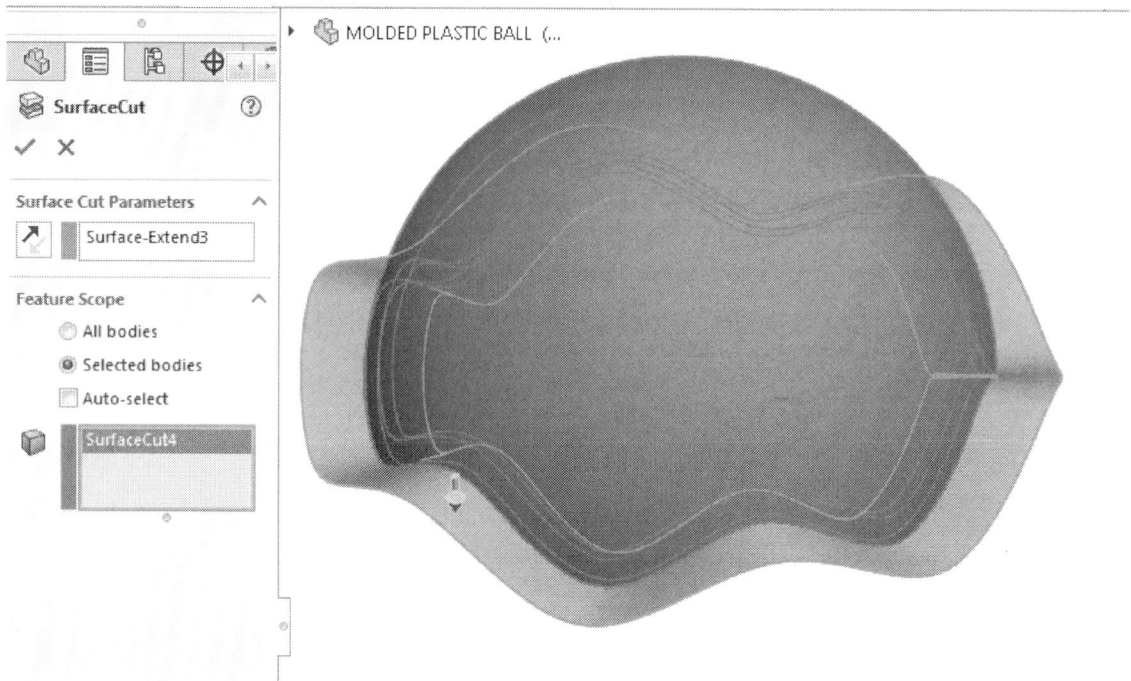

Figure 368 - Using Cut With Surface

Click OK.

Your part should now look as shown in the following image.

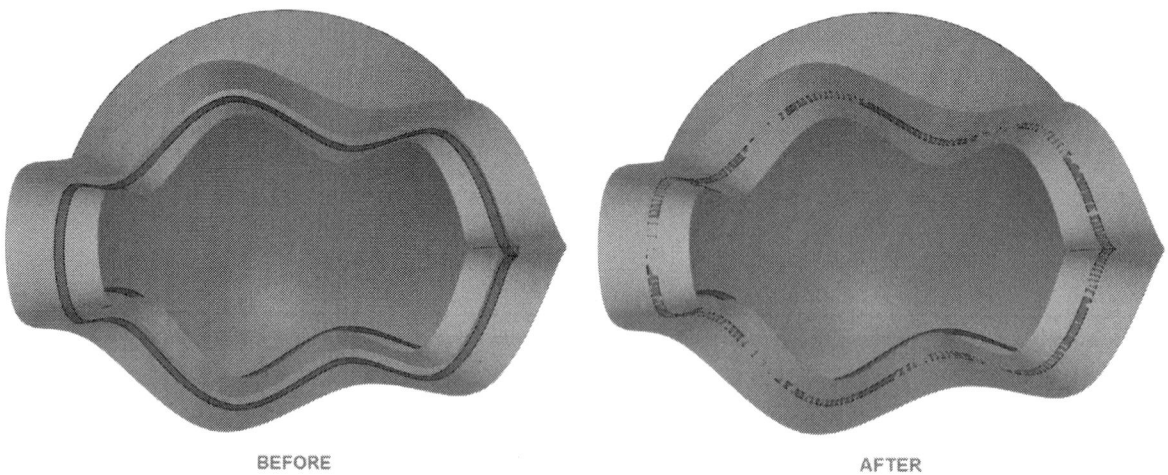

BEFORE

AFTER

Figure 369 - Before and After Cut With Surface Operation

Hide all surface bodies and show the hidden solid body - Red Part (Bottom Part).

The cross sectional view of your part about the Right Plane should now look as shown in the following image.

238

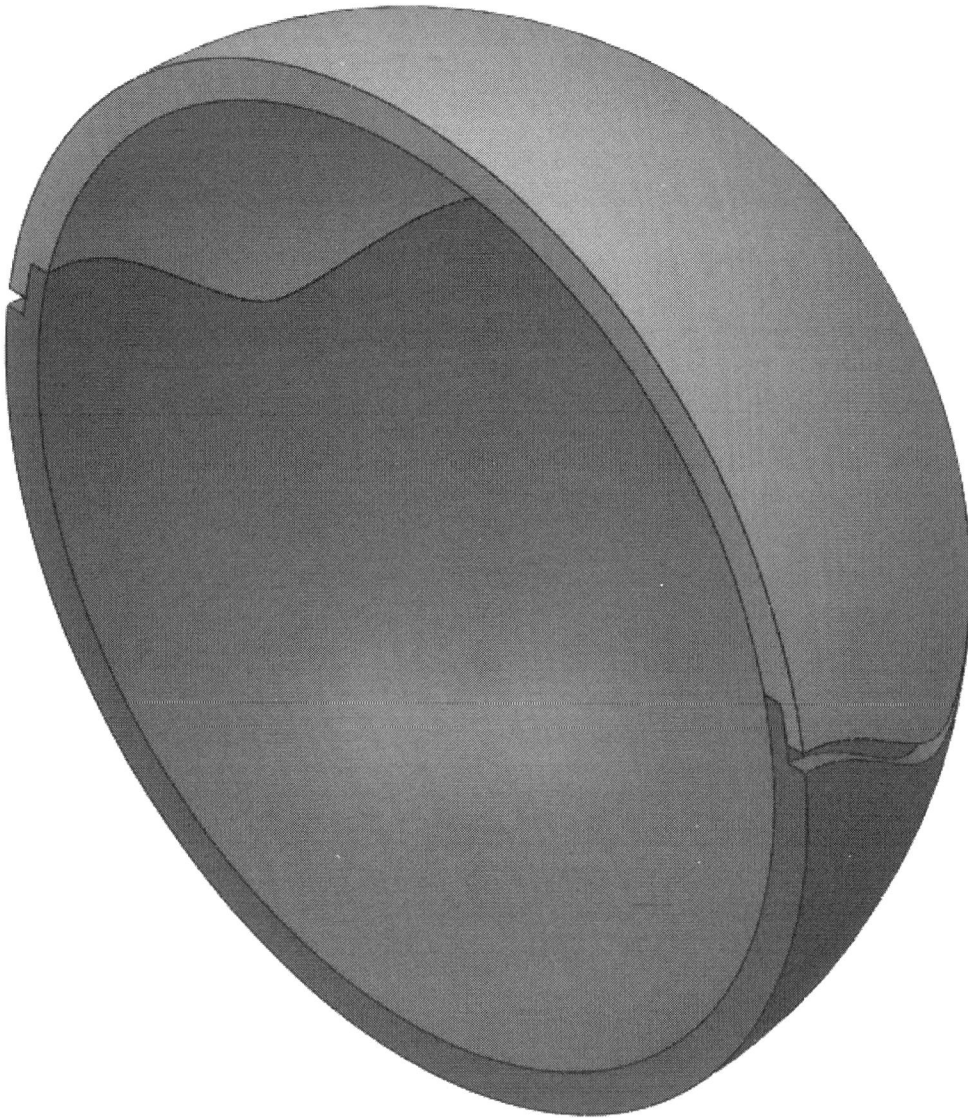

Figure 370 - Cross Sectional View after the Cut With Surface Operation

Save You Part.

MEASURING SURFACE AREA

Click Measure (Evaluate toolbar) or Tools > Evaluate > Measure.

Select the top surface in the Graphics Area as shown in the following image - the answer is thus 6208.09 square millimeters.

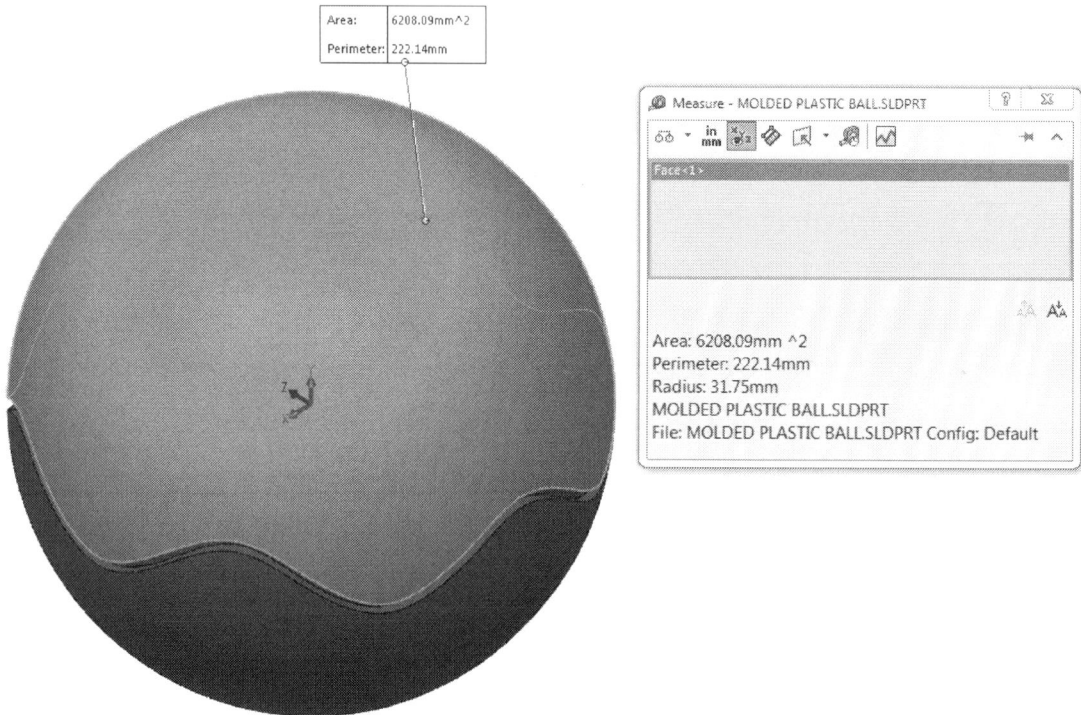

Figure 370 - Measuring surface area

Close the Measure Tool.

Save your part.

Congratulations, you have reached the end of the book. Thank you and wish you all the best.

Any review, comments, compliments or complaints on the site where you bought this book would be greatly appreciated as it would encourage the Author to continue writing more books as well as improve the quality of future books. May you also remember you may send an email to the Author directly at cswpasmebook@gmail.com - include the title of the Book in your email subject since the Author has other books on CSWPA Exams is busy writing more books.

INDEX

Printed in Great Britain
by Amazon